GUANIDINO COMPOUNDS IN BIOLOGY AND MEDICINE

# Guanidino Compounds in Biology and Medicine

*Edited by*

JOSEPH F. CLARK

*Department of Neurology*
*University of Cincinnati Medical Center*
*45267-0536, Cincinnati, OH*
*USA*

Reprinted from *Molecular and Cellular Biochemistry*, Volume 244 (2003)

Kluwer Academic Publishers
Dordrecht / Boston / London

**Distributors for North, Central and South America:**
Kluwer Academic Publishers
101 Philip Drive
Assinippi Park
Norwell, Massachusetts 02061 USA
Telephone (781) 871-6600
Fax (781) 681-9045
E-mail <kluwer@wkap.com>

**Distributors for all other countries:**
Kluwer Academic Publishers Group
Distribution Centre
Post Office Box 322
3300 AH Dordrecht, The Netherlands
Telephone 31 78 6392 392
Fax 31 78 6546 474
E-mail <services@wkap.nl>

Electronic Services <http://www.wkap.nl>

**Library of Congress Cataloging-in-Publication Data**

A C.I.P. Catalogue record for this book is available from the Library of Congress

ISBN 1-4020-7297-X



*Printed on acid-free paper*

Printed in The Netherlands

# Molecular and Cellular Biochemistry:

## An International Journal for Chemical Biology in Health and Disease

**CONTENTS VOLUME 244, Nos. 1 & 2, February 2003**

GUANIDINO COMPOUNDS IN BIOLOGY AND MEDICINE
Dr. Joseph F. Clark

*Molecular and Cellular Biochemistry* **244:** 1, 2003.


# Preface

Guanidino compounds comprise creatine, arginine, and the guanidines. In the past two years there have been over 2000 published articles with the names of these compounds in the title. One can go to any nutrition or health food store and buy these as supplements because it is believed they improve health and athletic performance.

In this special issue of *Molecular and Cellular Biochemistry* the International Guanidino Compounds in Biology and Medicine Society has invited these authors to present their findings which include an up-to-date summary of the scientific and clinical aspects of essentially all the biologically active Guanidino compounds. The articles summarize the current scientific knowledge of these compounds with reference to relevant clinical conditions. The reader will find a variety of articles discussing the chemical, biological and clinical functions of these compounds. This translation of science and clinical applications make this a valuable resource for learning the latest on the subject of Guanidino Compounds in Biology and Medicine.

The compilation of this important work has been made easier by assistance from Muscle Tech and Avicena as well as the Japanese Guanidino Compounds in Biology and Medicine Society and Children's Hospital Medical Center Cincinnati. The contributors and independent referees have provided an invaluable service in producing this work.

Joseph F. Clark
Department of Neurology, University of Cincinnati
Cincinnati, OH, USA
E-mail: joseph.clark@uc.edu

*Molecular and Cellular Biochemistry* **244:** 3–9, 2003.


# Inhibition by heparin of protein kinase C activation and hydroxyl radical generation in puromycin aminonucleoside treated isolated rat hepatocytes

Kazumasa Aoyagi,[1] Yutaka Kuzure,[2] Siranoush Shahrzad,[3] Aki Hirayama,[3] Sohji Nagase[3] and Atsushi Ueda[3]

[1]*Department of Internal Medicine, College Clinic, Tsukuba College of Technology, Tsukuba, Ibaraki;* [2]*Bio-System Research, University of Tsukuba, Tsukuba;* [3]*Institute of Clinical Medicine, University of Tsukuba, Tsukuba, Japan*

## Abstract

Heparin has been reported to have many actions similar to calcium-dependent protein kinase (PKC) inhibitors. We have found that puromycin aminonucleoside (PAN) increases hydroxyl radical generation and this was prevented by H-7, a PKC inhibitor in isolated rat hepatocytes. In this study, we investigate the effect of heparin on the increased hydroxyl radical generation as well as PKC activation by PAN in isolated rat hepatocytes. To estimate the amount of hydroxyl radical generation, we measured methylguanidine (MG) and creatol which are the products from the reaction of creatinine and hydroxyl radical. Synthetic rate of MG plus creatol in isolated rat hepatocytes incubated in Krebs-Henseleit bicarbonate buffer containing creatinine and tested reagents were recorded. This rate with or without PAN was 231 ± 11 or 112 ± 5.6 nmol/g wet cells/4 h (mean ± S.E., n = 5), respectively. Heparin concentrations of 3.3, 6.6 and 10 U/ml inhibited MG plus creatol synthesis in the presence of PAN by 30, 38 and 39%, and without PAN by 8.4, 27 and 34%, respectively. Statistical significance was observed except for 3.3 U/ml without PAN. The ratio of PKC in membrane/cytoplasmic fraction, an indicator of PKC activation, increased 2.8- and 3-fold that of the 0 time after 60 and 120 min incubation with PAN while heparin at 10 U/ml almost completely suppressed this increase in the ratio of PKC. The PKC ratio of the membrane/cytoplasmic fraction obtained from hepatocytes with heparin alone or without PAN and heparin was almost unchanged during the tested period. Variation of PKC levels in membrane fraction is similar to that of PKC ratio of the membrane/cytoplasmic fraction. Increased creatol synthesis by PAN and its inhibition by heparin were observed in the same samples as those used for the PKC study.

These results indicate that heparin inhibits the increase in hydroxyl radical generation induced by PAN through inhibition of PKC activation in isolated rat hepatocytes. (Mol Cell Biochem **244**: 3–9, 2003)

*Key words*: heparin puromycin aminonucleoside, protein kinase C, reactive oxygen

## Introduction

Heparin has been widely used as an anticoagulant in hemodialysis and cardiopulmonary bypass operation. Many papers report favorable effects of heparin on kidney diseases such as glomerulonephritis [1], renal failure [2], diabetic vascular complications [3, 4] and chronic puromycin aminonucleoside (PAN) nephrosis [5]. PAN has been known to induce heavy proteinuria [6] as well as an increase in the glomerular messangial matrix resembling the features of focal glomerular sclerosis, an intractable nephrosis [7]. Therefore, many investigators have used this reagent to create animal models for the study of nephrosis. Besides its anticoagulant activity, there are many inhibitory effects of heparin on the following: smooth cell proliferation [8], tissue factor expression [9] c-fos and c-mic mRNA expression [10], transcription of tissue-type plasminogen activator [11] induction of matrix metalloproteinase [12] PDGF stimulated MAP kinase

*Address for offprints*: K. Aoyagi, College Clinic, College of Tsukuba College of Technology, Tsukuba, Ibaraki, 305-0821, Japan
(E-mail: aoyagi@k.tsukuba-tech.ac.jp)

activation [13] and endothelin-1 synthesis in endothelial cells [14]. In most of these reports, activation of protein kinase C (PKC) and/or inhibitions of these phenomena by PKC inhibitors were reported.

Recently, increased reactive oxygen species (ROS) generation has been implicated as the cause of the toxicity induced by PAN [15, 16]. We, also, reported that PAN increases generation of hydroxyl radical, the most reactive ROS in isolated rat hepatocytes [17–19] as well as glomerulus [20]. As for the molecular mechanism for the toxicity of PAN, we noted that 1-(5-isoquinoline sulfonyl)-2-methylpiperazine dihydrochloride (H-7), a PKC inhibitor, inhibits this increases of hydroxyl radical generation induced by PAN [21]. We also found that PAN activates PKC in isolated rat hepatocytes (in submission). This increased hydroxyl radical generation is indicated by an increased synthesis of methylguanidine (MG) and creatol in isolated rat hepatocytes within an hour of the addition of PAN [19]. Since we showed that MG, a uremic toxin, is formed from creatinine through a hydroxyl radical adduct of creatinine [17–19] that was identified as creatol later [22, 23], it has become apparent that the synthetic rate of MG and/or creatol from a certain level of creatinine can be used as a marker for the rate of synthesis of the hydroxyl radical [24–26].

In the present study, we investigated the effects of heparin on the activation of PKC together with the hydroxyl radical generation in PAN treated-isolated rat hepatocytes.

## Materials and methods

### *Materials*

Puromycin aminonucleoside was purchased from Sigma Chemical Co., St. Louis, USA. Heparin was purchased from Novo Co. Ltd. The creatol standard was kindly donated by Dr. Ienaga of the Nippon Zoki Pharmaceutical Co. Ltd. Monoclonal antibodies specific against $\alpha$, $\beta$ and $\gamma$ isozymes were purchased from Seikagaku Kogyo Co. (Tokyo, Japan). The cell lysate for the PKC standard was purchased from the Transduction Laboratory (Lexington, KY, USA).

### *Preparation of isolated rat hepatocytes*

Male Wistar rats weighing 300–350 g were used in all experiments. The rats were allowed free access to water and laboratory chow containing 25% protein. Isolated hepatocytes were prepared essentially according to the method of Berry and Friend [27] as described previously [28, 29]. We calculated that $9.8 \times 10^7$ cells correspond to 1 g of wet hepatocytes.

### *Incubation of cells for MG and creatol synthesis*

Hepatocytes were incubated in 6 ml of Krebs-Henseleit bicarbonate buffer containing 3% bovine serum albumin, 10 mM sodium lactate, 16.6 mM creatinine, indicated substances and with or without 1.9 mM PAN. The incubation mixture was shaken at 60 cycles/min in a 30-ml conical flask with a rubber cap under 95% oxygen and 5% carbon dioxide at 37°C. Equilibration of the buffer was repeated every hour. Incubation was arrested by the addition of 0.6 ml of 100% (w/v) trichloroacetic acid.

### *Determination of MG and creatol*

After sonication, the supernatant of cells and medium was obtained by centrifugation at 1700 × g for 15 min at 0°C, and 0.2 ml of the extract was used for MG and creatol measurements. MG was determined by high-performance liquid chromatographic analysis using 9,10-phenanthrenequinone for the post-labeling method as described previously [24]. Dimethylformamide for fluorometrical use was purchased from Wako Co., Japan. Creatol was separated by a cation exchange resin column larger than that for MG analysis and was converted to MG by heating at 125°C under strong alkaline condition using the HPLC system modified from the MG determination apparatus [30].

### *Subcellular fractionation of PKC from isolated hepatocytes*

Hepatocytes (1.5 g wet cells) were incubated in 35 ml of the same buffer as used for MG and creatol synthesis with or without PAN. An aliquot (5 ml) of the cell suspension was taken at 0, 15, 30, 60 and 120 min and the cells were collected by centrifugation. The collected cells were homogenized with a Dounce homogenizer (100 strokes) in 2 ml of 50 mM Tris-HCl pH 7.6 containing 0.25 M sucrose, 2 mM EDTA, 0.2 mM phenylmethanesulfonyl fluoride and 0.1 mM leupeptin. After an initial centrifugation (600 × g for 10 min) to remove nuclei, the supernatant was centrifuged at 12000 × g for 10 min to remove mitochondria and the supernatants were separated into cytoplasmic and insoluble fractions by centrifugation (105,000 × g for 45 min). The insoluble (membrane) fractions were homogenized in 200 ml of the buffer used for homogenization and were kept on ice for 30 min after the addition of 1% Triton X-100 for solubilization. The cytoplasmic and the insoluble fractions were boiled for 5 min in the presence of 2% sodium dodecyl sulfate (SDS), and both fractions were kept at –80°C. Protein in both fractions was determined by the Bicinchoninic acid protein assay method [31].

*Determination of PKC*

Both fractions were boiled for 3 min with 1% β-mercaptoethanol, 5% glycerol and 0.006% bromophenol blue. Proteins (110 μg protein) in the above two fractions were separated by electrophoresis on 8% SDS-polyacrylamide gels (14 × 14 cm). As a positive control of PKC (1 μg protein), rat brain lysate obtained from Transduction Laboratories was used. The proteins were transferred to nitrocellulose filters. Immunoreactive proteins were detected using the mixture of antibodies specific against α, β and γ isozymes. These were diluted 100 times and mixed (1:1:1). Enhanced chemiluminescence (ECL) western blotting detection reagents (RPN 2106, Amersham Pharmacia Biotech, NJ, USA) were used for detecting PKCs. The bands of PKCs were quantified by densitometry.

*Statistics*

Significance of experimental results between two groups was evaluated using the unpaired Student's *t*-test. For comparisons involving more than two groups, ANOVA was applied. Results were expressed as mean ± S.E. The differences were considered statistically significant when the calculated P value was less than 5%.

## Results

*Inhibition by heparin of MG and creatol synthesis in isolated rat hepatocytes*

Effect of heparin on the synthesis of creatol and/or MG in isolated rat hepatocytes with or without PAN was investigated. The synthetic rate of MG in isolated rat hepatocytes with and without PAN were 177 ± 12 and 91.3 ± 3.2 nmol/g of wet cells/4 h (mean ± S.E.). Heparin at the concentrations of 3.3, 6.6 and 10 U/ml significantly inhibited MG synthesis by 38, 39 and 41% in the presence of PAN and by 24, 31 and 45% in the absence of PAN, respectively as shown in Fig. 1a. The MG plus creatol synthetic rate with and without PAN were 231 ± 11 and 111.6 ± 5.6 nmol/g of wet cells /4 h (mean ± S.E.), respectively. Heparin at 3.3, 6.6 and 10 U/ml inhibited MG plus creatol synthesis with PAN by 30, 3 8 and 39%, and without PAN by 8.4, 26.5 and 34% as shown in Fig. 1b. MG plus creatol synthesis was significantly inhibited by heparin at any tested concentration except at 3.3 U/ml without PAN.

The inhibition rate of 30% by heparin of 3.3 U/ml in the presence of PAN is much higher than that of 8.4% in the absence of PAN. The inhibition rate did not increase proportional to the heparin concentration. This indicates that the inhibition is not caused by a simple hydroxyl radical scavenging activity of heparin.

*Inhibition by heparin of PKC activation in isolated rat hepatocytes*

The translocation of PKC from the cytoplasm to the membrane fraction has been reported as the most reliable indicator of the activation of the conventional type of PKC in cells [32–34]. Bands of PKC in the membrane fractions were detected by the Western blotting method and their densities obtained by densitometry from isolated rat hepatocytes incubated in control solution (without PAN and heparin ), with PAN, with PAN and heparin, or with heparin alone respec-

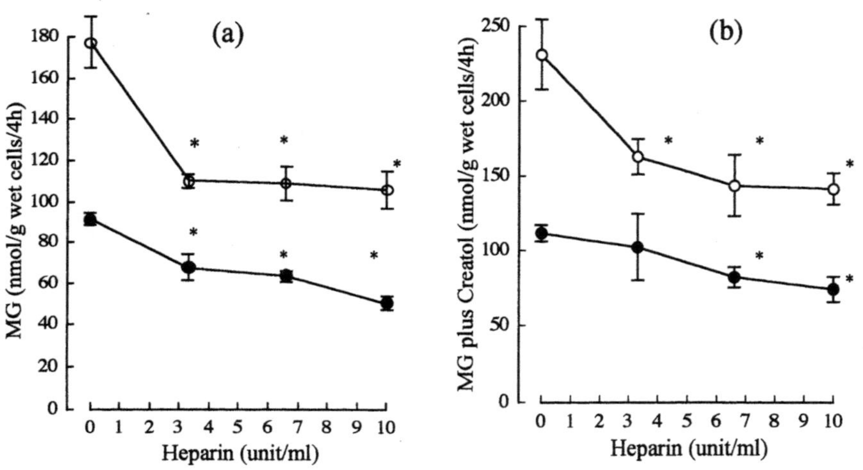


*Fig. 1.* Inhibition of MG or MG plus creatol synthesis by heparin in isolated rat hepatocytes. Isolated rat hepatocytes (0.1 g wet cells) were incubated with various concentrations of heparin with (O) or without 1.9 mM PAN (●) as described in 'Materials and methods'. Values are expressed as the mean ± S.E. (n = 5). *Indicates a p value less than 0.05 vs. the control value (without heparin). Data were subjected to ANOVA with pairwise comparison by the Bonferroni method.

tively. PKC in the membrane fractions increased at 60 and 120 min incubations in PAN-treated hepatocytes. As for the PKC bands from hepatocytes incubated with PAN plus heparin, inhibition of the increase of PKC in the membrane fractions at 60 and 120 min was noted. Since the total amount of PKC is more than 10 times of that of the membrane fraction, activation of PKC will not affect the PKC amount within a short period. Thereafter, cytoplasmic PKC can be used to correct the variation in density caused by the different exposures to luminescence by the Western blot method. The PKC ratio of the membrane/cytoplasmic fraction of the isolated rat hepatocytes determined after 0, 15, 30, 60 and 120 min incubations with or without 1.9 mM PAN were as follows. In the presence of PAN, the PKC ratio in the membrane/cytoplasmic fraction after 60 and 120 min incubation was 2.8 and 3.0 times higher than that of the 0 time control (Fig. 2a). On the other hand, the PKC ratio of the hepatocytes incubated with heparin and PAN, with heparin alone and without PAN or heparin did not change or changed little during the tested incubation period (Fig. 2a). PKC levels in the membrane fraction after adjusting for the densities based on assumption that 0 time density is a constant is shown (Fig. 2b). There are no major differences between the results using cytoplasmic PKC or 0 time membrane PKC for the correction. Creatol was determined in the same samples as those used for testing PKC activation. Creatol levels in the presence of PAN at 30, 60 and 120 min incubation increased to 3.2, 3.0 and 2.3 times of that seen without PAN (Fig. 2c). Heparin almost completely inhibited this increased synthesis of creatol.

These results indicate that heparin inhibits both PKC activation induced by PAN and MG generation in isolated rat hepatocytes.

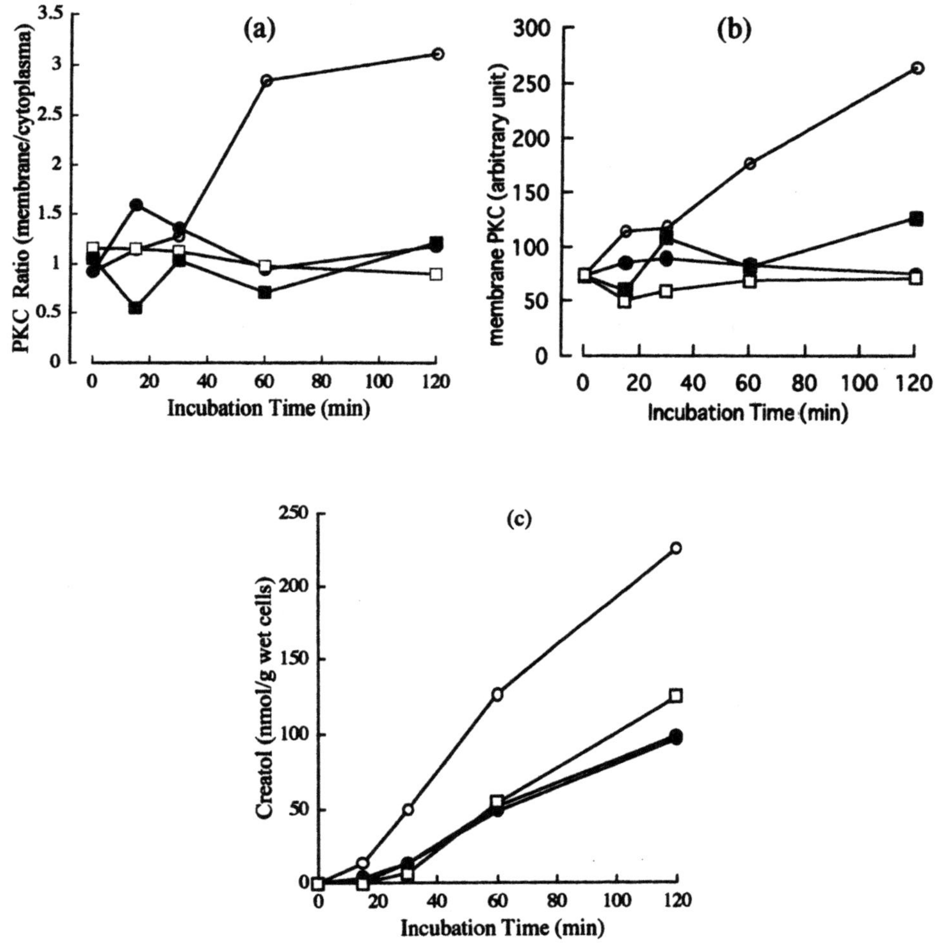


*Fig. 2.* Inhibition of PKC activation in PAN-treated hepatocytes by heparin. The ratio of PKC in the membrane/cytoplasmic fraction is shown in (a). The PKC level in the membrane fraction after adjusting the level based on the idea that 0 time density is a constant is shown (b). Creatol synthesis in isolated hepatocytes in the same sample as that for PKC study under various conditions are shown in (c). The symbols □ represents PAN and heparin, O represents PAN alone, ● represents no additives and ■ represents heparin alone.

## Discussion

In this study, we demonstrate that heparin inhibits the PKC activation as well as MG synthesis in PAN-treated isolated rat hepatocytes. The heparin concentration of 3.3 U/ml that significantly inhibited MG synthesis in PAN-treated hepatocytes is lower than that in the blood of the patients undergoing hemodialysis therapy or cardiopulmonary bypass surgery [35]. The inhibition rate of MG synthesis did not increase at concentrations of heparin higher than 6.6 and 10 U/ml. These data suggest that the inhibition dose is not dependent on the ROS scavenging activity of heparin, since the ROS scavenging activity of heparin is relatively low [36]. With regard to the heparin effect on ROS, it has been reported that heparin inhibit ROS generation in leukocytes [37] and in rheumatoid synovial cell A (macrophage-like) and B (fibroblast-like) stimulated by substance P [38]. It is reported that heparin reduces hepatic ischemia-reperfusion injury with decreased lipid peroxide *in vivo* [39]. However, in this report, anticoagulant role of heparin was not excluded as a cause of this favorable effect. This report may be the second instance in which heparin inhibits ROS generation in non-inflammatory cells following synovial B cells.

Favorable effects of heparin on glomerulonephritis [40] as well as chronic PAN-nephrosis [5] has been reported at concentrations insufficient to inhibit coagulation. Similar favorable effects of heparin or non-coagulant heparin like substances on the vascular complications in diabetes mellitus are also reported [4]. As the reason for these favorable effects of heparin, inhibition of mesangial cell proliferation and expression of matrix protease have been suggested [41]. These effects of heparin are attributed to the action of PKC inhibition. PKC plays an important role in many biological processes, especially in cell proliferation and the events following the inflammation [32]. Considering that H-7, a PKC inhibitor, prevented an increase of hydroxyl radical generation by PAN [21], inhibition of PKC activation by heparin in PAN-treated hepatocytes should lead to the suppression of hydroxyl radical generation. This report directly demonstrates that heparin inhibits the translocation of PKC in non-inflammatory cells. Similar reports prior this of the inhibition of translocation of PKC by heparin in non-inflammatory cells could not be found.

In addition to the kidney, PAN is also known to affect the liver. Aside from our studies [15–17], findings in the liver of PAN-treated rats include increased liver weight [42], inhibited phenobarbital induction of microsomal enzymes [43] and increased lipoperoxidation and glutathione peroxidase [44]. Activation of PKC in liver caused cholestasis [45], stimulation of apical exostosis [46], mitogen-activated protein kinase activation [47] and phospholipase D activation [48]. Among these effects, activation of phospholipases may lead to an increase in hydroxyl radical generation through arachidonic acid metabolism which is known to generate hydroxyl radicals in gallbladder muscle cells [49].

As a mechanism of inhibition of PKC activation, it was reported that heparin acts as an inositol 1,4,5-triphosphate receptor antagonist in receptor coupled G protein-mediated contraction of isolated intestinal muscle cells [50] and decreases intracellular $Ca^{2+}$ concentration [51].

Thus, heparin, which both scavenges ROS and inhibits PKC activation or PKC activity, represents a potential treatment for many diseases in which PKC activation is a major element.

## Acknowledgements

This study was supported in part by the Scientific Research Funds of the Ministry of Education, Science and Culture of Japan (C-05670942, C-06671126), Heparin Conference, and Grant for Scientific Research Expenses for Health and Welfare Programs, Ministry of Health and Welfare, and Japan Society for the Promotion of Science (P99168). We thank Professor Burton D. Cohen (Department of Medicine, Albert Einstein College of Medicine) for his critical comments. We thank Dr. Kenji Nomoto, Tsukuba Institute of the Eisai Pharmaceutical Co. Ltd. for this technical advice in the PKC study.

## References

1. Freedman P, Meister HP, De la Paz A, Ronaghy H: The clinical, functional, and histologic response to heparin in chronic renal disease. Invest Urol 7: 398–409, 1970
2. Barsotti G, Cupisti A, Gervasi GB, Bartoli C, Barsotti M, Pasquariello A, Moriconi L, Giovannetti S: Effects of oral administration of heparan sulphate in the rat remnant kidney model. Nephron 81: 10–316, 1999
3. Yokoyama H, Myrup B, Oturai P, Deckert T: Heparin, a possible therapy for diabetic complications: The effect on mesangial and myomedial cells *in vivo* and *in vitro*, especially in relation to extracellular matrix. J Diabetes Comp 9: 97–103, 1995
4. Myrup B, Hansen PM, Jensen T, Kofoed-Enevoldsen A, Feldt-Rasmussen B, Gram J, Kluft C, Jespersen J, Deckert T: Effect of low-dose heparin on urinary albumin excretion in insulin-dependent diabetes mellitus. Lancet 345: 421–422, 1995
5. Diamond JR, Karnovsky MJ: Nonanticoagulant protective effect of heparin in chronic aminonucleoside nephrosis. Renal Physiol 9: 366–374, 1986
6. Frenk S, Antonowicz I, Chaig JM, Metecoff J: Experimental nephrotic syndrome induced in rats by aminonucleoside renal lesion and body electrolyte composition. Proc Soc Exp Biol Med 89: 424–427, 1955
7. Glasser RJ, Velosa JA, Michael AF: Experimental model of focal sclerosis. 1. Relationship to protein excretion in aminonucleoside nephrosis. Lab Invest 36: 527–534, 1977
8. Castellot JJ Jr, Pukac LA, Caleb BL, Wright TC Jr, Karnovsky MJ: Heparin selectively inhibits a protein kinase C-dependent mechanism of cell cycle progression in calf aortic smooth muscle cells. J Cell Biol 109: 3147–3155, 1989

9. Xuereb JM, Herbert JM, Sie P, Boneu B, Constans J: Effects of heparin and related sulfated polysaccharides on tissue factor expression induced by mitogenic and non-mitogenic factors in human vascular smooth muscle cells. Thromb Haemost 81: 151–156, 1999
10. Pukac LA, Castellot JJ Jr, Wright TC Jr, Caleb BL, Karnovsky MJ: Heparin inhibits c-fos and c-myc mRNA expression in vascular smooth muscle cells. Cell Reg 1: 435–443, 1990
11. Au YP, Kenagy RD, Clowes AW: Heparin selectively inhibits the transcription of tissue-type plasminogen activator in primate arterial smooth muscle cells during mitogenesis. J Biol Chem 267: 3438–3444, 1992
12. Kenagy RD, Nikkari ST, Welgus HG, Clowes AW: Heparin inhibits the induction of three matrix metalloproteinases (stromelysin, 92-kD gelatinase, and collagenase) in primate arterial smooth muscle cells. J Clin Invest 93: 1987–1993, 1994
13. Adachi T, Nakashima S, Saji S, Nakamura T, Nozawa Y: Mitogen-activated protein kinase activation in hepatocyte growth factor-stimulated rat hepatocytes: Involvement of protein tyrosine kinase and protein kinase C. Hepatology 23: 1244–1253, 1996
14. Imai T, Hirata Y, Emori T, Marumo F: Heparin has an inhibitory effect on endothelin-1 synthesis and release by endothelial cells. Hypertension 21: 353–358, 1993
15. Diamond JR, Bonventre JV, Karnovsk MJ: A role for oxygen free radicals in aminonucleoside nephrosis. Kidney Int 29: 478–483, 1986
16. Srivastava RN, Diviven S, Kalia A, Travis LB, Ansari NH: Increased glomerular and urinary malondialdehyde in puromycin aminonucleoside-induced proteinuria in rats. Pediatr Nephrol 9: 48–51, 1995
17. Aoyagi K, Nagase S, Narita M, Tojo S: Role of active oxygen on methylguanidine synthesis in isolated rat hepatocytes. Kidney Int 22: S229–S233, 1987
18. Aoyagi K, Nagase S, Sakamoto M, Narita M, Tojo S: Puromycin aminonucleoside stimulates the synthesis of methylguanidine: A possible marker of active oxygen generation in isolated rat hepatocytes. In: A. Mori, B.D. Cohen, H. Koide (eds). Guanidines 2. Plenum Press, New York, 1989, pp 71–77
19. Aoyagi K, Akiyama K, Kuzure Y, Takemura K, Nagase S, Ienaga K, Nakamura K, Koyama A, Narita M: Synthesis of creatol, a hydroxyl radical adduct of creatinine and its increase by puromycin aminonucleoside in isolated rat hepatocytes. Free Radic Res 29: 221–226, 1998
20. Aoyagi K, Akiyama K, Tomida C, Gotoh M, Hirayama A, Takemura K, Ueda A, Nagase S, Koyama A, Narita M: Imaging of hydroperoxides in a rat glomerulus stimulated by puromycin aminonucleoside. Kidney Int 71(suppl): S153–S155, 1999
21. Aoyagi K, Takemura K, Tomida C, Goto M, Koyama A, Narita M: Inhibition of methylguanidine synthesis in isolated rat hepatocytes by protein kinase C inhibitor and calmodulin antagonist. In: P.P. De Deyn, B. Maeacau, A. Qureshi, A. Mori (eds). Guanidino Compounds in Biology and Medicine 2. John Libbey & Co. Ltd., London, 1997, pp 311–316
22. Nakamura K, Lenaga K, Yokozawa T, Fujitsuka N, Oura H: Production of methylguanidine from creatinine via creatol by active oxygen species: Analyses of the catabolism *in vitro*. Nephron 58: 42–44, 1991
23. Ienaga K, Nakamura K, Yamakawa H, Yokozawa T, Oura H, Nakano K: The use of $^{13}$C-labeling to prove that creatinine is oxidized by mammals into creatol and 5-hydroxy-1-methylhydantonin. J Chem Soc Chem Commun: 509–510, 1991
24. Aoyagi K, Nagase S, Koyama A, Narita M, Tojo S: Products of creatinine with hydroxyl radical as a useful marker of oxidative stress *in vivo*. Meth Mol Biol 108: 157–164, 1998
25. Sakamoto M, Aoyagi K, Nagase S, Ishikawa T, Takemura K, Narita M: Methylguanidine synthesis by reactive oxygen species from human leukocytes. Japanese J Nephrol 31: 851–858, 1998 (in Japanese)
26. Takemura K, Aoyagi K, Nagase S, Sakamoto M, Ishikawa T, Narita M: Effect of hyperbaric therapy on urinary methylguanidine excretion in normal human and patients with renal failure. In: P.P. De Deyn, B. Marescau, V. Stalon, I.A. Qureshi (eds). Guanidino Compounds in Biology and Medicine. John Libbey & Co. Ltd., London, 1992, pp 301–330
27. Berry MN, Friend DS: High-yield preparation of isolated liver cells. J Cell Biol 43: 506–520, 1969
28. Zahlten RN, Stratman FW, Lardy HA: Regulation of glucose synthesis in hormone-sensitive isolated rat hepatocytes. Proc Natl Acad Sci USA 70: 3213–3218, 1973
29. Aoyagi K, Ohba S, Narita M, Tojo S: Regulation of biosynthesis of guanidinosuccinic acid in isolated rat hepatocytes and *in vivo*. Kidney Int 24: S224–S228, 1983
30. Nakamura K, Ienaga K, Nakano K, Nakai M, Nakamura Y, Hasegawa G, Sawada M, Kondo M, Mori H, Kanatsuna T: Creatol, a creatinine metabolite, as a useful determinant of renal function. Nephron 66: 140–146, 1994
31. Smith PK, Krohn RI, Hermanson GT, Malia AK, Gartner FH, Provenzano MD, Goeke EK, Olson BJ, Klenk DC: Measurement of protein using bicinchoninic acid. Anal Biochem 150: 76–85, 1985
32. Nishizuka Y: Intracellular signaling by hydrolysis of phospholipids and activation of protein kinase C. Science 258: 607–614, 1992
33. Dekker LV, Parker PJ: Protein kinase C – a question of specificity. Trends Biochem Sci 19: 73–77, 1984
34. Nishizuka Y: Protein kinase C and lipid signaling for sustained cellular responses. FASEB J 9: 484–496, 1995
35. Hardy JF, Belisle S, Robitaille D, Perrault J, Roy M, Gagnon L: Measurement of heparin concentration in whole blood with the Hepcon/HMS device does not agree with laboratory determination of plasma heparin concentration using a chromogenic substrate for activated factor X. J Thoracic Cardiovasc Surg 112: 154–161, 1996
36. Lapenna D, Mezzetti A, de Gioia S, Ciofani G, Marzio L, Di Ilio C, Cuccurullo F: Heparin: Does it act as an antioxidant *in vivo*? Biochem Pharmacol 44: 188–191, 1992
37. Laghi-Pasini F, Pasqui AL, Ceccatelli L, Di Perr T: Heparin inhibits FMLP and Con-A dependent activation of human polymorphonuclear leucocytes *in vitro*. Int J Tissue React 5: 145–151, 1983
38. Tanabe T, Otani H, Mishima K, Ogawa R, Inagaki C: Mechanisms of oxyradical production in substance P stimulated rheumatoid synovial cells. Rheumatol Int 16: 159–167, 1996
39. Liu W, Pitcher DE, Morris SL, Pugmire JE, Shores JT, Ingram CE, Glew RH, Morris DM, Fry DE: Differential effects of heparin on the early and late phases of hepatic ischemia and reperfusion injury in the pig. Shock 12: 134–138, 1999
40. Burg M, Ostendorf T, Mooney A, Koch KM, Floege J: Treatment of experimental mesangioproliferative glomerulonephritis with non-anticoagulant heparin: Therapeutic efficacy and safety. Lab Invest 6: 505–516, 1997
41. Wang ZQ, Liang KH, Pahl MV, Vaziri ND: Effect of heparin on mesangial growth and gene expression of matrix proteins. Nephrol Dialysis Transplant 13: 3052–3057, 1998
42. Hoffman A, Levy G: Kinetics of drug action in disease states. XXIX. Effect of experimental nephrotic syndrome on the pharmacodynamics of eptabarbital: Implications of severe hypoalbuminemia. J Pharmacol Exp Ther 249: 117–122, 1989
43. Sidhu JS, Omiecinski CJ: Protein synthesis inhibitors exhibit a nonspecific effect on phenobarbital-inducible cytochrome P450 gene expression in primary rat hepatocytes. J Biol Chem 273: 4769–4775, 1998
44. Pedraza-Chaverri J, Arevalos AE: Tissue lipoperoxidation and glutathione peroxidase activity in puromycin aminonucleoside injected rats. Int J Biochem 26: 1139–1145, 1994
45. Roma MG, Orsler DJ, Coleman R: Canalicular retention as an *in vitro*

assay of tight junctional permeability in isolated hepatocyte couplets: Effects of protein kinase modulation and cholestatic agents. Fund Appl Toxicol 37: 71–81, 1997

46. Bruck R, Nathanson MH, Roelofsen H, Boyer JL: Effects of protein kinase C and cytosolic $Ca^{2+}$ on exocytosis in the isolated perfused rat liver. Hepatology 20: 1032–1040, 1994
47. Adachi T, Nakashima S, Saji S, Nakamura T, Nozaw Y: Mitogen-activated protein kinase activation in hepatocyte growth factor-stimulated rat hepatocytes: Involvement of protein tyrosine kinase and protein kinase C. Hepatology 23: 1244–1253, 1996
48. Adachi T, Nakashima S, Saji S, Nakamura T, Nozawa Y: Phospholipase D activation in hepatocyte growth factor-stimulated rat hepatocytes mediates the expressions of c-jun and c-fos: Involvement of protein tyrosine kinase, protein kinase C, and $Ca^{2+}$. Hepatology 24: 1274–1281, 1996
49. Yu P, Chen Q, Xiao Z, Harnett K, Biancani P, Behar J: Signal transduction pathways mediating CCK-induced gall bladder muscle contraction. Am J Physiol 275: G203–G211, 1998
50. Murthy KS, Grider JR, Makhlouf GM: Receptor-coupled G proteins mediate contraction and $Ca^{2+}$ mobilization in isolated intestinal muscle cells. J Pharmacol Exp Ther 260: 90–97, 1992
51. Tanabe T, Otani H, Mishima K, Ogawa R, Inagaki C: Mechanisms of oxyradical production in substance P stimulated rheumatoid synovial cells. Rheumatol Int 16: 159–167, 1996

*Molecular and Cellular Biochemistry* **244:** 11–15, 2003.


# Inhibition of arginine synthesis by urea: A mechanism for arginine deficiency in renal failure which leads to increased hydroxyl radical generation

Kazumasa Aoyagi

*Department of Internal Medicine, College Clinic, Tsukuba College of Technology, Tsukuba, Ibaraki, Japan*

## Abstract

We have reported that (1) the synthesis of GSA, a uremic toxin, increases depending on the urea concentration and (2) GSA is formed from argininosuccinic acid (ASA) and the hydroxyl radical or SIN-1 which generates superoxide and NO simultaneously. However, an excess of NO, which also serves as a scavenger of the hydroxyl radical, inhibited GSA synthesis. We also reported that arginine, citrulline or ammonia plus ornithine, all of which increase arginine, inhibit GSA synthesis even in the presence of urea. To elucidate the mechanism for increased GSA synthesis by urea, we investigated the effect of urea on ASA and arginine, the immediate precursor of NO.

Isolated rat hepatocytes were incubated in 6 ml of Krebs-Henseleit bicarbonate buffer containing 3% bovine serum albumin, 10 mM sodium lactate, 10 mM ammonium chloride and with or without 36 mM of urea and 0.5 or 5 mM ornithine at 37°C for 20 min. *In vivo* experiments, 4 ml/100 g body weight of 1.7 M urea or 1.7 M NaCl were injected intra-peritoneally into 5 male Wistar rats. Two hours after the intra-peritoneal injection of urea or 1.7 M NaCl, blood, liver and kidney were obtained by the freeze cramp method and amino acids were determined by an amino acid analyzer (JEOL:JCL-300).

ASA in isolated hepatocytes was not detected with or without 36 mM (200 mgN/dl) urea, but the arginine level decreased from 36 to 33 nmol/g wet cells with urea. Ornithine which inhibits GSA synthesis, increased ASA markedly in a dose dependent manner and increased arginine. At 2 h after the urea injection the rat serum arginine level decreased by 42% (n = 5), and ornithine and citrulline levels increased significantly. Urea injection increased the ASA level in liver from 36–51 nmol/g liver but this was not statistically significant.

We propose that urea inhibits arginine synthesis in hepatocytes, where the arginine level is extremely low to begin with, which decreases NO production which, in turn, increases hydroxyl radical generation from superoxide and NO. This may, also, be an explanation for the reported increase in oxygen stress in renal failure. (Mol Cell Biochem **244**: 11–15, 2003)

*Key words*: urea, arginine, reactive oxygen, uremia, guanidinosuccinic acid

## Introduction

Synthesis of guanidinosuccinic acid (GSA) increases in patients with renal failure and has been recognized as a uremic toxin [1] because of its various biological activities, such as its strong inhibition of platelet aggregation [2], induction of hemolysis [3], induction of generalized clonic and tonic convulsions [4] and activation of the N-methyl-D-aspartate (NMDA) receptor which generates NO in neurologic systems [5]. The synthesis of GSA in patients with renal failure increases depending on the urea level [6, 7]. Furthermore, we have demonstrated urea concentration-dependent GSA synthesis using isolated rat hepatocytes [8]. In addition, we have shown that GSA is formed from the reaction of arginino-

*Address for offprints*: K. Aoyagi, College Clinic, College of Tsukuba College of Technology, Tsukuba, Ibaraki, 305-0821, Japan (E-mail: aoyagi@k.tsukuba-tech.ac.jp)

succinic acid (ASA) and reactive oxygen species, particularly the hydroxyl radical [9,10]. ASA is an intermediate of the urea cycle and also a precursor of arginine. In *in vitro* experiments, we have reported that GSA is formed from ASA and SIN-1 which generates superoxide and NO simultaneously. Since urea is a known inhibitor of the enzymatic conversion of ASA to arginine we looked for an increase in ASA, a precursor of GSA [11], as a cause of this urea stimulated GSA synthesis. In addition, we previously reported that arginine, citrulline or ornithine plus ammonia, all of which increase arginine, inhibit GSA synthesis in the presence of urea, and ornithine without ammonia, which does not increase arginine, did not inhibit GSA synthesis [8, 12]. From the *in vitro* experiments using SIN-1, we propose that scavenging the hydroxyl radical by the excesses of NO arising from large amounts of arginine is a possible cause of this inhibition of GSA synthesis. Therefore, we investigated both the level of ASA, a substrate for GSA, and the level of arginine, a substrate of NO synthetase, as well as other urea cycle members in isolated rat hepatocytes and in living rats to elucidate the mechanism of urea stimulated GSA synthesis and its inhibition by arginine.

## Materials and methods

Isolated rat hepatocytes prepared from male Wistar rats were incubated in Krebs-Henseleit bicarbonate buffer as described previously [8]. Cells (0.2 g wet cell) were incubated in 6 ml of Krebs-Henseleit bicarbonate buffer containing 3% bovine serum albumin, 10 mM sodium lactate, 10 mM ammonium chloride and with or without 36 mM of urea and 0.5 or 5 mM ornithine at 37°C for 20 min. Incubation was arrested by the addition of 0.6 ml of 100% (wt/vol) trichloroacetic acid. After sonication, the supernatant of cells and medium was obtained by centrifugation at 1700 g for 15 min at 0°C, and 0.2 ml of the extract was used for amino acid analysis and GSA determination. GSA was determined by high-performance liquid chromatography using 9,10-phenanthrenequinone, which reacts specifically with mono-substituted guanidino compounds, for postlabeling after separation on a cation exchange column as described previously [8].

In the *in vivo* experiments, 4 ml/100 g body weight of 1.7 M urea or 1.7 M NaCl were injected intra-peritoneally into 5 male Wistar rats weighing from 255–274 g. Two hours after the intra-peritoneal injection of urea or 1.7 M NaCl, blood, the livers and kidneys were collected after ether anesthesia. Tissues were obtained by the freezed cramp method using liquid nitrogen and homogenized by Polytron in 2 volume of 10% trichroloacetic acid (TCA). The supernatant after centrifugation was used for amino acid analysis using an amino acid analyzer (JEOL:JCL-300).

## Results

### *Effect of urea and/or ornithine on the arginine and ASA levels in isolated rat hepatocytes*

Ornithine in the presence of ammonium chloride inhibited the urea-stimulated GSA synthesis in a dose-dependent manner was shown in Fig. 1. Chromatograms of ASA, a substrate for GSA, obtained by the amino acid analyzer were shown in Fig. 2. In the presence of 36 mM urea, a notch was recognized as ASA, but did not increase significantly; 0.5 mM ornithine increased ASA somewhat and 5 mM ornithine increased ASA to 16 mM. It is apparent that the ASA level is not the rate limiting factor for GSA synthesis, because 5 mM ornithine almost completely inhibited GSA synthesis. The arginine level in isolated rat hepatocytes was decreased by urea from 36 to 33 nmol/g wet cells, as shown in Fig. 3, but this was not statistically significant. On the contrary, the combination of ornithine plus ammonia that inhibited GSA synthesis, increase arginine. Arginine concentration increased from 0.03–0.05 mM in the presence of 0.5 mM ornithine and 0.03–0.22 mM with 5 mM ornithine. Citrulline concentration of control was 0.6 mM and increased by urea to 0.64 mM, and ornithine at 0.5 and 6 mM increased citrulline to 0.86 and 2.1 mM, respectively.

### *Effect of urea on the level of arginine, ASA and urea cycle members* in vivo

The concentration of urea in serum, liver and kidney at 2 h after the intraperitoneal injection of urea or NaCl are shown in Fig. 4. The urea concentration of serum and liver increased by 70 and 45 mM at 2 h after the urea injection. The serum

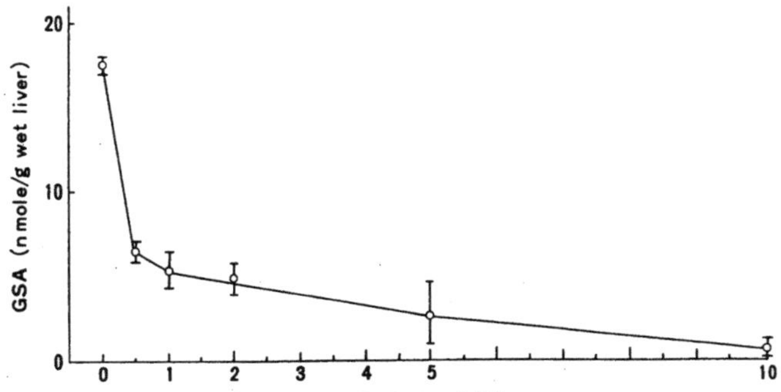


*Fig. 1.* Inhibitory effect of ornithine on guanidinosuccinic acid synthesis stimulated by urea in isolated rat hepatocytes. Cells (0.08 g of wet cells) were incubated as described in 'Materials and methods'. Each point represents the mean of duplicate incubation. Bars indicate the range of each determination.

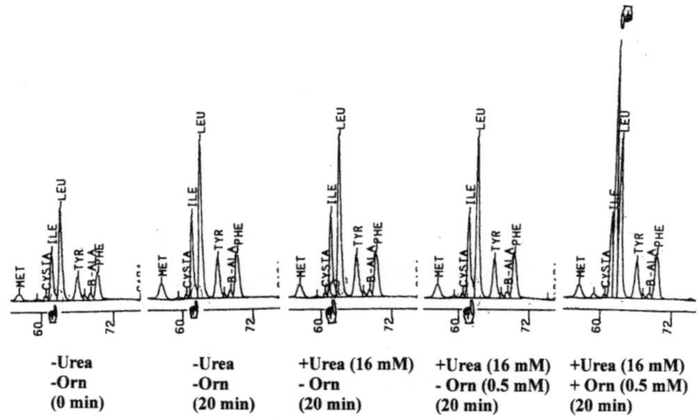


*Fig. 2.* Chromatograms of ASA from various conditions were obtained by the amino acid analyzer as described in 'Materials and methods'. Pointer indicates ASA peak.

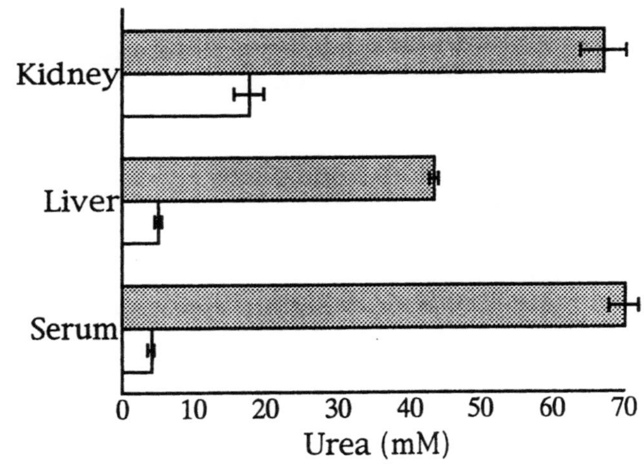


*Fig. 4.* Effect of urea on urea level in serum, liver and kidney. Urea concentration without urea was represented by open column and with urea was represented by dotted column. Values were obtained from 5 rats.

concentration of arginine significantly decreased by 43% after urea injection, and citrulline and ornithine increased as shown in Fig. 5. The arginine level in liver did not change as shown in Fig. 6. Among the urea cycle members, ornithine and citrulline increased significantly in the liver. In the kidney, the arginine level did not vary in response to the urea injection, but ornithine increased. The ASA level was increased from 35 ± 3.7 to 51 ± 8.8 nmol/g liver by urea injection but this was not statistically significant. The increased ornithine and citrulline and slight increase of ASA in the liver and similar amino acid change in the serum suggests that urea inhibits both ASA synthesis from citrulline and arginine synthesis from ASA.

## Discussion

It has been reported that the arginine levels decrease and citrulline levels increase in serum from patients with renal failure. This change in arginine and citrulline has been explained by the loss of renal tubular cells which form arginine from citrulline [14]. We show that urea decreases serum arginine and increases citrulline and ornithine even in normal rats. Therefore, urea may play an important role in this abnormal metabolism of arginine and citrulline in patients with renal failure. We have already reported that arginine, citrulline and ornithine plus ammonia, all of which increase arginine, an NO precursor, inhibit GSA synthesis in the presence of urea [13]. Urea decreases arginine which may lead to limited NO synthesis in hepatocytes because the arginine level is already

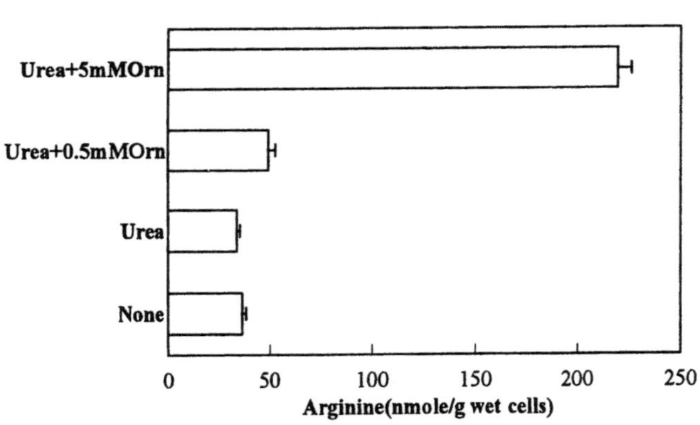


*Fig. 3.* Effect of urea and/or ornithine on arginine level in isolated rat hepatocytes. Hepatocytes were incubated for 20 min as described in 'Materials and methods'.

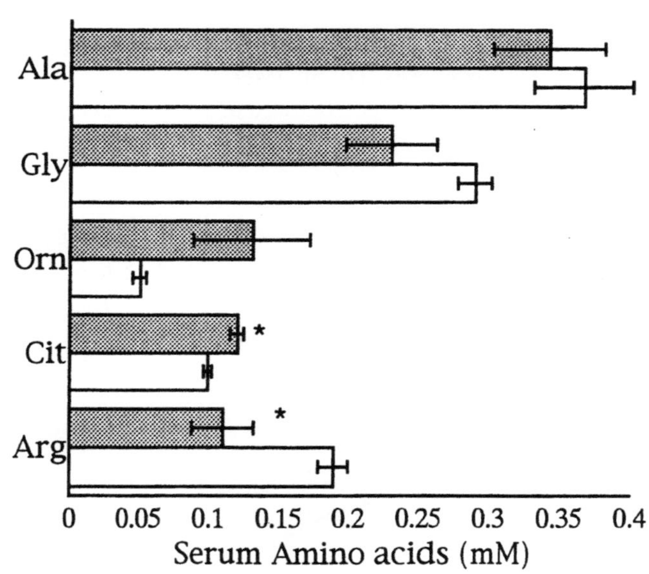


*Fig. 5.* Effect of urea on serum amino acids level. Serum amino acids without urea was represented by open column and with urea was represented by dotted column. Values were obtained from 5 rats. *Represented p less than 0.05.

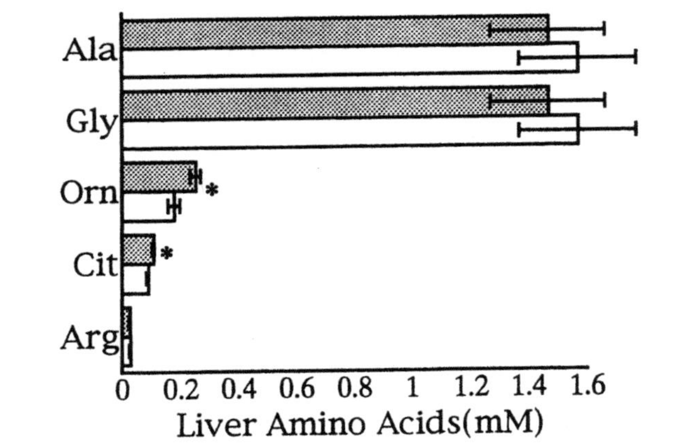


*Fig. 6.* Effect of urea on liver amino acids level. Liver amino acids without urea was represented by open column and with urea was represented by dotted column. Values were obtained from 5 rats. *Represented p less than 0.05.

low as a result of the activity of arginase. We showed that GSA is formed from ASA and the hydroxyl radical or SIN-1, which generates NO and superoxide simultaneously. However, SIN-1 at higher concentrations inhibited GSA synthesis from ASA [11]. Excess of NO, compared with superoxide anion, has been reported at the higher concentrations of SIN-1 [15]. The reason for the reduced GSA synthesis at higher concentrations of SIN-1 may be explained as a result of the scavenging of the hydroxyl radical by NO resulting in nitrous acid. Therefore, it is suggested that hydroxyl radical generation from the superoxide anion and NO may be regulated by the amount of NO generated from arginine in hepatocytes. Both the complete lack of NO, which does not generate the hydroxyl radical from the superoxide anion, and the excess of NO, which scavenges the hydroxyl radical, may lead to decreased GSA synthesis. The highest hydroxyl radical generation is observed at a specific ratio of NO to superoxide.

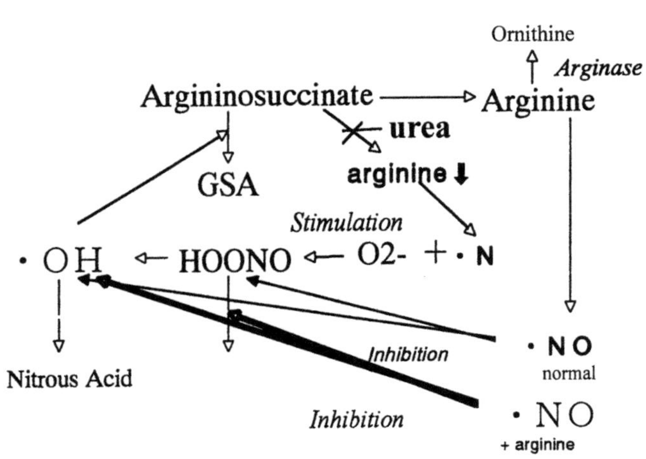


*Fig. 7.* A possible mechanism of urea stimulated GSA synthesis through arginine.

Urea may decrease the NO level at conditions suitable to the generation of the hydroxyl radical. These process is shown in Fig. 7. Patients with renal failure, who have high levels of urea, are felt to be exposed to increased oxidative stress [16, 17]. The mechanism for this increased oxidative stress has not been elucidated. Inhibition of arginine synthesis by urea may explain at least a part of this oxidative stress in the patients with renal failure. We reported that removal of urea by hemodialysis reduced this oxidative stress [18]. Moreover, it has been reported that a single session of hemodialysis changes the oxidized form of albumin to the reduced form [19]. Restriction of protein intake leads to both a decrease in urea levels and a decrease of arginine supply. Our study suggests that the supply of enough arginine as a source of NO in addition to protein restriction is a more effective way to prevent the progression of renal failure than dietary restriction alone. The most important finding of this study is that GSA may be a useful indicator of the hydroxyl radical generation *in vivo* and it can be detected using urine in both healthy and uremic subjects.

## Acknowledgements

We thank Professor Burton D. Cohen (Department of Medicine, Albert Einstein College of Medicine) for his critical comments.

## References

1. Bonas J, Cohen B, Natelson S: Separation and estimation of certain guanidino compounds: Application to human urine. Microchem J 7: 63–77, 1963
2. Horowitz HI, Stein IM, Cohen BD, White JG: Further studies on the platelet-inhibitory effect of guanidinosuccinic acid and its role in uremic bleeding. Am J Med 49: 336–345, 1970
3. Giovannetti G, Cioni L, Balestri OL, Biagini M: Evidence that guanidines and some related compounds cause haemolysis in chronic uremia. Clin Sci 34: 141–148, 1968
4. D'Hooge R, Pei YQ, Marescau B, De Deyn PP: Convulsive action and toxicity of uremic guanidino compounds: Behavioral assessment and relation to brain concentration in adult. J Neurol Sci 112: 96–105, 1992
5. D'Hoog R, Raes A, Lebrun P, Diltoer M, Van Bogaert PP, Manil J, Colin F, De Deyn PP: N-methyl-D-aspartate receptor activation by guanidinosuccinate but not by methylguanidine: Behavioural and electrophysiological evidence. Neuropharmacology 35: 433–440, 1996
6. Stein IM, Cohen BD, Kornhouser RS: Guanidinosuccinic acid in renal failure, experimental azotemia and in bone errors of the urea cycle. N Eng J Med 280: 926–930, 1969
7. Kopple JD, Gordon SI, Wang M, Swendseid ME: Factors affecting serum and urinary guanidinosuccinic acid level in normal and uremic subjects. J Lab Clin Med 90: 303–311, 1977
8. Aoyagi K, Ohba S, Narita M, Tojo S: Regulation of biosynthesis of guanidinosuccinic acid in isolated rat hepatocytes and *in vivo*. Kidney Int 24(supple16): S224–S228, 1983

9. Aoyagi K, Nagase S, Tomida C, Takemura K, Akiyama K, Koyama A: Synthesis of guanidinosuccinate from argininosuccinate and reactive oxygen *in vitro*. Enzyme Prot 149: 199–204, 1996
10. Aoyagi K, Nagase S, Gotoh M, Akiyama K, Satoh M, Hirayama A, Koyama A: Role of reactive oxygen and argininosuccinate in guanidinosuccinate synthesis in isolated rat hepatocytes. Enzyme Prot 49: 205–211, 1996
11. Aoyagi K, Akiyama K, Shahrzad S, Tomida C, Hirayama A, Nagase S, Takemura K, Koyama A, Ohba S, Narita M: Formation of guanidinosuccinic acid, a stable nitric oxide mimic, from argininosuccinic acid and nitric oxide-derived free radicals. Free Radic Res 31: 59–65, 1999
12. Menyhart J, Grof J: Urea as a selective inhibitor of argininosuccininate lyase. Eur J Biochem 75: 405–409, 1977
13. Aoyagi K, Shahrzad S, Iida S, Tomida C, Hirayama A, Nagase S, Takemura K, Koyama A, Ohba S, Narita M, Cohen BD: Role of nitric oxide in the synthesis of guanidinosuccinic acid, an activator of the N-methyl-D-aspartate receptor. Kidney Int 78(suppl): S93–S96, 2001
14. Swendseid ME, Wang M, Vyhmeister I, Chan W, Siassi F, Tam CF, Kopple JD: Amino acid metabolism in the chronically uremic rat. Clin Nephrol 3: 240–246, 1975
15. Hamaguchi N, Hayashi K, Endoh H, Niki E: Dual action of nitric oxide on lipid oxidation and antioxidant. Seikagaku 68: (abstr), 1015, 1996
16. Kuroda M, Asaka S, Tofuku Y, Takeda R: Serum antioxidant activity in uremic patients. Nephron 413: 293–298, 1985
17. Aoyagi K, Nagase S, Narita M, Tojo S: Role of active oxygen on methylguanidine synthesis in isolated rat hepatocytes. Kidney Int 22(suppl): S229–S233, 1987
18. Nagase S, Aoyagi K, Hirayama A, Gotoh M, Ueda A, Tomoda C, Kanemaki T, Nagai Y, Kikichi H, Koyama A: Favorable effect of hemodialysis on decreased serum antioxidant activity in hemodialysis patients demonstrated by electron spin resonance. J Am Soc Nephrol 8: 1157–1163, 1997
19. Sogami M, Era S, Nagaoka S, Kuwata K, Kida K, Miura K, Inouye H, Suzuki E, Hayano S, Sawada S: HPLC-studies on nonmercapt-mercapt conversion of human serum albumin. Int J Pept Prot Res 25: 398–402, 1985

*Molecular and Cellular Biochemistry* **244:** 17–30, 2003.


# Smooth muscle and NMR review: An overview of smooth muscle metabolism

Shinsuke Nakayama[1] and Joseph F. Clark[2]

[1]*Department of Cell Physiology, Nagoya University Graduate School of Medicine, Nagoya, Japan;* [2]*Department of Neurology, University of Cincinnati Medical Center, Cincinnati, OH, USA*

## Abstract

Nuclear magnetic resonance (NMR) is a non-invasive technique which allows us to examine the biochemical, physiological and metabolic events occurring inside living tissue; such as vascular and other smooth muscles.

It has been found that the smooth muscle metabolism is compartmented such that mitochondrial function fuels contraction and that much glycolytic ATP production is used for membrane pumps. Using NMR we have been able to observe the ATP and phosphocreatine (PCr) concentrations and estimate the ADP concentration, as well as flux through the creatine kinase (CK) system. It has also been found that the smooth muscle metabolism is able to maintain ATP concentration in the absence of mitochondrial function (cyanide inhibition). Therefore, the vessels are able to adapt to metabolic demands as necessary.

NMR is versatile in the information it can provide because it has also yielded important contributions with regard to the intracellular pH and ionic status. For example, the intracellular free $Mg^{2+}$ ($[Mg^{2+}]_i$) can be measured with NMR simultaneously with ATP concentrations and NMR has shown us that the $[Mg^{2+}]_i$ is highly protected in the muscle (within confined range), but also responds to the environment around it.

In this review we conclude that NMR measurements of smooth muscle research is a useful technique for assessing chronic and acute changes that occur in the tissue and during diseases. (Mol Cell Biochem **244**: 17–30, 2003)

*Key words*: smooth muscle, energetics, creatine kinase, $Mg^{2+}$, pH

*Abbreviations*: PPA – phenylphosphonic acid; PSS – physiological saline solution; GPC – glycerol phosphocholine; PCr – phosphocreatine; Cr – creatine; NMR – nuclear magnetic resonance; CK – creatine kinase; Mi-CK – mitochondrial creatine kinase; PME – phosphomonoesters; NTP – nucleoside triphosphates; $NAD^+$/NADH – nicotinamide adenine dinucleotide; TMSPS – 3-(Trimethylsilyl) propane-sulfonic acid; β-GPA – β-guanidinopropionic acid; FBP – fructose 1,6 bisphosphate; DOG-6P – 2-deoxy-glucose-6-phosphate; $K_D$'MgATP' – apparent dissociation constant of MgATP, $K_D$'MgATP'; HPLC – high-performance liquid chromatography; IAA – iodoacetate

## 1. Preface

Smooth muscle is generally considered to have a slower metabolism and lower concentration of metabolites compared to striated muscles such as the heart and skeletal muscle. This distinction has sometimes resulted in smooth muscle studies being conducted after striated muscle studies. However, this does not mean the smooth muscle metabolism is not novel or dynamic. For example, the metabolism of vascular smooth muscle is functionally compartmented with what appears to be differential control for aerobic and anaerobic glycolysis as well as oxidative phosphorylation [1–6]. NMR studies of living tissues sometimes assumes that the cell's metabolites will be well mixed during the duration of the experiment. Although smooth muscle cells are small compared to most striated muscles, the smooth muscle metabolites and energetic processes have been shown to be highly organized and structured [3, 5, 6]. Therefore the smooth muscle system provides a unique environment with multiple factors contributing to its metabolism and metabolic control, such as homeostatic

*Address for offprints*: J.F. Clark, Department of Neurology, University of Cincinnati Medical Center, Cincinnati, OH 45267-0525, USA (E-mail: joseph.clark@uc.edu)

control and ionic regulation. We hope this review is useful for better understanding not only in smooth muscle metabolism itself, but also in other related tissues and organs, for example brain with respect of creatine kinase.

## 2. Metabolism studies

### 2.1. $^{31}P$ *NMR spectrum*

In almost all smooth muscles measured with $^{31}$P-NMR, there are six major peaks easily resolvable depending on their chemical shifts: PME (phosphomonoesters), Pi (inorganic phosphate), PCr (phosphocreatine), and γ-, α- and β-atoms of NTP (nucleoside triphosphates). The right hump in the α-NTP peak arises from $NAD^+$/NADH (nicotinamide adenine dinucleotide). In most cases the PME peak can be divided into two subpeaks: PME1 and PME2. From their chemical shifts, these two peaks are considered to mainly arise from PEt (phosphorylethanolamine) and PCh (phosphorylcholine), respectively. The observations of the PDE (phosphodiester) peaks varies greatly from tissue to tissue [7].

The peak area corresponds to the concentration of each compound after correcting for each saturation factors. The γ- and α-NTP peaks also contain contributions of NDPs (nucleotide diphosphates), such contribution would become significant especially under metabolic inhibition [8]. In this review the b-NTP peak is treated as the ATP concentration ([ATP]) in the cytoplasm, although it contains 20–30% of other NTP, such as GTP etc. [9, 10].

The ratio of [PCr] to [ATP] ([PCr]/[ATP]) is often used to an index of cellular energy level. As an overview of smooth muscle metabolism, it is easy to point out that [PCr]/[ATP] in various smooth muscles ranged widely even under control conditions. Interestingly, [PCr]/[ATP] seems to be related with physiological functions of smooth muscles [10]: Fast phasic smooth muscles such as guinea-pig taenia caeci and pig urinary bladder have high [PCr]/[ATP] (1.4–1.7), while it is quite low (0.5–0.7) in tonic smooth muscles, such as pig carotid artery and sheep aorta; In the slow phasic smooth muscles of guinea-pig stomach, [PCr]/[ATP] (0.9) is between fast phasic and tonic smooth muscles (Fig. 1, Table 1). Since this wide variation of [PCr]/[ATP] is seen in resting conditions (even when [Pi] is low,) mechanisms to regulate the cellular creatine (Cr) level (e.g. creatine transporters) may be linked with physiological functions of smooth muscle.

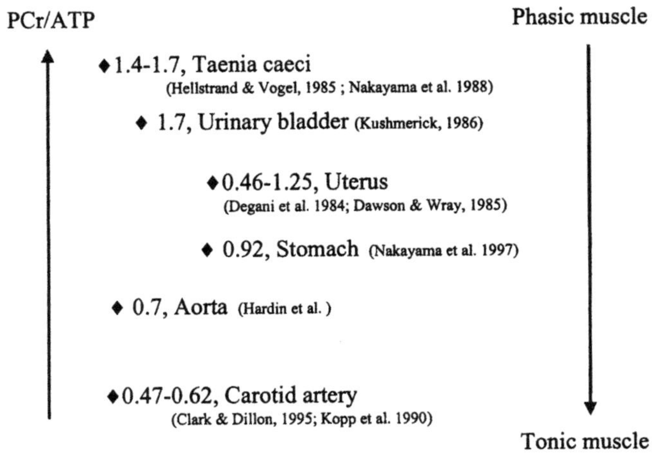


*Fig. 1.* Apparent relationship between PCr:ATP and physiological functions of smooth muscle. In This figure we see the relative relationship between the energy metabolism of the smooth muscle and its functional characteristic of phasic versus tonic. The rapid contracting and relaxing taenia caeci has a relative high PCr:ATP ratio and is a phasic smooth muscle. Whereas the tonic carotid artery has a much lower PCr:ATP ratio. Thus, the NMR measurements of smooth muscle metabolism reflect the function of the tissue.

### 2.2. *Kinetic analysis (saturation transfer)*

#### 2.2.1. *Technical aspects*

Magnetic resonance technology can measure, quantitatively and qualitatively, exchange of subgroups between molecules. For example, in muscle energetics the exchange of phosphate between ATP and PCr via creatine kinase (Eq. 1) can be quantified using saturation transfer. Whereas this same exchange can qualitatively be demonstrated using 2D nuclear-overhauser exchange (NOE) measurements.

$$[PCr] + [ADP] \longleftrightarrow [ATP] + [Cr] + [H^+]. \quad (1)$$

In Fig. 2 a representative series of spectra obtained during a saturation transfer experiment is shown. The * represents the position of specific irradiation to 'saturate' the peak of interest. This saturation occurs because a net randomization of nuclear spins caused by the irradiation. It must be emphasized that randomization of nuclear spins does not mean that the molecule is changed or that the enzyme kinetics are altered. The nuclei have simply been 'labeled' with a change in spin. Because the phosphate of PCr or ATP retains the randomization of spins when it is exchanged to the other molecule via CK, the height of the peak is concomitantly decreased. The decrease in peak height is due to the phosphate resonance now being represented by both the saturated spin and the non-saturated species. The duration of the saturation is determined by the T1 relaxation time of the species. Therefore the T1 is quantitatively changed according to the change in T1, and this change in T1 is due to the native T1 relaxation time and the contribution to the T1 from the other species. Therefore, the pseudo first order rate constant of exchange can be determined with these data.

*Table 1.* Intracellular concentrations of phosphorus compounds (mM) and pH in smooth muscles measured with $^{31}$P-NMR and chemical analyses

| | [PCr] | [ATP (NTP)] | [Pi] | [PME] | [ADP] | $pH_i$ | Condition |
|---|---|---|---|---|---|---|---|
| Kopp *et al.*, 1990 [68] | | | | | | | (37°C) |
| Pig carotid artery | 0.62# | 1# | 0.58# | 2.02# | | 7.02 | $^{31}$P-NMR measurement of chemical extract |
| Clark and Dillon, 1995 [60] | | | | | | | |
| Pig carotid artery | (0.65) | (1.39) | | | 0.031* | | (37°C) |
| Hardin *et al.*, 1992 | | | | | | | |
| Sheep aorta [81] | (0.14) | (0.20) | | | | 7.05 | Room temperature |
| Kushmerick *et al.*, 1986 [9] | | | | | | | |
| Rabbit urinary bladder | 1.3# | 1# | 1.20# | 1.60# | | 7.10 | (25°C) |
| | (1.64) | (0.97) | (2.02) | (1.09) | (0.23) | 7.10 | |
| Uterus | (1.29 | (1.19) | (2.82) | | (0.28) | 7.01 | |
| Nakayama and Smith (unpublished observation) | | | | | | | |
| Guinea-pig urinary bladder | 1.7# | 1# | | | | | (32°C) |
| Degani, 1984 [15] | | | | | | | |
| Rat uterus | 2.1 | 4.6 | 0.8 | 5.4 | 0.04* | | (4°C) |
| Dawson and Wray, 1985 [20] | | | | | | | |
| Rat uterus | 3.0 | 2.4 | 1.5 | 6.3 | | 7.09 | Non-pregnant (37°C) |
| | 4.2 | 3.1 | 1.6 | 6.6 | | 7.08 | Pregnant |
| | 2.3 | 2.3 | 2.5 | 9.8 | | 6.83 | Post-partum |
| Vogel et al., 1983 [19] | | | | | | | |
| Guinea-pig taenia caeci | 1.5–2.0# | 1# | | | | 7.0 | (23°C) |
| Hellstrand and Vogel, 1985 [25] | | | | | | | |
| Rabbit urinary bladder | 1.9 | 1# | | | | 7.0 | (23°C), $Ca^{2+}$-free medium |
| | (1.76) | (0.79) | (1.30) | | (0.15) | | |
| taenia caeci | 1.4 | 1# | | | | 7.0 | (23°C), $Ca^{2+}$-free medium |
| | (1.07) | (0.61) | (0.48) | | (0.18) | | |
| Nakayama *et al.*, 1988 [26] | | | | | | | |
| Guinea-pig taenia caci | 1.71# | 1# | 0.58# | | | 7.05 | (32°C) |
| Nakayama *et al.*, 1995 [7] | | | | | | | |
| Guinea-pig taenia caeci | | | | 1.59 | | | (32°C), $^{31}$P-NMR measurements of chemical extract |
| Nakayama *et al.*, 1997 [10] | | | | | | | |
| Guinea-pig stomach | 0.92# | 1# | 0.54# | 3.09# | | 7.08 | (32°C) |
| Yoshizaki *et al.*, 1987 [13] | | | | | | | |
| Bull frog stomach | 4.8 | 3.6 | 2.4 | 3.0 | | 7.27 | (20–25°C) |
| | (3.1) | (2.6) | | | | | |

Asterisks indicate [ADP] calculated assuming equilibrium for the creatine kinase reaction. In some NMR measurements, the concentrations of phosphorus compounds are expressed relative to that of ATP (indicated by #). In the NMR measurements of Dawson and Wray, the concentrations of phosphorus compounds are expressed in mM. In other NMR measurements and chemical analyses, the values are expressed in μmoles/g wet weight.

Quantification of this exchange is achieved by applying the following formula: Kf = –(1/T1)(1/T1*). Where T1 is the relaxation time and T1* is the relaxation time during saturation and Kf is the pseudo first order rate constant in s-1. the pseudo first order rate constant can be used for determining flux because Kf times the concentration of the representative

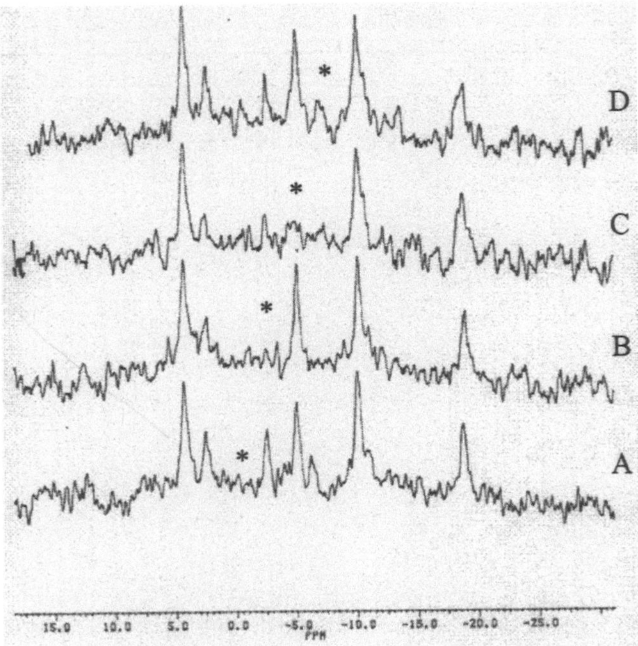


*Fig. 2.* In this figure we see a representative series of NMR spectra showing the sequence of a typical saturation transfer experiments. The bottom trace (A) is the initial $^{31}P$ NMR spectrum and recalling the information presented in Fig. 1 the PCr:ATP ratio is consistent with that previously reported. There is a saturation pulse being applied up filed of the PCr peak to control for non-specific saturation. In spectrum B there has been a continuous wave saturation applied to the PCr resonance. Thereby making the NMR relatively unable to receive the PCr's signal. Spectrum C has the saturation applied to the γ-ATP peak. Again the peak is essentially not 'visible' to the NMR and we also see a concomitant decrease in the PCr resonance. Finally, in Spectrum D there is a saturating pulse applied between the γ-ATP and β-ATP to again control for non-specific saturation effects. The data from these experiments can be used to calculate the molecular exchange between ATP and PCr via the creatine kinase reaction.

metabolite is equal to the flux in mM/s. For example, if Kf for PCr to ATP flux was 0.1 and the concentration of PCr ([PCr]) was 5 mM, the flux would be 0.5 mM/s: $0.1 \times 5 = 0.5$.

Saturation transfer is not the only technique capable of quantifying molecular exchange between molecules. Inversion recovery is also capable of measuring the kinetics of exchange. In this technique we selectively invert a peak of interest with a selective 180° pulse. After the selective inversion, the peak recovers according to its T1 relaxation kinetics. However when there is exchange between molecules, the peak it exchanges with falls and then recovers [16]. This 'recovery' occurs due to its relaxation kinetics and the amount of exchange between the two molecules.

There are two main techniques used for saturating a peak of interest. The most commonly used technique is continuous wave (CW) saturation. This is when a continuous, narrow band of rf irradiation is applied to the peak of interest. Generally this needs to be applied for a length of time 5× longer than the T1 to ensure complete saturation. It is also advisable for the saturation transfer experiment to be performed while using a deuterium lock to avoid filed drift which could lead to incomplete saturation.

An alternate method for saturating a peak of interest is the DANTE pulse sequence. The mechanism for achieving this saturation is beyond the scope of this article but a more complete discussion can be found in Clark *et al.* [11].

Qualitative assessment of the presence of kinetic exchange between molecular species can be obtained using 2D NMR. 2D NMR spectroscopy generally obtains spectroscopic information in 2 frequency domains. For determining exchange of species such as phosphate between PCr and ATP 2D NOE is used. NOE exchange is not as quantitative as saturation transfer techniques, but it has an advantage over saturation methods because it can detect *multiple* exchanges simultaneously. This is important in a system with multiple peaks and unknown exchanges, but its utility for smooth muscle exchanges is less clear.

### 2.2.2. *Creatine kinase activity and compartmentation*

Using magnetization transfer techniques Clark and Dillon [11] found that the vascular smooth muscle CK reaction is largely at equilibrium with little relative contribution of ATPase activity at rest; resulting in a 2 site exchange mechanism. However, during pathological activation using an *in vitro* model of cerebral vasospasm [12], we have found that the ATPase contribution can increase strikingly (unpublished data of J.F. Clark *et al.*). This is not surprising because the dynamic energy demands of the vessels would suggest that ATPase activity could approach that of CK activity but might not surpass the CK buffering capacity thereby avoiding an induction of a rigor like state with excess ADP.

Yoshizaki *et al.* [13] performed NMR exchange experiments on smooth muscle of the bullfrog stomach. They determined that the flux through CK to be 0.77 mmol/g ww/s and that this flux was 100 times faster than the ATP turnover rate. They concluded that the CK reaction was at equilibrium because the oxygen consumption measured in the bullfrog stomach was 100 fold slower than the CK flux. They also reported a discrepancy in the forward (PCr → ATP) and reverse (ATP → PCr) fluxes of 1.7. This discrepancy was attributed to compartmentation of ATP pools or other reactions involving ATP. Compartmentation of nuclear species contributing to the relaxation calculations would cause a systematic decrease in the determination of T1 (transverse relaxation time). This effect would also be dependent upon the relative and total size of the metabolite pool effected.

Under anaerobic conditions [PCr] and [ATP] fell in the bullfrog stomach, Yoshizaki *et al.* [13] concluded that the rate of CK was too rapid to be shifted from equilibrium to allow for the simultaneous fall in [ATP]. They again suggested that their results could be explained by compartmentation of adenine nucleotides in the bullfrog stomach muscle. Thus implying that the CK reaction was buffering one ATP compartment at the expense of a fall in [ATP] at another compartment.

Degani *et al.* [14] used steady state saturation transfer (SSST) and DANTE for inversion transfer (IT) to study the kinetics through the CK reaction in the immature rat uteri. After injection of estrogen (24 h) there was an increase in PCr and an increase in the flux between PCr and ATP. Flux from ATP to PCr could not be observed in these experiments because no exchange of magnetization was seen in the ATP resonance when PCr was saturated. Nonetheless, they assumed equilibrium of the CK reaction to estimate the flux. Interestingly, there was a significant decrease in the PME resonance after estrogen injection compared to control.

Oxygen consumption was 1.5 ml/mg dry weight/h for control and injected uteri. This is not surprising because it would require a substantial synthesis of mitochondrial oxidative enzymes to see an increase in oxygen consumption. There was no increase in mitochondrial CK but there was however synthesis of BB-CK. An increase in total CK activity as well as BB-CK was observed by Clark *et al.* [5] and is consistent with the increased flux from PCr to ATP seen in this relatively short time course.

The conclusion drawn from these studies is that flux through the CK reaction increase only 24 h after injection of estrogen in parallel with an increase in CK activity. There was no change in mitochondrial CK or oxidative phosphorylation. Therefore the CK system is not responding in parallel to estrogen stimulation. But, the increase in BB-CK does parallel the increase in flux observed between PCr and ATP.

The results of Degani *et al.* [14, 15] were interesting in that they produced an increase in CK activity with an increase in flux from PCr to ATP. The isoenzyme that increased in this experiment is predominantly BB-CK. BB-CK has been found to be the isoenzyme localized at the contractile proteins in the guinea pig myometrium [5] In this paper the CK reaction was assumed to be in equilibrium to estimate the ATP to PCr flux. On the other hand, Yoshizaki et al. [13] reported a discrepancy between the flux of PCr → ATP compared to ATP → PCr of 1.7. Thus the assumption of equilibrium for estimating the reverse flux must be made with caution. ATP to PCr exchange might be due to the action of Mi-CK and the absence of Mi-CK in the immature rat uteri may result in a loss of visible exchange from ATP → PCr. Indeed the Mi-CK system greatly increases in the myometrium during gestation [5]. If the Mi-CK and oxidative systems increase with gestation, and if Mi-CK is the dominant isoenzyme producing PCr, than a lack of visible exchange may be due to the low Mi-CK activity. Therefore, when the Mi-CK increases with gestation the ATP → PCr exchange may increase as well.

Another explanation regarding the fluxes of the CK reaction may be due to inhomogeneous distribution of adenine nucleotides in the smooth muscle cell [13]. However, Clark and Dillon [16] performed conventional saturation transfer (CST) and multisite saturation transfer (MST) on the porcine carotid artery. They found that CK was indeed at equilibrium and that even under stimulated conditions, the ATPase was not rapid enough to be observed in the exchange experiments.

In early $^{31}$P-NMR studies, in guinea-pig taenia caeci, Verume and Nicolay [17] observed a large fall in [PCr] during contraction with relatively small changes in [ATP]. The [Pi] varied inversely with the change in [PCr] and there was significant acidification, down to a pH of 6.5 during stimulation. The [ADP] calculated from the Keq of CK was 0.7 mmol/g during stimulation. This is the highest reported value of ADP from NMR data we have found in the literature [18]. During contraction both [PCr] and [ATP] decreased though the [ATP] was buffered because its fall was less than that of PCr. They attributed this fall in [ATP] (despite of the presence of PCr) to the relatively low specific activity of CK in the guinea pig taenia caeci. However [ATP] fell only when [PCr] fell to 'very low levels'. Despite the striking fall in [PCr] and [ATP], the tissue metabolite concentrations, as well as function, returned to control levels upon relaxation. On the other hand, Vogel *et al.* [19] studied the effect of inhibition with cyanide on the taenia coli at 23°C. They found that the tissue was indeed stable for 24 h in the absence of perfusion Pi. They were not able to determine the pHi because intracellular Pi was not observable. The tissue remained viable with a [PCr] to [ATP] ratio of 1.5–2.0 throughout the experiments. They attempted to quantify [ADP] by subtraction of β-ATP peak area from γ-ATP peak area, and found that [ADP] was below the resolution of their system. They did however observe a striking fall in [PCr] with the $^{31}$P-NMR during contraction.

### 2.3. *Specific smooth muscle studies*

#### 2.3.1. *Uterine smooth muscle*

Several studies from several groups have used NMR to study uterine energetics [14, 20]. In the rat uterus, Dawson and Wray [20] have found that PCr is increased by 40% during gestation while [ATP] is up 30%. During gestation there is also an increase in total CK activity [5]. The increase in CK, [PCr] and [ATP] is concomitant with the increased flux through the CK reaction as measured by magnetization transfer experiments [14]. Magnetization transfer is a technique by which one can determine the rate of molecular exchange through a chemical reaction and is discussed in more detail below. Gestation is characterized by a host of metabolic and energetic changes [5] and NMR is well suited for studying these changes.

During gestation, the activity of CK increases and [PCr] increases as well. Degani *et al.* [15] examined estrogen induced changes in the immature (25 day old) rat uteri. They found that there was an increase in free $Mg^{2+}$ (0.2–0.4 mmol/Kg ww) in the estrogen (17β-estrodiol) treated uteri and a parallel increase in [PCr]/[ATP] (4.6 ± 0.4 and 2.1 ± 0.2). The greatest change in the high energy phosphates occurred between 6 and 24 h after injection of estrogen. Interestingly at 4°C the [PCr] to [ATP] ratio was the greatest. All of their experiments were thus performed at 4°C.

Dawson and Wray [20] studied the effects of pregnancy as well as parturition in the rat uteri. They found similar values for [ATP] as reported by Degani *et al.* [15], but more [PCr] compared to the immature rat uteri at 4°C. There was an increase in [ATP] and [PCr] during gestation with no change in [PME] and [Pi]. Post partum, however, [Pi] and [PME] increased, while [ATP] and [PCr] went down. It is possible that [PME] changes reflect the increased membrane metabolism (Table 1).

Uterine energetics and $pH_i$ have been studied with $^{31}$P-NMR during metabolic inhibition. Cyanide caused intracellular acidification in the pregnant and non-pregnant rat uterus. Inhibition of oxidative phosphorylation also caused an abolition of [PCr] in the non-pregnant rat with a decrease in [ATP] while in the pregnant uterus there was a large decrease in [PCr] and a fall in [ATP] [22].

Smooth muscle contractility may be resistant to $pH_i$ changes because Wray *et al.* [23] found that with cyanide (CN) inhibition in the pregnant uterus when trimethylamine (TMA), (added to prevent $pH_i$ changes caused by CN) there was a large fall in [PCr] and increased Pi with no change in $pH_i$. However upon reperfusion there was significant alkalinization and a recovery of [PCr]. Treatments with CN nonetheless consistently caused a decrease in contractility regardless of the $pH_i$. Contractility was thus correlated to energetic status rather than $pH_i$.

Significantly more lactate was produced in the pregnant compared to non-pregnant uteri (0.43 ± 0.07 and 0.25 ± 0.09 mmol/g/min respectively: [22]). The rate of lactate production however was not significantly different between the two with cyanide treatment (about 2.0 mmol/g/min). Interestingly, intracellular acidification of the uterus during inhibition of lactate efflux with α-cyano-4-hydroxycinnamate (CHC), was significantly increased. CN and CHC treatments caused an even grater acidification (down to pH 6.8 in the non-pregnant uterus), suggesting an important role for lactate efflux in maintaining $pH_i$.

Stimulating the uterus with 70 mM KCl caused a marked fall in [PCr] and a small decrease in [ATP]. [Pi] increased concomitantly as PCr falls, while $pH_i$ decreased from 7.18 to 6.78. In experiments where glycolysis was inhibited with iodoacetate (IAA), NMR observable PCr and ATP was abolished and this effect was not reversible [22]. This is not surprising because an oxidative substrate such as acetate or pyruvate was not supplied while glucose could not pass through glycolysis. Therefore one can not conclude that glycolysis is essential for uterine function without repeating these experiments in the presence of an oxidative substrate.

Indeed the opposite may be correct because Boehm *et al.* [24] found that in the glucose or pyruvate perfused porcine carotid artery there was a decrease in ATP with no change in NMR observable [PCr]. However, without OAA the contribution from glycogen's glycosyl groups to glycolysis has yet to be determined. Nonetheless, it appears that uterine smooth muscle utilizes oxidative phosphorylation for normal contractile function.

#### 2.3.2. *Gastrointestinal smooth muscle*

In the guinea-pig taenia caeci NMR was used to investigate relationship between high-$K^+$-induced contracture and energy metabolism [25, 26]. The magnitude of high-$K^+$-induced contraction is decreased by repeating application of $Ca^{2+}$ in $Ca^{2+}$-free and substrate-free solutions [27]. This reduction of tension development is accompanied by a decrease in oxygen consumption. Subsequent application of either glucose or β-hydroxybutyrate restores both high-$K^+$-contraction and oxygen consumption. β-Hydroxybutyrate is catabolized by TCA cycle not via glycolytic pathway. Chemical analyses using aequorin fluorescence (reported by Ashoori *et al.* [27]) showed that ATP and PCr contents decrease during repeated high-$K^+$-contractions, and that the subsequent application of either metabolic substrate restore both of them. The authors pointed out a positive correlation between tension development and ATP content. On the other hand, Nakayama *et al.* [26] showed using $^{31}$P-NMR that prolonged exposures to a high-$K^+$ solution in the absence of metabolic substrate gradually decreased both high energy phosphates (ATP and PCr). The subsequent application of glucose caused only a small recovery in [PCr], and the recovery of [ATP] was negligible. This NMR study suggested that the magnitude of tension development does not directly correlate with total ATP content itself, but it is possibly controlled by the local concentrations of these compounds via changes in ATP turnover rate. Also, in this NMR study [ATP] was decreased to less than 50% by exposures to high-$K^+$-solutions in hypoxia (glucose remained). [ATP] was only partially restored by reapplication of normal solution, and the subsequent addition of adenosine completely restored it [26]. This suggested that the drastic reduction of [ATP] observed in $^{31}$P-NMR involves a significant contribution of the reduction of adenosine pool.

In guinea-pig stomach $^{31}$P-NMR has recently been applied to investigate the relationship between metabolic inhibition and slow wave (spontaneous electrical and mechanical) activities [10]. Guinea-pig stomach has spontaneous oscillations (3–6 cycles/min) of the membrane potential, and the contraction is associated with its depolarization phase. When the isolated stomach smooth muscle was exposed to CN, slow waves immediately suppressed and the resting membrane potential is slightly depolarized. Applications of IAA also suppressed slow wave activity, but it took longer (for 10–20 min). The duration of the depolarizing phase (action potential duration: APD) reduced gradually during its applications, and eventually slow waves ceased. The resting membrane potential was not significantly changed throughout. Furthermore, the amplitude of the tonic potential evoked by electric current application was not changed throughout IAA application, and that the reduction of APD was still observed even in the presence of glibenclamide [10], an inhibitor of ATP-sensitive $K^+$ channels. These results suggest that unlike cardiac muscle, the decrease in the APD is not caused by activation of ATP-sensitive $K^+$ channels [10], although in the same smooth muscle tissue the presence of these channels and hyperpolarization upon their activation have been shown using cromakalim [28] which is known to activate smooth muscle type ATP-sensitive $K^+$channels [29, 30].

During applications of IAA, the concentrations of high energy phosphates measured by $^{31}$P-NMR slowly decreased in stomach smooth muscle [10]. After 25 min, at which time slow wave activity would cease, [PCr] and [ATP] decreased to 60 and 80% of the control, respectively, while [Pi] fell below a detectable level. On the other hand, applications of CN decreased [PCr] and [ATP] to 30 and 80%, respectively. The fact that high levels of [ATP] still remained, supports the above hypothesis that no significant contribution of ATP-sensitive $K^+$ channels upon suppression of slow wave activity during metabolic inhibition described above. Also, it has been suggested that nucleoside diphophates (NDP) activate smooth muscle type ATP-sensitive $K^+$ channels [29–31]. Assuming the chemical equilibrium of the creatine kinase reaction, the $^{31}$P-NMR observation by Nakayama *et al.* [10] of significant decrease in [Pi] following IAA application, however, deduces decrease in ADP. Thus, it seems unlikely that the NDP-dependent activation of the $K^+$ channel occurs in this case. It is concluded that slow wave activity in guinea-pig stomach is eliminated by metabolism-dependent mechanism(s) other than ATP-sensitive $K^+$ channels during metabolic inhibition, especially upon exposure to IAA. During any metabolic inhibition, this unique mechanism would preserve cell viability of smooth muscles which possess spontaneous contraction.

2.3.3. *Vascular and other smooth muscles*

Adams and Dillon [32] determined the glucose dependence of energetics and contractility upon the porcine carotid artery. They found a dependence upon glucose for generating tension when the carotid arteries were stimulated with norepenephrine (NE). Application of high $K^+$-solution however was capable of producing maximal contraction even under glucose-free conditions when NE had failed to produce a response. Despite the inability to contract without glucose via NE stimulation, during application of NE for 3 h the porcine carotid artery still had normal levels of high energy phosphate even in the absence of glucose. This study showed that there is a substrate dependence for normal excitation contraction coupling which is independent upon total tissue levels of ATP and PCr. This evidence supports the well documented compartmentalization of metabolism in vascular smooth muscle [1].

Clark and Dillon [33] used $^{31}$P-NMR to determine the toxicity of 3-(Trimethylsilyl) propane-sulfonic acid (TMSPS) in porcine carotid arteries. TMSPS is an analogue of tetramethyl silane (TMS) and might be used as an $^1$H or $^{13}$C NMR marker for experiments in aqueous solution. They showed that [PCr] decreased at concentrations of TMSPS above 3 mM. In parallel experiments 3 mM TMSPS was the upper limit for its action on force maintenance of the carotids. They concluded that up to 3 mM TMSPS could be used as an $^1$H or $^{13}$C NMR marker in the porcine carotid artery and possibly other tissues because of its relatively low toxicity.

Fisher and Dillon [34] demonstrated that phenylphosphonic acid (PPA) was an effective pHe indicator and measure of extracellular volume in the isolated perfused rabbit urinary bladder. They showed that [PPA] from 1–20 mM had no detectable effect on the $^{31}$P NMR visible metabolites and no effect on charbachol induced contractions. PPA was an effective pHe marker with a linear response to pH changes from 6.40–7.60. According to its ability to quickly load, and be rinsed out of the vasculature of the rabbit bladder, they concluded that the PPA was remaining extracellular [34]. PPA has proven useful as a marker of extracellular volume as well as pHe in tissues other than the bladder.

Fisher and Dillon [34] used $^{31}$P-NMR to study the effects of insulin on the isolated perfused rabbit urinary bladder. They perfused the rabbit urinary bladder with a physiological saline solution containing red blood cells with 7 mM glucose and 0.15 mM pyruvate as substrate. They found that perfusion with 0.1 mU/ml insulin caused a 16% increase in PCr and a 12% decrease in [Pi]. There was also an intracellular alkalinization with $pH_i$ going from $6.90 \pm 0.01$ to $7.14 \pm 0.06$ while $pH_e$ (extracellular pH) was not changed. Extracellular volume increased, however, by $54 \pm 15\%$ as seen by the change in the PPA peak area. Free $Mg^{2+}$ however remained constant at about 0.52 mM before and during the insulin treatment. This increased ratio of extracellular volume is probably related to vascular smooth muscle dilation due to insulin perfusion causing hyperpolarization of the vascular smooth muscle. The insulin perfusion, facilitated glucose uptake and increased glycolysis could then produce the increased energy status reported. The energy status was increased by an increase in [PCr] and increased [PCr]/[ATP], because there was no significant change in [ATP]. However no change in free [ADP] was observed in this study. The authors concluded that insulin induced significant changes in the intracellular concentration of phosphorylated metabolites during 1 h of perfusion in the rabbit bladder as well as the extracellular volume and cytoplasmic alkalinization. Part of these changes may be attributed to increased perfusion achieved as evidenced by increased tissue volume, as well as the increased glucose uptake, caused by the insulin treatment. It is doubtful that the increased tissue volume could be explained by insulin causing PPA somehow to enter the cells of the urinary bladder because the PPA still had a rapid, monoexponential washout after insulin treatment.

Boehm *et al.* [24] studied the relationship between substrate and metabolism in the porcine carotid artery using $^{31}$P-NMR. They superfused the carotid arteries with a modified Krebs solution containing glucose or pyruvate as the substrate under fully oxygenated conditions. They found that the creatine analogue β-guanidinopropionic acid (β-GPA) was phosphorylated more rapidly with glucose as the substrate. The rate of phosphorylation was not dependant upon the [ATP], because CN (cyanide) caused only a slight decrease in ATP but a striking decrease in β-GPA-phosphate production. This observation is taken to indicate that oxidative metabolism controls phosphocreatine concentration, while glycolytic metabolism controls ATP. Part of this conclusion is supported by the report of Clark *et al.* [35] where it was reported that β-GPA is not phosphorylated by mitochondrial creatine kinase. They concluded that ATP was compartmentalized in the smooth muscle such that all of it was not readily available for phosphorylating β-GPA.

### 2.4. $^{13}$C and $^{1}$H studies

Carlier *et al.* [36] studied vascular smooth muscle metabolism during renovascular hypertension using multinuclear NMR. Renovascular hypertension was produced using the one kidney one clip method. Septic shock was also used as an experimental model in dogs. $^{31}$P-NMR measurements showed that the [ADP]/[ATP] ratio increased in the hypertensive animals. The ADP concentration in these experiments was estimated by subtracting β-ATP from α-ATP. The ATP/PCr ratio remained constant at about 3.

Aortas isolated from dogs with septic shock were used to acquire $^{13}$C-NMR spectra and they found resonances from CH3-CH2, CH2-CH2-CO, CH2-CH=CH, CH2n, CH3-CH2-CO, and CH=. $^{1}$H-NMR from lyophilized aortas of the septic dog showed resonances from valine, isoleucine, alanine, lactate, glutamate, creatine/PCr, choline, carnitine and taurine. They showed that increased energy demand with norepinephrine stimulation can be observed with $^{31}$P-NMR as a loss of observable PCr and decreased [ATP]. There was also an increase in the PME region, possibly due to the increased glycolytic metabolism with concomitant increase in 'sugar phosphates'. Despite continued stimulation, there is a recovery of the high energy phosphates to prestimulation levels. They speculated, as do others, that vascular smooth muscle energetic changes maybe involved in the pathology of cardiovascular hypertension and that NMR may be useful to follow and analyze such pathological changes.

Exogenously administered fructose 1,6 bisphosphate (FBP) has been reported to protect various tissues and aid in recovery following hypoxia and surgery. The mechanism by which FBP has this action is not known, it may act by being taken directly into the cell or have secondary effects through biophysical actions. Hardin and colleagues [37–41] investigated the effects of exogenous administration of FBP to the porcine carotid artery and concluded that it was metabolised in the cell. They used [1,3-$^{13}$C]FBP and monitored the $^{13}$lactate produced by the artery in the superfusate during hypoxia. The labelled FBP was converted to lactate by the carotid artery during hypoxia and that at high doses of FBP (20 mM) there was a significant protection of vascular tone. They also observed two peaks at 68.7 ppm and 70.1 ppm which have yet to be identified. Therefore they conclude that FBP is metabolised by the cytosolic glycolytic enzymes and has striking benefits to the vessels metabolism and function.

## 3. Ion analyses

### 3.1. $pH_i$ studies

Since Moon and Richards [42] had reported, the Pi peak position ($\delta_{op}$) has been frequently used for estimation of $pH_i$. The $pH_i$ is estimated from the known pH-dependence of the chemical shift of Pi. using a Henderson-Hasselbalch type equation:

$$pH_i = pK + \log_{10}((\delta_{op} - \delta_1)/(\delta_2 - \delta_{op})), \tag{2}$$

where pK is –log of the second dissociation constant of $H^+$ and Pi, and $\delta_1$ and $\delta_2$ are the chemical shifts of $H_2PO_4^-$ and $HPO_4^{2-}$, respectively. Because of large changes in the chemical shift of Pi over the physiological pH range, this method provides a high resolution for $pH_i$ estimation, although the estimated $pH_i$ is averaged over the entire tissue. Another advantage to the use of NMR in estimating $pH_i$ is that this technique is non-invasive. Under normal conditions the $pH_i$ is in various smooth muscles ranges around 7.0–7.3.

The $pH_i$ can be also estimated from the chemical shifts of PME peaks [7] and from the chemical shifts of γ- and β-ATP peaks [43]. The latter method simultaneously provides $[Mg^{2+}]_i$. Although the resolutions of these methods are lower compared to the case of Pi, these methods are useful when the Pi peak is below detectable levels, e.g. when contraction is eliminated in $Ca^{2+}$-free solutions [7, 44, 45], $Na^+$-free solutions [43], and high-$Mg^{2+}$ solutions [45].

In vascular smooth muscle (rat aorta), the $pH_i$ estimated from the Pi peak (NMR study) showed that intracellular acidosis caused by removal of ammonium chloride was associated with transient constriction [46]. On the other hand, in uterine smooth muscle intracellular acidosis caused by the same treatment suppressed spontaneous contraction [47, 48]. This study was done by the same group, but using fluorescent $pH_i$ indicators. These results suggest that the effect of $pH_i$ on contractility varies among smooth muscles.

In uterine smooth muscle, using $^{31}$P-NMR intensive work has been done to investigate the relationship between contractility and changes in $pH_i$ associated with different metabolic conditions. First of all, it has been shown that the $pH_i$ changes depending on the functional state of the uterus: The $pH_i$ of the rat uterus significantly decreases during and just after parturition, and eventually recovers [20]. However, this transient fall of $pH_i$ was not a consequence of the mechanical work itself, because Caesarean-sectioned uteri showed similar changes in $pH_i$. The changes in $pH_i$ seen depending on the functional state were presumed to be due to combined changes in many endocrine levels, i.e. only effects of estrogen [15] did not mimic the changes in $pH_i$ seen during the course of pregnancy and parturition.

Numerous metabolic inhibitors were applied to the uterus, as described above (see Section 2.3.1 for more detailed experiments). Applications of CN, a representative inhibitor for oxidative phosphorylation caused a significant intracellular acidification accompanied by a marked increase in the lactate production rate. Furthermore, when lactate efflux was blocked by α-cyano-4-hydroxycinnamate (CHC), a significant intracellular acidification was observed. This acidification was pronounced by additional applications of CN [22]. These results suggest that the lactate production rate makes a great contribution in decreasing $pH_i$. The author also showed that irrespective of the presence of CN, a glycolytic inhibitor, iodoacetic acid caused acidic changes, suggesting some other mechanisms (and/or factors) to decrease $pH_i$.

To further investigate the effects of $pH_i$ on uterine contractility during metabolic inhibition, a weak base (trimethylamine) was applied in the presence of CN. This treatment nullified intracellular acidosis, but did not restore the contractions, which suggests that changes in the metabolic condition depressed the contraction [23].

Effects of ischemia on uterine contraction was also investigated using a surface-coil. The $^{31}$P-NMR spectrum and uterine pressure were simultaneously measured from intact rat uteri. During a 20 min occlusion of the uterine artery the $pH_i$ dropped significantly accompanied by significant decreases in high energy phosphates (in both [ATP] and [PCr]). These changes were reversed by the subsequent reperfusion of blood flow after 30 min [49, 50]. Further, an *in vivo* NMR study using a surface-coil showed that $pH_i$ decreased by about 0.1 units during a single contraction in rat uterus [51]. This fall of $pH_i$ is thought to cause relaxation via a negative feed back mechanism.

In gastro-intestinal smooth muscles the $pH_i$ was estimated using the chemical shifts of Pi. In guinea-pig stomach [10] application of CN caused intracellular acidosis (approximately 0.2 pH units). On the other hand, IAA caused a transient alkalinization followed by gradual decrease in $pH_i$. However, as the Pi peak was often decreased to undetectable levels, the decrease in $pH_i$ was not confirmed in all preparations. Also, due to inhibition of glycolytic pathway by IAA, PMEs other than PEt and PCh were accumulated in the tissue, thus the PME peaks were not appropriate to estimate pHi in this case. When glycolysis was inhibited with 2-deoxyglucose, the 2-deoxy-glucose-6-phosphate (DOG-6-P) peak appeared in the PME region and gradually became larger during the exposures. However, the DOG-6-P peak was clearly isolated from the other two PME peaks. The chemical shift of DOG-6-P provides $pH_i$ estimation as accurately as Pi over physiological $pH_i$ range. The observation that this PME peak shifted down field, also indicated a gradual decrease in $pH_i$ during prolonged inhibition of glycolysis. This paradoxical intracellular acidosis may be ascribed to that hydrogen ion is released correlating with breakdown of ATP [52] and/or that some $pH_i$-regulatory mechanisms may be energy metabolism-dependent, e.g. ATP-dependence of $Na^+$-$H^+$ exchanger [53].

Several groups reported the effects of high-$K^+$-induced contracture on $pH_i$ in longitudinal smooth muscle of the taenia caeci. Apparent discrepancies among the works may have caused by the experimental condition used by each group. Vermue and Nicolay [17] reported that the $pH_i$ decreased in high-$K^+$ solutions at a room temperature. In contrast, Hellstrand *et al.* [19, 25] reported that $pH_i$ was relatively stable at a room temperature, and using a higher perfusion rate (10–20 ml/min) than that of Vermue and Nicolay [17]. Nakayama *et al.* [26] also examined the effects of high-$K^+$-induced contractures using a similar flow rate (12 ml/min at 32°C). They found a significant decrease in $pH_i$ (to approximately 6.6) after 50 min of exposure to a high-$K^+$-containing with a substrate-free solution. The $pH_i$ partially recovered (to 6.8) after 150 min, may be due to a decrease in lactate production brought about by depletion of glycogen. The subsequent readmission of glucose resulted in a significant decrease in $pH_i$, again. The experiments described for the taenia caeci were carried out in the presence of $Ca^{2+}$. As for the case of other tissues (e.g. skeletal muscle [54]), $Ca^{2+}$ seemed to be an important factor for the lactate production, because under $Ca^{2+}$-free conditions the $pH_i$ was little changed by exposure to high-$K^+$-solutions even in hypoxia. Using a fluorescent pH indicator Obara *et al.* [55] has demonstrated a similar result that hypoxia did not cause any significant decrease in $pH_i$ under a $Ca^{2+}$-free and high-$K^+$ condition.

NMR studies in the guinea-pig taenia caeci provided several lines of supporting evidence for the presence of $pH_i$-regulatory mechanisms. Removal of extracellular $Na^+$ caused a slow, but significant decrease in $pH_i$ [43]. This phenomenon suggest that $Na^+$-dependent mechanisms ($Na^+$-$H^+$ exchange and $Na^+$-$HCO_3^-$ co-transport, reviewed by Aickin, [56] are important in regulating $pH_i$ in smooth muscle. Also, in the presence of amiloride (a blocker for $Na^+$-$H^+$ exchanger), removal of bicarbonate significantly decreased $pH_i$, suggesting an importance of bicarbonate-dependent mechanism(s): $Na^+$-$HCO_3^-$ co-transport and/or $Cl^-$-$HCO_3^-$ exchange.

Contribution of proton consumption in the creatine kinase reaction upon breakdown of PCr (as shown in Eq. 1) has been clearly demonstrated in cardiac [57] and skeletal muscles [58] under metabolic inhibition. However, this effect on $pH_i$ is not significant in smooth muscles, including the guinea-pig taenia caeci, although the creatine kinase reaction does occur during metabolic inhibition. This could be explained by the combination of the following reasons: (1) relatively small amount of PCr (e.g. Butler and Siegman [18]), (2) low time resolution of $^{31}$P-NMR measurement and (3) low creatine kinase activity in smooth muscle [59, 60].

### 3.2. *Magnesium and other ions*

The chemical shift of ATP peaks, especially β-ATP, depends on the ratio of free to metal-bound ATP [61]. Since the apparent dissociation constants of CaATP ($K_D^{\text{'CaATP'}}$) and MgATP ($K_D^{\text{'MgATP'}}$) are nearly the same [61], the effect of $Ca^{2+}$ binding to ATP is considered negligible in this $Mg^{2+}$ measurement under normal intracellular ionic environments because the $Mg^{2+}$ concentration is much greater than the $Ca^{2+}$ concentration. This is an advantage of $^{31}$P-NMR to measure $[Mg^{2+}]_i$ compared to fluorescent dye techniques, i.e. The dissociation constant of mag-fura for $Ca^{2+}$ is ~ 80 times smaller than that for $Mg^{2+}$ so that contribution of intracellular $Ca^{2+}$-dependent fluorescence is not negligible in $Ca^{2+}$-containing solutions [62, 63]. In red blood cells, Gupta *et al.* [64] estimated the intracellular free $Mg^{2+}$ concentration ($[Mg^{2+}]_i$ ) from the observed chemical shift of ATP ($\delta_{oATP}$) using the following equation:

$$[Mg^{2+}]_i = K_D^{\text{'MgATP'}} (\delta_{oATP} - \delta_{fATP}) / (\delta_{bATP} - \delta_{oATP}), \qquad (3)$$

where $\delta_{fATP}$ and $\delta_{bATP}$ are the chemical shifts of free and $Mg^{2+}$-binding ATP, respectively. The observed chemical shift of β-ATP [9, 65] and observed separation between the α- and β-ATP peaks are usually used because the chemical shift of β-ATP is most $Mg^{2+}$-specific, and that of the α-ATP peak is least modulatable with $Mg^{2+}$ [44]. On the other hand, the chemical shift of γ-ATP is affected by both $Mg^{2+}$ and pH, thus, this peak is not preferable to estimate $[Mg^{2+}]_i$ in ordinary NMR experiments.

The chemical shift of the β-ATP peak is affected by pH to some extent. For an accurate estimation of $[Mg^{2+}]_i$, $K_D^{\text{'MgATP'}}$, $\delta_{fATP}$ and $\delta_{bATP}$ should be corrected by $pH_i$, because these parameters are also pH-sensitive [43, 66]. The chemical shifts of PME peaks as well as Pi peak can be used in the correction with $pH_i$ [43, 45, 65]. As described above, the chemical shift of Pi has the greatest resolution in $pH_i$ estimation, but the PME peaks are also useful, for example, under $Ca^{2+}$-free conditions, often used in smooth muscle NMR studies. Using the pH-dependency of the above three parameters, Nakayama *et al.* [43] described simultaneous equations for $[Mg^{2+}]_i$ and $pH_i$, with respect of β- and γ-ATP (Eq. 4β and Eq. 4γ, respectively):

$$[Mg^{2+}]_i = K_D^{\text{'MgATP'}} (pH_i) (\delta_{obATP} - \delta_{fbATP} (pH_i))/(\delta_{bbATP} (pH_i) - \delta_{obATP}), (4\beta)$$

$$[Mg^{2+}]_i = K_D^{\text{'MgATP'}} (pH_i) (\delta_{ogATP} - \delta_{fgATP} (pH_i))/(\delta_{bgATP} (pH_i) - \delta_{ogATP}), (4\gamma).$$

By solving the simultaneous equations derived from the observed chemical shifts of γ- and β-ATP, $[Mg^{2+}]_i$ and $pH_i$ are calculated at the same time [43]. They successfully have demonstrated changes in $pH_i$ as well as $[Mg^{2+}]_i$ during removal and readmission of $Na^+$, as expected that these ions are regulated by $Na^+$-dependent transporters.

The estimated 'normal' $[Mg^{2+}]_i$ using the methods described above ranged between 0.3–0.5 mM in various smooth muscles [9, 15, 43–45, 67, 68]. These values are substantially smaller than those in skeletal muscles estimated with NMR [69, 70]. Also, it is noteworthy that in any type of muscles the estimated normal value of $[Mg^{2+}]_i$ is much lower than the value expected from the electro-chemical gradient across the plasma membrane. Furthermore, it has been pointed out that small changes in $[Mg^{2+}]_i$ around its normal concentration would affect smooth muscle contraction via changes in $Mg^{2+}$-dependent intracellular enzyme activity [71].

Pathological and hormonal changes in $[Mg^{2+}]_i$ were measured in several smooth muscles. In rat uterine smooth muscle [15], it has been reported that applications of estrogen increased $[Mg^{2+}]_i$ by approximately 25% after 1.5 h. Also, the increase in $[Mg^{2+}]_i$ is accompanied by an decrease in [PCr]/[ATP]. These results are consistent with the fact that the equilibrium constant of CK is $Mg^{2+}$-dependent in the concentration range of the normal $[Mg^{2+}]_i$ [72].

$[Mg^{2+}]_i$ regulation was also investigated in many tissues including smooth muscle. In the porcine carotid artery, it has been reported that significant decreases in $[Mg^{2+}]_i$ was observed when the substrate was switched from glucose to pyruvate [71, 73]. Insulin is an important hormonal factor regulating glucose uptake and energy metabolism. In rabbit urinary bladder, insulin had however, little effect on the $[Mg^{2+}]_i$, although it induced small increases in $pH_i$ and [PCr] [74].

As described in Section 2, metabolic inhibition causes significant decrease in [ATP] which is one of major intracellular $Mg^{2+}$ buffering agents. In porcine carotid artery it has been shown that under ischemic conditions the increase in $[Mg^{2+}]_i$ is much less than the estimated release of $Mg^{2+}$ brought about by a decrease in ATP content [9]. This suggests that some mechanisms regulating $[Mg^{2+}]_i$ are operated in smooth muscle, even under metabolic inhibition. In porcine carotid artery, [ADP] was increased by hypoxia and the ADP peaks became visible and were successfully separated from the ATP peaks [8]. The chemical shift of β-ADP can also be used to estimate $[Mg^{2+}]_i$, as long as $[Mg^{2+}]_i$ was kept below 1 mM, i.e. when $[Mg^{2+}]_i$ is above 1 mM, the chemical shifts are not great enough to separate.

In the guinea-pig taenia caeci, the underlying mechanisms for maintaining the low $[Mg^{2+}]_i$ level was intensively investigated. $[Mg^{2+}]_i$ was reversibly increased by removal of extracellular $Na^+$ which would eliminate $Na^+$ gradient across the plasma membrane [43, 45]. The increase in $[Mg^{2+}]_i$ during $Na^+$ removal was potentiated by increasing the extracellular $Mg^{2+}$ concentration, while simultaneous removal of the extracellular $Na^+$ and $Mg^{2+}$ attenuated the increase. During the removal and readmission of extracellular $Na^+$ changes in [ATP] did not correlate with those of $[Mg^{2+}]_i$. These results suggested that a $Na^+$-$Mg^{2+}$ exchange system pumps out $Mg^{2+}$ to maintain the low $[Mg^{2+}]_i$ level using the energy from the $Na^+$ gradient, and that the role of ATP-driven $Mg^{2+}$ pump, if present, is relatively minor under normal conditions. When ouabain was applied to eliminate the $Na^+$ gradient, $[Mg^{2+}]_i$ increased. Furthermore, amiloride, a non-specific blocker for $Na^+$ transporters reversibly increased $[Mg^{2+}]_i$ [65]. In these experiments with $Na^+$ transporter inhibitors, extracellular $Ca^{2+}$ was removed in order to enhance changes in $[Mg^{2+}]_i$. Under exposure to ouabain and amiloride, the activity of $Na^+$-$K^+$ pump was monitored with $^{87}$Rb uptake. These results support the role of $Na^+$-$Mg^{2+}$ exchange in smooth muscle of the guinea-pig taenia caeci.

In the guinea-pig taenia caeci, simultaneous removal of extracellular $Ca^{2+}$ and $Mg^{2+}$ decreased $[Mg^{2+}]_i$, irrespective of the presence of $Na^+$ ($K^+$-substitution) [45]. In addition, amiloride did not prevent the decrease in $[Mg^{2+}]_i$ during the simultaneous removal of divalent cations [65]. These results suggest the presence of another $Mg^{2+}$ pathway. This $Mg^{2+}$ pathway is blocked by $Ca^{2+}$, and through this pathway $Mg^{2+}$ is passively transported. The increase in $[Mg^{2+}]_i$ during $Na^+$ removal was potentiated in the absence of extracellular $Ca^{2+}$. This can be explained by inhibition of the passive $Mg^{2+}$ pathway by $Ca^{2+}$ and competition of $Mg^{2+}$ and $Ca^{2+}$ at the $Na^+$-$Mg^{2+}$ exchanger [43, 65].

From ion flux measurements, the presence of $Na^+$-$Mg^{2+}$ exchange system and its role have been proposed in various tissues [75], although there is some controversial results reported in several tissues and cells. $^{31}$P-NMR measurements obtained in smooth muscles reinforced the hypothesis of $Na^+$-

*Table 2.* Original and re-estimated $[Mg^{2+}]_i$

| | Original $[Mg^{2+}]_i$ (mM) | Kd used (μM) | Free ATP (%) | $pH_i$ | Temp (°C) | Re-est. $[Mg^{2+}]_i$ (mM) | Kd (μM) from Zhang |
|---|---|---|---|---|---|---|---|
| Degani *et al.*, 1984 [15] | | | | | | | |
| Rat uterus | 0.38 | 140 | 27 | (7.4) | 4 | 0.92 | 229 |
| Kushmerick *et al.*, 1986 [9] | | | | | | | |
| Rabbit urinary bladder | 0.46 | 77.6 | 14.1 | 7.10 | 25 | 0.79 | 133 |
| Uterus | 0.4 | 77.6 | 16.2 | 7.01 | 25 | 0.71 | 138 |
| Kopp *et al.*, 1990 [68] | | | | | | | |
| Pig carotid artery | 0.54 | 77.6 | 12.6 | 7.03 | 37 | 0.65 | 94 |
| Nakayama and Tomita, 1990 [44] | | | | | | | |
| Guinea-pig taenia caeci | 0.31 | 41 | 11.2 | 7.07 | 32 | 0.84 | 107 |
| Nakayama *et al.*, 1994 [43] | | | | | | | |
| Guinea-pig taenia caeci | 0.33 | 41 | 11.6 | 7.13 | 32 | 0.81 | 105 |

Original values of $[Mg^{2+}]_i$ are listed with $pH_i$; temperature and dissociation constant of MgATP (Kd) used in each reference. The free ATP level calculated from the data in each reference is also listed. The $[Mg^{2+}]_i$ values were re-estimated with standardised Kd values. Kd values (at 25 and 27°C) were calculated from $pH_i$ (observed in each experiment) using formulae described by Zhang *et al.* (1997). Subsequently, a Kd value for the temperature applied in each experiment was obtained using Van't Hoff isochore.

$Mg^{2+}$ exchange, and proved that the $[Mg^{2+}]_i$ is actually modulatable through $Na^+$-$Mg^{2+}$ exchange, but it takes a substantially longer time compared to other ions, such as $Ca^{2+}$. Using a fluorescent $Mg^{2+}$ indicator (mag-fura) under $Ca^{2+}$-free conditions, supporting evidence for $Na^+$-$Mg^{2+}$ exchange has been published in the same smooth muscle [76].

Comparison between control, Wister-Kyoto and Sprague-Dawley, and spontaneously hypertensive rats has revealed that $[Ca^{2+}]_i$ and $[Na^+]_i$ measured with $^{19}F$ and $^{23}Na$-NMR techniques respectively are significantly higher in hypertensive aorta [77]. The higher level of $[Ca^{2+}]_i$ in hypertensive aorta may be explained by $Na^+$-$Ca^{2+}$ exchange mechanism (an increase in $[Na^+]_i$ would increase $[Ca^{2+}]_i$). $[Mg^{2+}]_i$ estimated from the chemical shift of ATP was slightly, but not statistically significantly lower in hypertensive aorta than in control. Also, in numerous smooth muscles (arteries: [8, 32]; uterus and urinary bladder: [9]), it has been reported that mechanical stretch does not cause significant change in $[Mg^{2+}]_i$ (measured with $^{31}P$-NMR), being consistent with the comparison between hypertensive and normotensive rats. On the other hand, Kopp *et al.* [68] reported a discrepancy that mechanical stretch caused a large increase in $[Mg^{2+}]_i$ in porcine carotid artery. Using the same $^{31}P$-NMR technique, they found that $Mg^{2+}$ increased in the pressurized artery from 0.54 up to 0.99 mM while phosphate resonances remained unchanged. The [PCr]/[Pi] ratio was 1.3 and the pHi was stable at about 7.03. They suggest that the change in $[Mg^{2+}]_i$ caused by stretch may play a role in controlling vascular smooth muscle contractility. The authors did not however address how this change in $Mg^{2+}$ might cause an increase in [ADP] via a shift in the $Mg^{2+}$ dependence upon CK reaction. MgADP has been implicated as a possible regulatory mechanism in vascular smooth muscle contractility [78, 79].

As shown earlier, $[Mg^{2+}]_i$ estimation in $^{31}P$-NMR depends on the apparent dissociation constant of MgATP ($K_D$'MgATP'). Recently, Zhang *et al.* [80] re-evaluated the dissociation constant of MgATP as a function of pH and temperature. Applications of this new estimation to the control conditions of previous studies provide larger $K_D$'MgATP' values than those used previously, whereby the recalculated values of $[Mg^{2+}]_i$ correspondingly increase about by twice (0.7–0.9 mM, Table 2). Removal of extracellular $Ca^{2+}$ causes marginal increases in $[Mg^{2+}]_i$ in guinea-pig taenia caeci (e.g. Nakayama and Nomura [65]). Under this condition the recalculated $[Mg^{2+}]_i$ was 0.8–0.9 mM. This value is comparable to the value estimated using $Mg^{2+}$ dyes [63].

## Acknowledgements

This work was supported by research grants from Goho Life Science International Fund, and for Cardiovascular Diseases from the Ministry of Health and Welfare (Japan to SN), and the NIH HL67186 (JFC).

## References


1. Paul RJ: Functional compartmentalization of oxidative and glycolytic metabolism in vascular smooth muscle. Am J Physiol 244: C399–C409, 1983
2. Lynch RM, Paul RJ: Compartmentation of glycolytic and glycogenolytic metabolism in vascular smooth muscle. Science 222: 1344–1346, 1983
3. Lynch RM, Paul RJ: Compartmentation of carbohydrate metabolism in vascular smooth muscle. Am J Physiol 252: C328–C334, 1987
4. Ishida Y, Paul RJ: Evidence for compartmentation of high energy phosphagens in smooth muscle. Prog Clin Biol Res 315: 417–428, 1989
5. Clark JF, Khuchua Z, Kuznetsov AV, Saks VA, Ventura-Clapier R: Compartmentation of creatine kinase isoenzymes in myometrium of gravid guinea-pig. J Physiol 466: 553–573, 1993
6. Boehm EA, Clark JF, Radda GK: Metabolite utilization and possible compartmentation in porcine carotid artery: A study using β-guanidinopropionic acid. Am J Physiol 268: C628–C635, 1995
7. Nakayama S, Hachisuka T, Itoh K, Matsumoto T, Tomita T: Phosphomonoesters in the guinea-pig taenia caeci: pH-Dependency of the phosphomonoester peaks in $^{31}P$-NMR. Jpn J Physiol 45: 411–422, 1995
8. Fisher MJ, Dillon PF: Direct determination of ADP in hypoxic porcine carotid artery using 31 P NMR. NMR Biomed 1: 121–126, 1988
9. Kushmerick MJ, Dillon PF, Meyer RA, Brown TR, Krisanda JM, Sweeney HL: $^{31}P$ NMR spectroscopy, chemical analysis, and free $Mg^{2+}$ of rabbit bladder and uterine smooth muscle. J Biol Chem 261: 14420–14429, 1986
10. Nakayama S, Chihara S, Clark JF, Huang S-M, Horiuchi T, Tomita T: Consequences of metabolic inhibition in smooth muscle isolated from guinea-pig stomach. J Physiol 505: 229–240, 1997
11. Clark JF, Harris GI, Dillon PF: Multisite saturation transfer using DANTE and continuous wave. Magn Reson Med 17: 274–278, 1991
12. Pyne GJ, Cadoux-Hudson TAD, Clark JF: Force function relations in vascular smooth muscle during cerebral vasospasm. Biophys J 72: A256, 1998
13. Yoshizaki K, Radda GK, Inubushi T, Chance B: $^1H$ and $^{31}P$-NMR studies on smooth muscle of bullfrog stomach. Biochem Biophys Acta 928: 36–44, 1987
14. Degani H, Victor TA, Kaye AM: Effects of 17β-estradiol on high energy phosphate concentrations and the flux catalyzed by creatine kinase in immature rat uteri: $^{31}P$ NMR studies. Endocrinology 122: 1631–1637, 1988
15. Degani H, Shaer A, Victor TA, Kaye AM: Estrogen-induced changes in high-energy phosphate metabolism in rat uterus. Biochemistry 23: 2572–2577, 1984
16. Clark JF, Dillon PF: Porcine carotid artery creatine kinase kinetics using NMR saturation transfer. Biophys J 57: 156a, 1990
17. Verume NA, Nicolay K: Energetics of smooth muscle taenia caecum of guinea-pig: A $^{31}P$ NMR study. FEBS Lett 156: 293–297, 1983
18. Butler TM, Siegman MJ: High-energy phosphate metabolism in vascular smooth muscle. Annu Rev Physiol 47: 629–643, 1985
19. Vogel HJ, Lilja H, Hellstrand P: Phosphorus-$^{31}$NMR studies of smooth muscle from guinea pig taenia coli. Biosci Rep 3: 863–870, 1983
20. Dawson MJ, Wray S: The effects of pregnancy and parturition on phosphorus metabolites in rat uterus studied by $^{31}P$ NMR. J Physiol 368: 19–31, 1985
21. Iyengar MR: Creatine kinase as an intracellular regulator. J Muscle Res Cell Motil 5: 527–534, 1984

22. Wray S: The effects of metabolic inhibition on uterine metabolism and intracellular pH in the rat. J Physiol 423: 411–423, 1990
23. Wray S, Duggins K, Iles R, Nyman L, Osman VA: The effects of metabolic inhibition and intracellular pH on rat uterine force production. Exp Physiol 77: 307–319, 1992
24. Boehm EA, Clark JF, Radda GK: Metabolite utilization in the porcine carotid artery: A $^{31}$P NMR study using the creatine analogue β-guanidinopropionic acid. J Physiol 477: 40P, 1994
25. Hellstrand P, Vogel HJ: Phosphagens and intracellular pH in intact rabbit smooth muscle studied by $^{31}$P-NMR. Am J Physiol 248: C320–C329, 1985
26. Nakayama S, Seo Y, Takai A, Tomita T, Watari H: Phosphorous compounds studied by $^{31}$P nuclear magnetic resonance spectroscopy in the taenia of guinea-pig caecum. J Physiol 402: 565–578, 1988
27. Ashoori F, Takai A, Tokuno H, Tomita T: Effects of glucose removal and readmission on the potassium contracture in the guinea-pig taenia coli. J Physiol 356: 33–48, 1984
28. Katayama N, Huang S-M, Tomita T, Brading AF: Effects of cromakalim on the electrical slow wave in the circular muscle of guinea-pig gastric antrum. Br J Pharmacol 109: 1097–1100, 1993
29. Beech DJ, Zhang H, Nakao K, Bolton, TB: K channel activation by nucleotide diphosphates and its inhibition by glibenclamide in vascular smooth muscle cells. Br J Pharmacol 110: 573–582, 1993
30. Beech DJ, Zhang H, Nakao K, Bolton TB: Single channel and whole-cell K-currents evoked by levocromakalim in smooth muscle cells from the rabbit portal vein. Br J Pharmacol 110: 583–590, 1993
31. Kajioka S, Kitamura K, Kuriyama H: Guanosine diphosphate activates an adenosine-5′-triphosphate-sensitive $K^+$ channels in the rabbit portal vein. J Physiol 444: 397–418, 1991
32. Adams GR, Dillon PF: Glucose dependence of sequential norepinephrine contractions of vascular smooth muscle. Blood vessels 26: 77–83, 1989
33. Clark JF, Dillon PF: Mechanical and metabolic toxicity of 3-(trimethylsilyl) propanesulfonic acid to porcine carotid arteries. Biochim Biophys Acta 1014: 235–238, 1989
34. Fisher MJ, Dillon PF: Phenylphosphonate: A $^{31}$P-NMR indicator of extracellular pH and volume in the isolated perfused rabbit bladder. Circ Res 60: 472–477, 1987
35. Clark JF, Khuchua Z, Kuznetsov AV, Vassileva AV, Boehm E, Radda GK, Saks VA: Actions of the creatine analogue β-guanidinopropionic acid on rat heart mitochondria. Biochem J 300: 211–216, 1994
36. Carlier PG, Grandjean J, Michel P, D'Orio V, Rorive GL: Arterial metabolism as studied *in vitro* by NMR: Preliminary results in normotensive and hypertensive aortas. Arch Int Physiol Biochem 93: 107–118, 1985
37. Hardin CD, Roberts TM: Metabolism of exogenously applied fructose 1,6-bisphosphate in hypoxic vascular smooth muscle. Am J Physiol 267: H2325–H2332, 1994
38. Hardin CD, Roberts TM: Gluconeogenesis during hypoxia in vascular smooth muscle studied by $^{13}$C-NMR. Physiol Res 44: 257–260, 1995
39. Hardin CD, Roberts TM: Compartmentation of glucose and fructose 1,6-bisphosphate metabolism in vascular smooth muscle. Biochemistry 34: 1323–1331, 1995
40. Hardin CD, Roberts TM: Regulation of glycogen utilisation, but not glucose utilisation, by precontraction glycogen levels in vascular smooth muscle. Biochemistry 36: 6954–6959, 1997
41. Hardin CD, Finder DR: Glycolytic flux in permeabilized freshly isolated vascular smooth muscle cells. Am J Physiol 274: C88–C96, 1998
42. Moon RB, Richards JH: Determination of intracellular pH by $^{31}$P nuclear magnetic resonance. J Biol Chem 248: 7276–7278, 1973
43. Nakayama S, Nomura H, Tomita T: Intracellular-free magnesium in the smooth muscle of guinea pig taenia caeci: A concomitant analysis for magnesium and pH upon sodium removal. J Gen Physiol 103: 833–851, 1994
44. Nakayama S, Tomita T: Depletion of intracellular free $Mg^{2+}$ in $Mg^{2+}$- and $Ca^{2+}$-free solution in the taenia isolated from guinea-pig caecum. J Physiol 421: 363–378,1990
45. Nakayama S, Tomita T: Regulation of intracellular free magnesium concentration in the taenia of guinea-pig caecum. J Physiol 435: 559–572, 1991
46. Spurway NC, Wray S: A phosphorus NMR study of metabolites and intracellular pH in rabbit vascular smooth muscle. J Physiol 393: 57–71, 1987
47. Taggart M, Wray S: Simultaneous measurement of intracellular pH and contraction in uterine smooth muscle. Pflügers Arch 423: 527–529, 1993
48. Parratt J, Taggart M, Wray S: Abolition of contractions in the myometrium by acidification *in vitro*. Lancet 344: 717–718, 1994
49. Harrison N, Larcombe-McDouall JB, Wray S: A 31P NMR investigation into the effects of repeated vascular occlusion on uterine metabolites, intracellular pH and force. NMR Biomed 8: 28–32, 1995
50. Harrison N, Larcombe-McDouall JB, Earley L, Wray S: An *in vivo* study of the effects of ischaemia on uterine contraction, intracellular pH and metabolites in the rat. J Physiol 476: 349–354, 1994
51. Larcombe-McDoual J, Buttell N, Harrison N, Wray S: *In vivo* pH and metabolite changes during a single contraction in rat uterine smooth muscle. J Physiol 518: 783–790, 1999
52. Eisner DA, Nichols CG, O'Neill SC, Smith GL, Valdeolmillos M: The effects of metabolic inhibition on intracellular calcium and pH in isolated rat ventricular cells. J Physiol 411: 393–418, 1989
53. Demaurex N, Romanek RR, Orlowski J, Grinstein S: ATP dependence of $Na^+/H^+$ exchange: Nucleotide specificity and assessment of the role of phospholipids. J Gen Physiol 109: 117–128, 1997
54. Dawson MJ, Gadian DG, Wilkie DR: Contraction and recovery of living muscles studied by 31P nuclear magnetic resonance. J Physiol 267: 703–735, 1977
55. Obara K, Bowman PS, Ishida Y, Paul RJ: Effects of hypoxia on $[Ca^{2+}]_i$, $pH_i$ and myosin light chain phosphorylation in guinea-pig taenia caeci. J Physiol 503: 427–433, 1997
56. Aickin CC: Movement of acid equivalents across the mammalian smooth muscle cell membrane. In: Proton Passage Across Cell Membrane. Ciba, Foundation Symposium 139. John Wiley & Sons Ltd., Chichester, 1988, pp 3–22
57. Allen DG, Morris PG, Orchard CH, Pirolo JS: A nuclear magnetic resonance study of metabolism in the ferret heart during hypoxia and inhibition of glycolysis. J Physiol 361: 185–204, 1985
58. Yamada T, Kikuchi K, Sugi H: $^{31}$P nuclear magnetic resonance studies on the glycogenolysis regulation in resting and contracting frog skeletal muscle. J Physiol 460: 273–286, 1993
59. Clark JF, Kemp GJ, Radda GK: The creatine kinase equilibrium free [ADP] and myosin ATPase in vascular smooth muscle cross-bridges. J Theor Biol 173: 207–211, 1995
60. Clark JF, Dillon PF: Phosphocreatine and creatine kinase in energetic metabolism of the porcine carotid artery. J Vasc Res 32: 24–30, 1995
61. Nageswara Rao BD: Phosphorus-$^{31}$NMR of enzyme complexes. In: G.D. Gorenstein (ed). Phosphorus-$^{31}$NMR. Academic Press, Orlando, Florida, 1984, pp 57–103
62. Raju B, Murphy E, Levy LA, Hall RD, London RE: A fluorescent indicator for measuring cytosolic free magnesium. Am J Physiol 256: C540–C548, 1989
63. Tashiro M, Konishi M: Basal intracellular free $Mg^{2+}$ concentration in smooth muscle cells of guinea-pig taenia cecum: Intracellular calibration of the fluorescent indicator fraptra. Biophys J 73: 3358–3370, 1997
64. Gupta RK, Gupta P, Moore RD: NMR studies of intracellular metal ions in intact cells and tissues. Annu Rev Biophys Bioeng 13, 221–246, 1984
65. Nakayama S, Nomura H: Mechanisms of intracellular $Mg^{2+}$ regulation affected by amiloride and ouabain in the guinea-pig taenia caeci. J Physiol 488: 1–12, 1995

66. Bock JL, Wenz B, Gupta RK: Changes in intracellular Mg adenosine triphosphate and ionized $Mg^{2+}$ during blood storage: Detection by $^{31}P$ nuclear magnetic resonance. Blood 65: 1526–1530, 1985
67. Dillon PF: Free intracellular $Mg^{2+}$ in relaxed and contracted porcine carotid artery. J Physiol 371: 161P, 1986
68. Kopp SJ, Barron JT, Tow JP. Phosphatic metabolites, intracellular pH and free $[Mg^{2+}]$ in single intact porcine carotid artery segments studied by $^{31}P$ NMR. Biochim Biophys Acta 1055: 27–35, 1990
69. Gupta RK, Moore RD: $^{31}P$ NMR studies of intracellular free $Mg^{2+}$ in intact frog skeletal muscle. J Biol Chem 255: 3987–3993, 1980
70. Dawson MJ, Gadian DG, Wilkie DR: Muscular fatigue investigated by phosphorus nuclear magnetic resonance. Nature 274: 4271–4275, 1978
71. Dillon PF: $^{31}P$ Nuclear magnetic resonance spectroscopy In: M. Bárány (ed). Biochemistry of Smooth Muscle Contraction. Academic Press, San Diego, 1996, pp 393–404
72. Lawson JW, Veech RL, Effects of pH and free $Mg^{2+}$ on the Keq of the creatine kinase reaction and other phosphate hydrolyses and phosphate transfer reactions. J Biol Chem 254, 6528–6537, 1979
73. Dillon PF: Influence of cellular energy metabolism on contractions of porcine carotid artery smooth muscle. J Vasc Res 37: 532–539, 2000
74. Fisher MJ, Dillon PF: $^{31}P$ NMR study of insulin effects on the isolated perfused rabbit urinary bladder. Magn Reson Med 9: 53–65, 1989
75. Flatman PW: Mechanisms of magnesium transport. Annu Rev Physiol 53: 259–271, 1991
76. Tashiro M, Konishi M: $Na^+$ gradient dependent $Mg^{2+}$ transport in smooth muscle cells of guinea pig taenia cecum. Biophys J 73: 3371–3384, 1997
77. Jelick LA, Gupta RK: NMR measurement of cytosolic free calcium, free magnesium, and intracellular sodium in the aorta of normal and spontaneously hypertensive rat. J Biol Chem 265: 1391–1400, 1990
78. Nishiye E, Somlyo AV, Somlyo AP: The effects of MgADP on cross-bridge kinetics: A laser flash photolysis study of guinea pig smooth muscle. J Physiol 460: 247–271, 1993
79. Fuglsang A, Khromov A, Torok K, Somlyo AV, Somlyo AP: Flash photolysis studies of relaxation and cross-bridge detachment: Higher sensitivity of tonic than phasic smooth muscle to MgADP. J Muscle Res Cell Motil 14: 666–673, 1993
80. Zhang W, Truttmann AC, Lüthi D, McGugan JAS: Apparent $Mg^{2+}$-adenosine 5-triphosphate dissociation constant measured with $Mg^{2+}$ macroelectrodes under conditions pertinent to $^{31}P$ NMR ionized magnesium determinations. Anal Biochem 251: 246–250, 1997
81. Hardin CD, Wisemand RW, Kushmerick MJ: Tension responses of sheep aorta to simultaneous decreases in phosphocreatine inorganic phosphate and ATP. J Physiol 458: 139–150, 1992

*Molecular and Cellular Biochemistry* **244:** 31–36, 2003.


# Methyl group deficiency and guanidino production in Uremia

Burton D. Cohen
*Albert Einstein College of Medicine, Bronx, NY, USA*

## Abstract

Guanidinosuccinic acid (GSA) is one of the earliest uremic toxins isolated and its toxicity identified. Its metabolic origins have remained obscure until recently when a series of studies showed that it arose from the oxidation of argininosuccinic acid (ASA) by free radicals. The stimulus for this oxidation, occurring optimally in the presence of the failed kidney, is the rising level of urea which, through enzyme inhibition, results in a decline in hepatic levels of the semi-essential amino acid, arginine. It is further noted that concentrations of GSA in both serum and urine decline sharply in animals and humans exposed to the essential amino acid, methionine. In this review the argument is advanced that uremics suffer from a defective ability to generate methyl groups due to anorexia, dietary restrictions and renal protein leakage. This leads to the accumulation of homocysteine, a substance known to produce vascular damage. Even in healthy subjects intake of choline together with methionine is insufficient to satisfy total metabolic requirements for methyl groups. In end-stage renal disease, therefore, protein restriction contributes to the build-up of toxins in uremia. Replacement using specific amino acid mixtures should be directed toward identified deficiencies and adequacy monitored by following serum levels of the related toxins, in this case GSA and homocysteine. (Mol Cell Biochem **244**: 31–36, 2003)

*Key words*: guanidinosuccinic acid, argininosuccinic acid, methyl groups, betaine, methionine, uremic toxins

The management of uremia has a bizarre history involving treatments, ingeniously devised, long before we knew what we were treating. Engineers got the jump on the clinicians when dialysis was undertaken in the 1940's for the removal of substances which were totally unidentified except for the somewhat naïve assumption that they were probably excreted by normal kidneys. It is as if the discovery of penicillin predated the germ theory. This success in advance of understanding set off a mad search for the so-called uremic toxins.

The first of these presumed poisons isolated and identified was guanidinosuccinic acid or GSA [1] (Fig. 1). It was found to be a substance manufactured in the human liver in the presence of excess urea [2–4] and in animals following injections of large quantities of urea [3, 5]. It is reduced or eliminated from the circulation by dialysis [5, 6] or by injections or ingestion of methionine [7] (Fig. 2). Its cellular toxicity is demonstrable in a number of systems but most notably as an inhibitor of platelet factor 3 activation and, thus, is a major contributor to the bleeding diathesis of uremia [8–10]. In addition, it is implicated in red cell hemolysis [11], white cell activation [12, 13], myocardial contractility [14–16] and pancreatic beta cell responsiveness [17]. It participates, therefore, in uremic anemia, immunodeficiency, cardiopathy and azotemic pseudodiabetes.

The urea molecule itself was promptly shown not to be a source of GSA [8] and it was felt from the onset, largely due to structural analogy, to be a consequence of the aberrant cleavage of argininosuccinic acid or ASA [2] (Fig. 3). ASA is familiar as a component of the Krebs-Hensleit urea cycle (Fig. 4). Subsequent studies noted that the reaction with ASA is a P450 dependent oxidation [18] (see Table 1) which involves oxygen in the form of free radicals such as the superoxide and hydroxyl ions [19–21].

Several authors [22, 23] call attention to enzyme retroinhibition or feedback control within the urea cycle as a potential stimulus to the aberrant cleavage of ASA. Excess urea is known to suppress ASA lyase interrupting the flow to arginine and potentially creating an accumulation of ASA capable of shunting into GSA under the conditions of mass action. However, in the rare instances of congenital absence of ASA

*Address for offprints*: B.D. Cohen, 160 Crosby Village Road, Eastham, MA 02642, USA (E-mail: burt1071@aol.com)

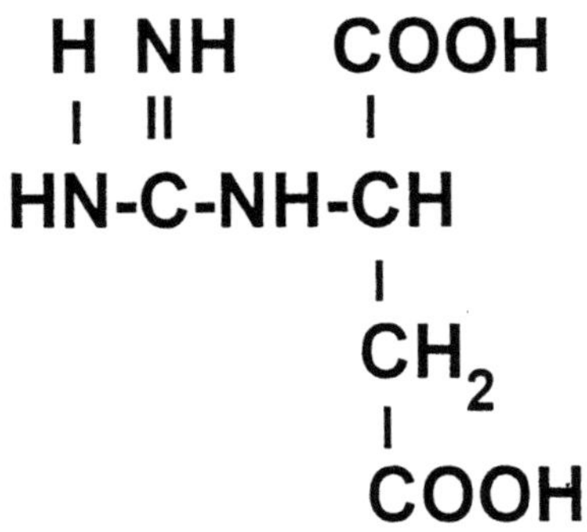

*Fig. 1.* Guanidinosuccinic acid or GSA.

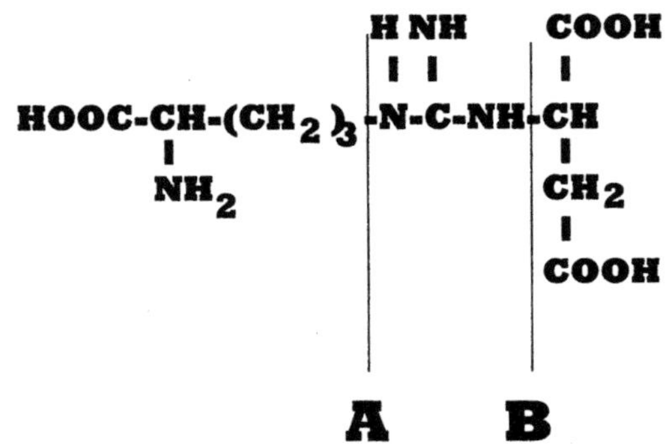


*Fig. 3.* Argininosuccinic acid or ASA. Point B is the normal cleavage point in the presence of ASA lyase. Point A is the proposed point of cleavage.

lyase, known as argininosuccinic aciduria, GSA is not found in the urine. It is similarly absent in all other reported cases of inborn errors of the urea cycle [5, 24, 25] (Fig. 5).

Can oxygen, or more specifically reactive oxygen species, represent the rate limiting factor in the GSA shunt? It has been proposed that uremia is a particularly oxidative state because of the loss of renal clearance of superoxide anions [23, 26].

Oxygen is toxic because it reacts with organic matter. Just about all metabolic processes produce an excess of electrons. When this excess overwhelms the capacity of the electron transport system, superoxides result. In the presence of either oxidized iron or nitric oxide they become hydroxyl radicals which are half of hydrogen peroxide. These free radicals or reactive oxygen species are fragments of molecules energetically predisposed to reconnect with their missing parts and they careen about inside cells like loose cannons damaging proteins, lipids and nucleic acids. All rapidly metabolizing tissues, such as phagocytes engulfing pathogens, produce superoxides. They are, therefore, ubiquitous and a most important insurance against superoxidation and the production of toxins such as GSA is the generation of free radical scavengers. One such protector against the accumulation of hydroxyl radicals is nitric oxide or NO.

In both the liver and the kidney arginine is a pivotal reactant. In addition to its role in protein synthesis, it is the sub-

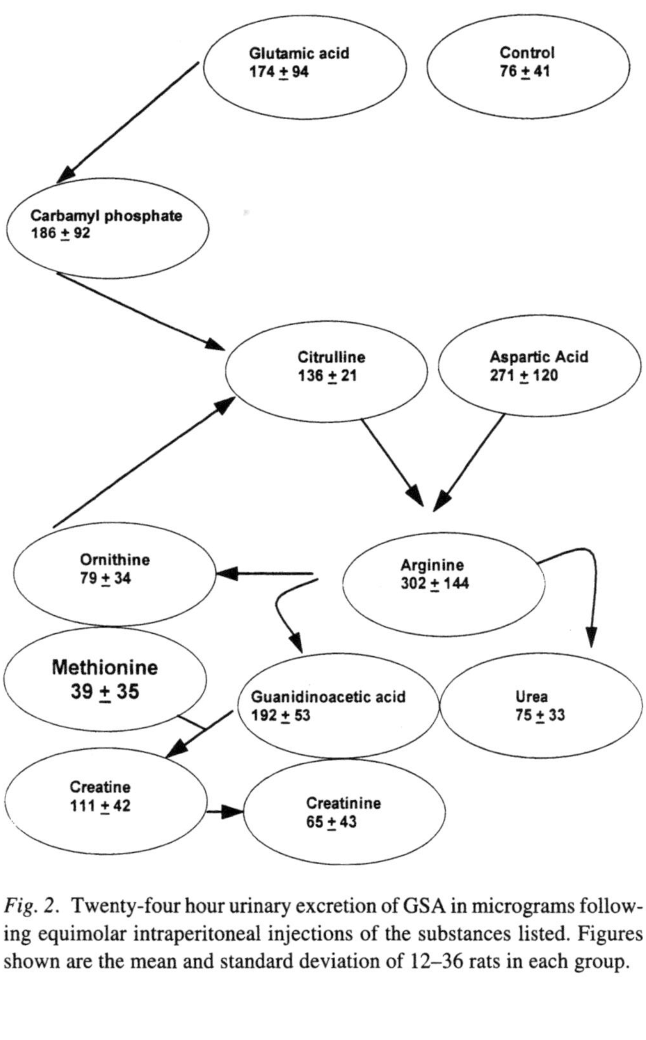


*Fig. 2.* Twenty-four hour urinary excretion of GSA in micrograms following equimolar intraperitoneal injections of the substances listed. Figures shown are the mean and standard deviation of 12–36 rats in each group.

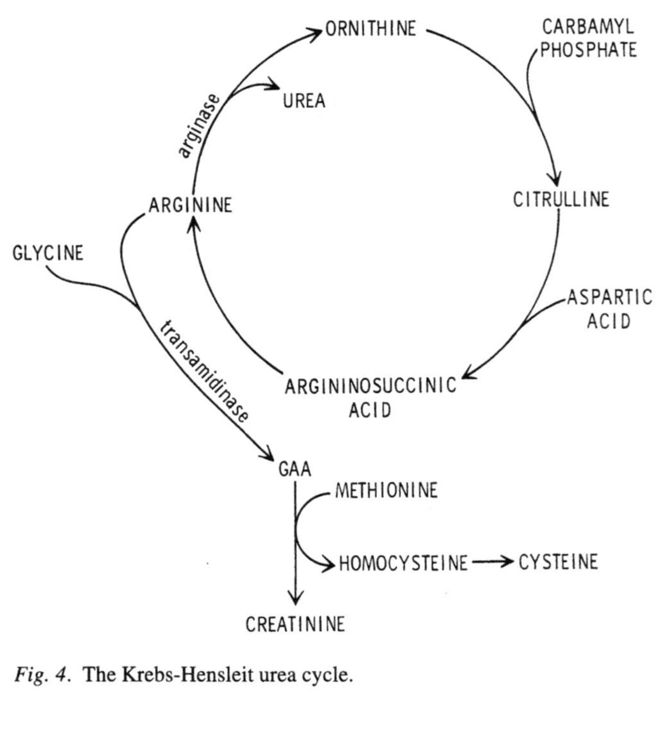


*Fig. 4.* The Krebs-Hensleit urea cycle.

*Table 1.* GSA output in rats in micrograms/mg CR/24 h following intraperitoneal injection as noted

| Controls | Urea | Urea + phenobarbitol |
|---|---|---|
| 9 | 21 | 12 |

Figures are the mean of 12 experiments in each group.

strate for transamidination of glycine to form creatine, the sole source of nitric oxide and the exclusive generator of urea. All three reactions, the enzymes for which are glycine-arginine transamidinase, nitric oxide synthase and arginase, compete and are substrate-dependent [27] with arginase accounting for the major share of activity. Substrate depletion leads to profound consequences. We have shown previously that glycine excess increases urinary guanidinoacetic acid and reduces production of NO presumably through competition for available arginine [28] (see Table 2). Aoyagi, meanwhile, has demonstrated that excess urea, while not increasing ASA, sharply reduces arginine and, thus, decreases NO leading to the hepatic accumulation of hydroxyl radicals [29, 30]. Through this complex sequence urea increases the generation of GSA.

The effect, therefore, is to create an alternate urea cycle in which the end-product is GSA rather than urea (Fig. 6). The free radical assault on ASA produces, in addition to GSA, the

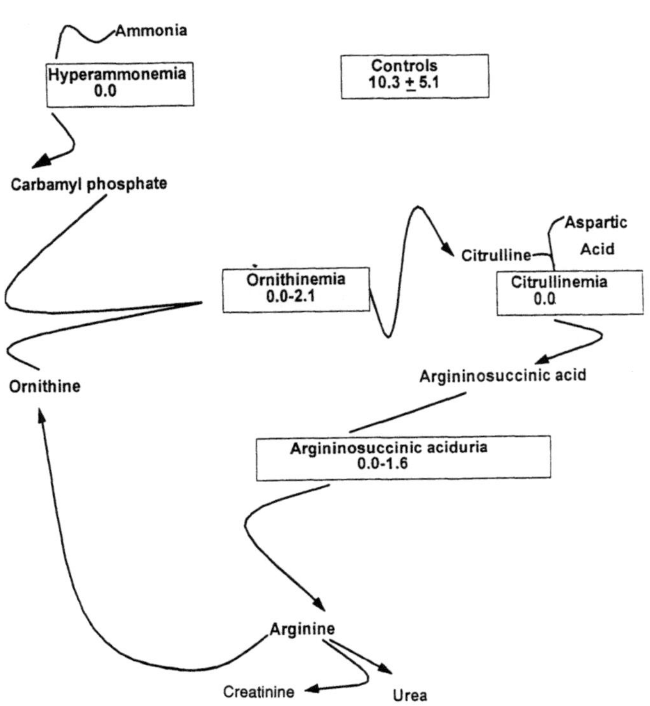


*Fig. 5.* Range of GSA (mg/g creatinine) in 8 untreated subjects known to have inborn errors of the urea cycle: 2 with ornithinemia, 3 with argininosuccinic aciduria and 1 each of the others. Controls represent the mean and standard deviation of 20 observations.

*Table 2.* The effect of glycine on the urinary output of the derivatives of arginine

| | NO | UGAA |
|---|---|---|
| Control | 0.051 | 228 |
| Glycine | 0.023 | 337 |
| Control | 0.003 | 509 |
| Glycine | 0.001 | 757 |
| Control | 0.007 | 501 |
| Glycine | 0.006 | 624 |

The results of 3 different experiments are shown: one involving 80 hospitalized subjects, 35 of whom received glycine 60 mmol daily in divided doses for periods ranging from 1–3 weeks, another involved 26 hospitalized subjects who served as their own controls receiving the same regimen for 2 days followed by 2 days off, a third group consisted in 5 normal controls who underwent the same regimen for 2 weeks bracketed by by a pretreatment and posttreatment control period of 2 weeks each. NO = nitric oxide, UGAA = urinary guanidinoacetic acid.

amino acid glutamine semialdehyde which undergoes transamination with a variety of amino acids to produce a ketoacid plus ornithine (Fig. 7). Together ornithine and the ketoacid combine to restore ASA and complete the GSA cycle.

GSA, however, unlike urea, is not an inert form of nitrogen capable of serving as waste. Relatively small amounts compared to urea, creatinine and, even, uric acid are found in urine. Instead GSA is transmethylated by methionine rendering it undetectable in our assays. Methionine, in the form of s-adenosylmethionine or SAMe, is the major metabolic source of methyl groups. Many substances are involved in this activity but a pivotal product is homocysteine (Fig. 8). There are numerous synthetic uses for the methyl group but its overwhelming biologic function is the production of creatine. Second only to ATPase, transmethylations are the most common enzyme reactions in biology and the use of the enzyme to create creatine consumes more SAMe than all other transmethylations combined [31].

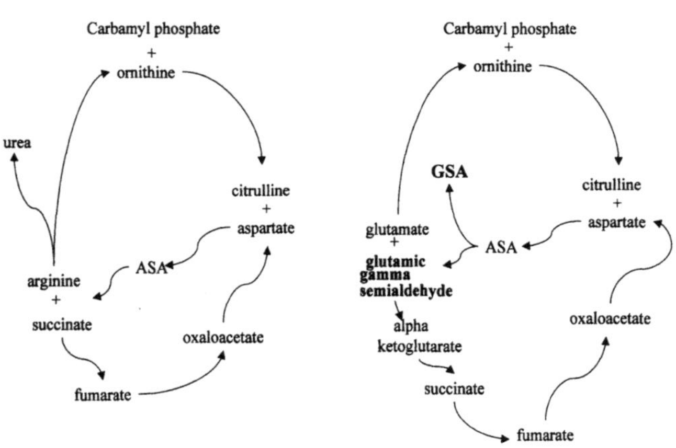


*Fig. 6.* The alternate urea cycle. The Krebs-Hensleit cycle is shown on the left and proposed GSA cycle on the right.

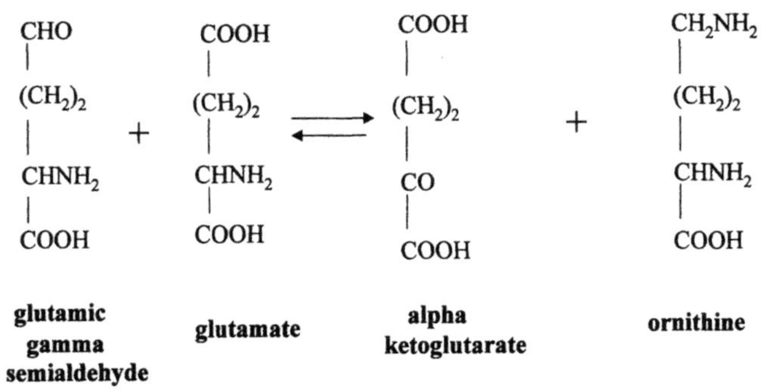


*Fig. 7.* Transamination pathway for the restoration of ornithine following the oxidation of ASA. The reaction has been demonstrated in liver and the specific enzyme partially purified from mitochondria.

**Adenosylmethionine + GAA ⟶**

**Homocysteine + Creatine + Adenosine**

**Adenosylmethionine + GSA ⟶**

**Homocysteine + MethylGSA + Adenosine**

*Fig. 9.* The normal path for creatine synthesis and the proposed GSA analogue.

The disappearance of GSA following treatment with methionine in a process chemically analogous to the formation of creatine from guanidinoacetic acid (Fig. 9) represents the most serious threat posed by GSA. To support this hypothesis we studied homocysteine in uremics [32–34], particularly attracted to the project by the clinical observation that atheromatous vasculitis, a well known consequence of homocysteinemia [35], is the most frequent cause of death in end-stage renal disease [36]. Both homocysteine and cysteine were, indeed, increased in uremic patients and probably contribute to the high incidence of atherogenesis [33, 34].

Methyl groups are obtained via three different routes, two of which are exogenous sources and the third is a path of *de novo* synthesis (Fig. 10). The major source of dietary methionine is meat protein and diets rich in methionine are an unfortunate burden on the failing kidney. Dietary choline becomes a methyl donor following oxidation to betaine and betaine can serve to restore methionine by methylation of homocysteine. In the presence of dietary deficiency, on the other hand, the de novo synthesis of methyl groups is stimulated [38]. This pathway uses methyltetrahydrofolate or methylTHF as a source. The methyltransferase, the enzyme responsible for the final transmethylation is vitamin B12 dependent. The methyl is derived from the beta carbon of serine which transfers its hydroxymethyl group to the hydrofolate which is then reduced to methylTHF.

Man and animals use more methyl groups than are consumed in their diets in the form of methionine or choline. The difference, about 50% of the daily need, is made up via the tetrahydrofolate-dependent pathway, synthesis of which increases if there is a decrease in methionine intake [39]. Uremics, with their poor appetites and recommended low protein diets, are under constant duress to produce enough methyl groups to sustain normal transmethylation needs and prevent the buildup of homocysteine.

Fifty years ago the focus of attention in managing uremia was slowing progression and postponing death which was rapid and inevitable. Severe restriction of fluids, minerals and protein was the backbone of treatment. With the advent of dialysis death from uremia was effectively delayed but cardiovascular disease and malnutrition became the focus of attention. Recent studies reveal an unanticipated irony. In end-stage renal disease high serum cholesterol, usually a poor prognostic finding, correlated positively with survival while low serum albumin reflecting protein malnutrition implied a poor outcome [40]. One of the most sensitive measures of undernutrition is an insufficiency of methyl groups as measured by increased serum GSA and homocysteine. Dietary supplementation in renal replacement therapy should include SAMe and betaine, in addition to increased folic acid [41–43], using serum GSA and homocysteine levels as markers.

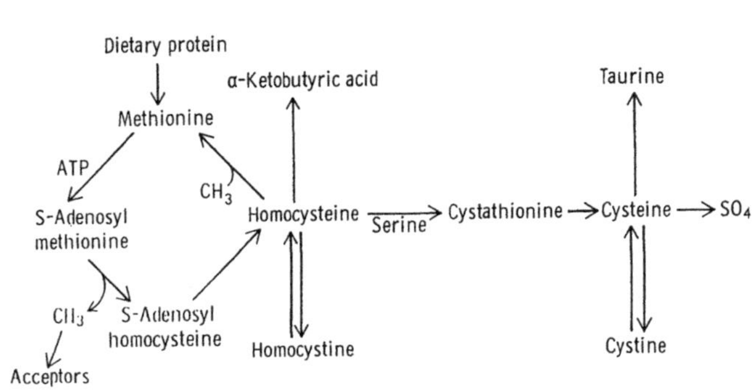


*Fig. 8.* The principal transmethylation pathway.

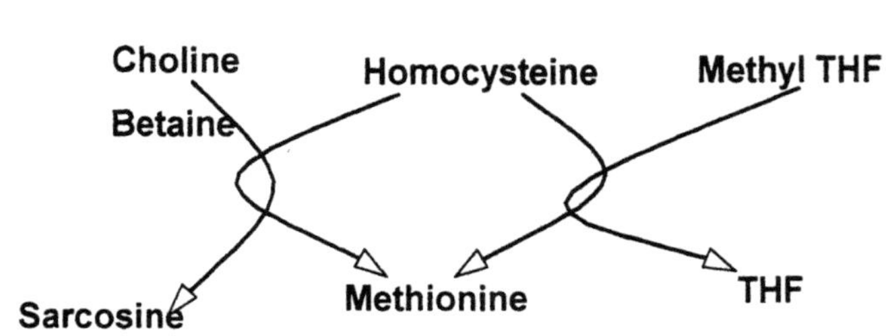


*Fig. 10.* Schematic illustrating three routes for methyl transfer.

# References


1. Bonas JE, Cohen BD, Natelson S: Separation and estimation of certain guanidino compounds. Application to human urine. Microchem J 7: 63–77, 1963
2. Cohen BD, Stein IM, Bonas JE: Guanidinosuccinic aciduria in uremia. Am J Med 45: 63–68, 1968
3. Aoyagi K, Ohba S, Narita M, Tojo S: Regulation of the biosynthesis of guanidinosuccinic acid in isolated rat hepatocytes and *in vivo*. Kidney Int 24: S224–S228, 1983
4. Kopple JD, Gordon SI, Wang M, Swendseid ME: Factors affecting serum and urinary guanidinosuccinic acid levels in normal and uremic subjects. J Lab Clin Med 90: 303–311, 1977
5. Stein IM, Cohen BD, Kornhauser RS: Guanidinosuccinic acid in renal failure, experimental azotemia and inborn errors of the urea cycle. N Engl J Med 280: 926–930, 1969
6. Perez G, Abraira C, Rey A, Horn M: Removal of guanidinosuccinic acid by hemodialysis. Nephron 19: 214–219, 1977
7. Cohen BD: Guanidinosuccinic acid in uremia. Arch Intern Med 126: 846–850, 1970
8. Dobelstein H, Edel HH, Schmidt M, Schubart G, Wenzierl M: Guanidinosuccinic acid and uremia: A clinical study. Klin Wochenschr 49: 348–357, 1971
9. Horowitz HI, Cohen BD, Martinez P, Papayyoanou MF: Defective ADP-induced platelet factor 3 activation in uremia. Blood 30: 331–340, 1967
10. Horowitz HI, Stein IM, Cohen BD, White JG: Further studies on the platelet inhibitory effect of guanidinosuccinic acid and its role in uremic bleeding. Am J Med 49: 336–345, 1970
11. Lonergan ET, Semar M, Sterzel RB, Tresor G, Needle MA, Voyles L, Lange K: Erythrocyte transketolase activity in dialyzed patients: A reversible metabolic lesion of uremia. N Engl J Med 284: 1399–1403, 1971
12. Slavin RD, Fitch CD: Inhibition of lymphocyte transformation by guanidinosuccinic acid, a surplus metabolite. Experientia 27: 1340–1341, 1971
13. Shainkin-Kestenbaum R, Winikoff Y, Dvilansky A, Chaimowitz C, Nathan I: Effect of guanidinopropionic acid on lymphocyte proliferation. Nephron 44: 295–298, 1987
14. Acquatella H, Perez-Rojas M, Burger B, Lozano JR: Experimental modification of myocardial contractility produced by a toxin retained in uremia: Guanidinosuccinic acid. Arch Inst Cardiol Mex 44: 624–639, 1974
15. Acquatella H, Perez-Rojas M, Burger B, Quinan-Baldo A: Left ventricular function in terminal uremia: A hemodynamic and echocardiographic study. Nephron 22: 160–174, 1978
16. Scheuer J, Stezoski SW: The effect of uremic compounds on cardiac function and metabolism. J Mol Cell Cardiol 5: 287–300, 1973
17. Cohen BD, Handelsman DG, Pai BN: Toxicity arising from the urea cycle. Kidney Int 7: S285–S287, 1975
18. Patel H, Cohen BD: The effect of inhibitors of cytochrome P-450 microsomal enzymes on the biosynthesis of potential uremic toxins. Ann NY Acad Sci 434: 468–470, 1984
19. Aoyagi K, Nagase S, Tomida C, Takemura K, Akiyama K, Koyama A: Synthesis of guanidinosuccinate from argininosuccinate and reactive oxygen *in vitro*. Enzyme Prot 49: 199–204, 1996
20. Aoyagi K, Nagase S, Gotoh M, Akiyama K, Safoh M, Hirayama A, Koyama A: Role of reactive oxygen and argininosuccinate in guanidinosuccinate synthesis in isolated rat hepatocytes. Enzyme Prot 49: 205–211, 1996
21. Aoyagi K, Shahrzad S, Iida S, Tomida C, Hirayama A, Nagase S, Cohen BD: Role of nitric oxide in the synthesis of guanidinosuccinic acid, an activator of the N-methyl-D-aspartate receptor. Kidney Int 59: S93–S96, 2001
22. Menyhart Y, Grof J, Somogyi J: Abstracts of the 9th Annu Meet Eur Soc Clin Invest 9: 37–38, 1975
23. Aoyagi K: Significance of reactive oxygen in kidney disease elucidated by uremic toxins: From uremia to nephrosis. In: T. Yoshikawa, S. Toyokuni, Y. Yamamoto, Y. Naito (eds). Free Radicals in Chemistry, Biology and Medicine, ch. 42. OICA International, London, 2000, pp 1–7
24. Bohles H, Cohen BD, Michalk D: Guanidinosuccinic acid excretion in argininosuccinic aciduria. Adv Exp Med Biol 153: 443–448, 1982
25. Cohen BD: Guanidinosuccinic acid in uremia. Arch Int Med 162: 846–850, 1970
26 Aoyagi K, Teraoka S: The artificial kidney for the treatment of redox state abnormalities in renal failure. J Artif Organs 4: 1–2, 2001
27. Palmer RMJ: The L-arginine nitric oxide pathway. Curr Opin Nephrol Hypertens 2: 122–128, 1993
28. Cohen BD, Patel H: Glycine in acute renal tubular necrosis. In: P.P. DeDeyn, B. Marescau, I.A. Qureshi, A. Mori (eds). Guanidino Compounds in Biology and Medicine 2. John Libbey, London, 1997, pp 273–277
29. Aoyagi K, Akiyama K, Shahrzad S, Tomida C, Hirayama A, Nagase S, Takemura K, Koyama A, Ohba S, Narita M: Formation of guanidinosuccinic acid, a stable nitric oxide mimic, from argininosuccinic acid and nitric oxide-derived free radicals. Free Radic Res 31: 59–65, 1999
30. Aoyagi K: Inhibition of arginine synthesis by urea: a new mechanism of arginine deficiency in renal failure which leads to increased hydroxyl radical generation. Mol Cell Biochem 39: (abstr), 2, 2001
31. Mudd SH, Levy, HI: Disorders of transulfuration. In: J. Stanbury (ed). The Metabolic Basis of Inherited Disease. McGraw-Hill, New York, 1978, 460 pp
32. Cohen BD, Patel H, Kornhauser RS: Alternate reasons for atherogenesis in uremia. Proc Dialysis Transplant Forum 7: 178–180, 1977
33. Cohen BD: Guanidino compounds: Implications in uremia. In: A. Mori, B.D. Cohen, A. Lowenthal (eds). Guanidines. Plenum Press, New York, 1985, pp 265–275
34. Cohen BD: Homocysteine and the guanidines: An alternate source of atherogenesis in uremia. In: J.F. Clark, A. Mori, I. Masashi (eds). Guanidino Compounds in Biology and Medicine. Blackwell Science, Japan, 1999, pp 259–262
35. McCully KS: Vascular pathology of homocysteinemia: Implications for the pathogenesis of arteriosclerosis. Am J Pathol 56: 111–123, 1969
36. Lindner A, Charra B, Sherrard DJ, Scribner BH: Accelerated atherosclerosisin prolonged maintenance hemodialysis. N Engl J Med 290: 697–701, 1974
37. Cohen BD, Patel H: Methionine and the control of creatine synthesis. Ann NY Acad Sci 494: 329–331, 1987
38. Mudd SH: Homocystinuria and homocysteine metabolism. In: W.L. Nyhan (ed). Heritable Disorders of Amino Acid Metabolism. Wiley, New York, 1974, 372 pp
39. Mudd SH: Methyl group transfer. In: W.H. Fishman (ed). Metabolic Conjugation and Metabolic Hydrolysis. Lippincott, Philadelphia, 1971, pp 297–343
40. Avram MM, Sreedhara R, Fein P, Oo KK, Chattopadhyay J, Mittman N: Survival on hemodialysis and peritoneal dialysis over 12 years with emphasis on nutritional parameters. Am J Kidney Dis 37: S77–S80, 2001

41. Van Guldener C, Janssen MJFM, DeMeer K, Donker AJM, Stehouwer CDA: Effect of folic acid and betaine on fasting and post methionine-loading plasma homocysteine and methionine levels in chronic hemodialysis patients. J Int Med 245: 175–183, 1999

42. Manns B, Hyndman E, Burgess E, Parsons H, Schaefer J, Snyder F, Scott-Douglas N: Oral vitamin B12 and high dose folic acid in hemodialysis patients with hyper-homocysteinemia. Kidney Int 59: 1103–1109, 2001

*Molecular and Cellular Biochemistry* **244:** 37–43, 2003.


# L-Canavanine as a radiosensitization agent for human pancreatic cancer cells

Aimee K. Bence,[1] Val R. Adams[2] and Peter A. Crooks[1]

[1]*Division of Pharmaceutical Sciences;* [2]*Division of Pharmacy Practice and Science, College of Pharmacy, University of Kentucky, Lexington, KY, USA*

## Abstract

This study evaluated the *in vitro* effect of L-canavanine on cell cycle progression in the two human pancreatic cancer cells lines PANC-1 and MIA PaCa-2. After 72 h of exposure to L-canavanine, the percentage of cells in the radiosensitive $G_2$/M phase of the cell cycle increased 6-fold in PANC-1 cells and 4-fold in MIA PaCa-2 cells, when compared to untreated cells. The capacity of L-canavanine to redistribute cells into the $G_2$/M phase of the cell cycle was both concentration- and time-dependent. Since many drugs that cause cells to accumulate in the $G_2$/M phase of the cell cycle are effective radiosensitization agents, the potential of L-canavanine to synergistically enhance the effects of ionizing radiation also was evaluated. The interaction between these treatment modalities was quantified using the median-effect equation and combination index analysis. L-Canavanine was found to be synergistic with radiation when either PANC-1 or MIA PaCa-2 cells were exposed to L-canavanine for 72 h prior to irradiation. These results suggest that L-canavanine in combination with radiation may have clinical potential in the treatment of pancreatic cancer. (Mol Cell Biochem **244**: 37–43, 2003)

*Key words*: L-canavanine, cell cycle analysis, ionizing radiation, pancreatic cancer, radiosensitization, median-effect analysis

## Introduction

Pancreatic cancer is the fourth leading cause of cancer death in the United States and it is estimated that in 2001 almost 29,000 individuals will die of this disease [1]. Cancer of the exocrine pancreas is rarely curable with the highest cure rate occurring only if the tumor is localized to the pancreas. Unfortunately, due to the non-specificity of the symptoms and the difficulty in early detection, less than 20% of patients present with disease confined to the pancreas, rendering surgical interventions ineffective. For patients with advanced cancer, the five-year survival rate is less than 3%, and most patients die within one year of initial assessment [2]. Conventional treatments of pancreatic cancer include surgery, radiation therapy, chemotherapy, and palliation of secondary symptoms. Despite the recent advances in both surgical techniques and chemotherapy regimens, the median survival for patients with metastatic disease is still less than six months and there is a clear need for more effective treatments for pancreatic cancer.

L-Canavanine (1), the natural occurring non-protein, δ-oxa analog of L-arginine (2), is found in a variety of higher plants [3] (Fig. 1). The anticancer properties of L-canavanine have been well described [4, 5]. L-Canavanine's usefulness as an anticancer agent is enhanced by its apparent cytoselective toxicity towards transformed cells [6, 7]. In particular, L-canavanine has demonstrated the capacity to inhibit the growth of pancreatic cancers both *in vitro* and *in vivo* [8]. L-Canavanine may have particular utility as a therapy for pancreatic cancer since it is selectivity taken up by the pancreas [9].

L-Canavanine is incorporated in place of L-arginine into newly synthesized proteins in a wide variety of organisms, resulting in the formation of non-functional proteins [3, 10, 11]. These non-functional proteins may be variously manifested as structural and functional defects, including morphological and developmental aberrations, altered protein

*Address for offprints*: P.A. Crooks, Division of Pharmaceutical Sciences, College of Pharmacy, University of Kentucky, 800 Rose Street, Lexington, KY 40536, USA

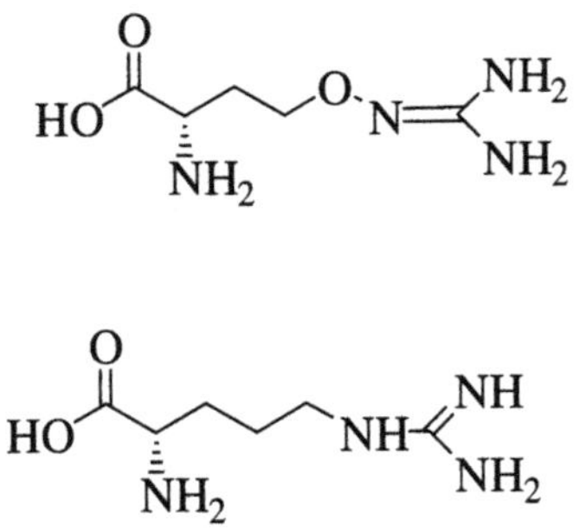

*Fig. 1.* Chemical structures of L-canavanine (1) and L-arginine (2).

conformation and structure, and impaired enzymatic activity, as well as decreased cellular tolerance to heat, radiation, and other stressors [12–14]. These structurally aberrant proteins have altered biological activity that may disrupt critical processes required for tumor growth. L-Canavanine also adversely affects DNA replication, histone and heat shock protein function, RNA synthesis, and RNA translation into protein [15–17].

Precise regulation of the cell cycle is vital for normal cell growth and proliferation. Many cytotoxic agents, including some anti-neoplastic agents, alter the complex molecular processes involved in the cell cycle, leading to a redistribution of cells within the cell cycle [18–20]. Despite its known anti-proliferative properties against pancreatic adenocarcinoma, the effect of L-canavanine on the cell cycle of pancreatic cancer cells has never been examined.

Ionizing radiation is a standard treatment modality for patients with pancreatic cancer. While this therapy does not have curative potential, radiation can reduce tumor size, prolong life, and provide palliative relief for secondary symptoms related to pancreatic cancer [21]. In 1983, Green *et al.* qualitatively evaluated the capacity of L-canavanine to enhance the effectiveness of the antiproliferative effects of radiation in human colorectal cancer cells [22]. Although the data indicated that in the colon cancer cell line HT-29/SP 1, L-canavanine enhanced the effects of radiation, no subsequent studies to quantify this effect were performed. Furthermore, no study has ever examined the radiosensitization potential of L-canavanine in pancreatic cancer cells.

Consequently, these studies were initiated to determine the effect of L-canavanine on the cell cycle of PANC-1 and MIA PaCa-2 cells. Additionally, the capacity of L-canavanine, in combination with ionizing radiation, to synergistically enhance the *in vitro* antiproliferative effects of these treatments in PANC-1 and MIA PaCa-2 was evaluated using median-effect and combination index analysis. The results of these studies indicate that L-canavanine may have potential as a radiosensitization agent in the treatment of pancreatic cancer, and that further *in vivo* studies are warranted to determine the clinical potential of this novel chemotherapeutic agent.

# Materials and methods

### *Materials*

The established human pancreatic cancer cell lines PANC-1 and MIA PaCa-2 were used in these experiments. All cells were obtained from American Type Cell Collection (Manassas, VA, USA). The cells were cultured in Dulbecco's Modified Eagle Medium (DMEM) (Gibco BRL, Gaithersburg, MD, USA) supplemented with 10% fetal bovine serum (Gibco), 4 mM glutamine (Atlanta Biologicals, Norcross, GA, USA), and a penicillin/streptomycin solution (100 units/ml and 100 µg/ml, respectively) (Gibco). The cells were maintained at 37°C in a humidified 5% $CO_2$ atmosphere for at least 72 h prior to each experiment. L-Arginine deficient media was prepared using Gibco's formulation for DMEM, but contained reduced amounts of L-arginine. MTS was obtained from Promega (Madison, WI, USA).

L-Canavanine was isolated from the seeds of the jack bean, *Canavalia ensiformis*, by repetitive recrystallization [23]. L-Canaline and L-indospicine were synthesized as previously reported [24, 25]; L-arginine, D-arginine, and L-homoarginine were obtained from Sigma (St. Louis, MO, USA). The flow cytometry experiments were conducted in the Flow Cytometry laboratory at the University of Kentucky, Lexington, KY, USA, using a Becton-Dickinson FACScan (Franklin Lakes, NJ, USA). Irradiation of the cells was performed using a Mark I $^{137}Cs$ irradiator from J.L. Shepard and Associates (Glendale, CA, USA). The computer program *Calcusyn* from Biosoft (Cambridge, UK) was used to calculate median-effect parameters, and combination index values. Results are expressed as the mean ± S.E.M, and a paired two-tailed Student's *t*-test was used to determine statistical significance. A p value of less than 0.05 was considered statistically significant.

### *Cell cycle analysis experiments*

PANC-1 and MIA-PaCa-2 cells were each plated in 20 ml of media in 100 mm dishes. The cells were allowed to grow for 24 h before the addition of L-canavanine. L-Canavanine was dissolved in 20 ml of DMEM and tested at a final concentration of 0.5, 1, 2, and 3 mM; L-arginine analogs were dissolved in 20 ml of DMEM and added to the cells to afford a final concentration of 3 mM. In the L-arginine-depleted experiments, the commercial media (L-arginine concentration 400 µM) was removed and replaced with L-arginine-depleted media (containing 0, 80, 160, 240, 320, or 400 µM of L-arginine). At 24, 48, or 72 h after drug treatment, the cells were harvested, pelleted, washed with cold phosphate buffered saline (PBS) and resuspended in 0.5 ml of cold PBS. While vortexing, 5 ml of cold 70% ethanol was added dropwise to fix the cells. After allowing the cells to stand for at least 15 min at 4°C, the cells were pelleted, washed with cold PBS, and resuspended in propidium iodide stain solution (0.01 M Tris base, 10 mM NaCl, 700 U RNAase, 0.075 mM propidium iodide stain, and Nonidet P-40) at a final concentration of $1 \times 10^6$ cells/ml. After incubating the cells at 37°C for 30 min, flow cytometry was used to analyze the propidium iodide staining and to determine the percentage of cells that were in the $G_1$, S, or $G_2$/M phase of the cell cycle.

*Radiosensitization cytotoxicity assays*

PANC-1 and MIA PaCa-2 cells were harvested and 100 µl of the cell suspension was plated in 96-well microtiter plates at a concentration of $8 \times 10^2$ cells/well. The experimental design used in these cytotoxicity assays included using a fixed ratio of L-canavanine (mM) to radiation (Gy). Consequently, the amount of L-canavanine added to each well was dependent on the amount of radiation that each well received. Each well was irradiated at one of the following irradiation energy levels: 24, 20, 16, 12, 8, or 4 Gy. L-Canavanine was dissolved in 100 µl of DMEM such that the final ratio of L-canavanine concentration (mM) to radiation energy (Gy) in each well was 1:1, 1:5, 1:10, 1:20, or 1:40. For example, wells exposed to 20 Gy received 20 mM (1:1), 4 mM (1:5), 2 mM (1:10), 1 mM (1:20) or 0.5 mM (1:40) of L-canavanine. Four replicate wells were used for the control wells and for each concentration of L-canavanine and radiation.

After the cells had adhered, 100 µl of the specific concentration of L-canavanine for the proper L-canavanine to radiation ratio was added to the wells. After 72 h, L-canavanine-containing media was replaced with 200 µl of media. The cells were irradiated (at either 24, 20, 16, 12, 8, or 4 Gy) and allowed to grow for an additional 72 h before the cell viability was assessed in the MTS assay [26].

*Median effect analysis*

The median-effect equation and combination index analysis were used to calculate the interaction between the treatment modalities, and to determine if the effect was antagonistic, additive, or synergistic. Both the median effect-equation and combination index analysis are considered generally applicable in biological systems, and their use has been extensively reviewed [27, 28]. The algebraic expression of the median-effect equation is: $f_a/f_u = (D/D_m)^m$, where $f_a$ is the fraction affected by the dose, $f_u$ is the fraction unaffected by the dose, D is the dose of the drug, $D_m$ is the dose of drug required to produce a median-effect (the dose of drug required to inhibit the growth of 50% of the cells), and *m* is a coefficient signifying the sigmoidicity of the dose-effect curve.

The logarithmic form of the median effect equation linearizes dose-response curves, regardless of whether they are hyperbolic curves ($m = 1$) or sigmoidal curves ($m = 2$), and can be plotted to generate the median-effect plot. When log (D) is plotted vs. log ($f_a/f_u$), the x-intercept yields log ($D_m$) from which the $IC_{50}$ can be determined, and the slope is the m value. The *m* and $D_m$ values obtained from the median-effect plot can be used to calculate the isoeffective dose for any effect level (i.e. $IC_{25}$, $IC_{50}$, or $IC_{75}$).

The combination index (CI), used to determine synergism or antagonism, is based on the multiple-drug effect equation. A CI value of one indicates that the effects of the two drugs are additive, while a CI value of greater than one is defined as antagonism between the drugs. A CI of less than one is defined as synergism. The equation used to calculate CI is: $(D)_1/(D_x)_1 + (D)_2/(D_x)_2 + \alpha\ (D)_1(D)_2/(D_x)_1(D_x)_2$, where $(D_x)_1$ and $(D_x)_2$ are the doses for x% inhibition by drug 1 and drug 2 alone. These values can be obtained from the median-effect equation. $(D)_1$ and $(D)_2$ are the doses in combination that also inhibit cell growth by ×%. In this experiment, it was ambiguous as to whether these drugs were acting in a mutually exclusive ($\alpha = 0$) or mutually non-exclusive ($\alpha = 1$) manner.

# Results

*Cell cycle analysis*

Propidium iodide and flow cytometry were used to assess whether L-canavanine's growth inhibition disrupted the cell cycle of PANC-1 and MIA PaCa-2 cells. The effect of both L-canavanine dose and exposure times were analyzed in these experiments; four concentrations of L-canavanine (0.5, 1, 2, and 3 mM) were analyzed at 24, 48, and 72 h after the addition of drug. At 24 and 48 h after treatment with 0.5, 1, and 2 mM of L-canavanine, the cells exposed to L-canavanine showed no statistically significant difference in the percentage of cells in each phase of the cell cycle, when compared to untreated cells. In MIA PaCa-2 cells, exposure to 3 mM of L-canavanine, however, caused an enhanced accumulation of cells in the $G_2$/M phase at both 24 h (25.5 ± 1.3% ) and 48 h (35.8 ± 3.3%). A similar profile was seen in the PANC-1 cells treated with L-canavanine. In cells treated with 3 mM L-canavanine 37.6 ± 0.8% (24 h) and 38.1 ± 1.8% (48 h) of the cells were in the $G_2$/M phase (Fig. 2).

As seen in Fig. 3, at 72 h after drug treatment, 2 mM of L-canavanine resulted in almost a 2-fold increase in cells in the $G_2$/M phase when compared to control (12.1 ± 1.4% to 22.1 ± 3.0%, respectively) in MIA PaCa-2 cells, while treatment with 3 mM L-canavanine caused over a 4-fold increase in the percentage of cells in the $G_2$/M phase when compared to control cells (12.1 to 53.8 ± 7.3%, respectively). There was

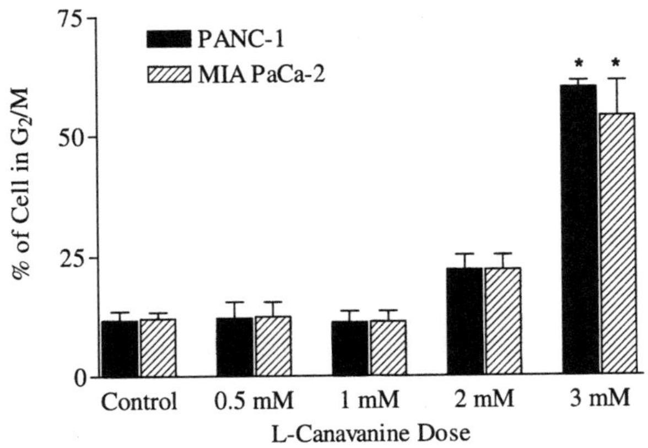


*Fig. 2.* The percentage of PANC-1 or MIA PaCa-2 cells in the $G_2$/M phase of the cell cycle after 72 h of exposure to L-canavanine. Cells were stained with propidium iodide and flow cytometric analysis was performed as described in 'Materials and methods'. A result was considered statistically significant (*) when $p \leq 0.05$.

a concomitant decrease in the percentage of cells in the $G_0$/$G_1$ phase from 55.3 ± 4.6 to 22.9 ± 8.1%. At 72 h after PANC-1 cells were exposed to L-canavanine, the percentage of cells in the $G_2$/M phase was 22.1 ± 3.0% (2 mM) and 59.7 ± 1.9% (3 mM), compared to only 11.7 ± 1.8% for untreated cells. A concurrent decrease in the percentage of cells in the $G_0$/$G_1$ phase was observed with the control cells having 57.3 ± 4.4% of cells in the $G_0$/$G_1$, and the cells treated with 3 mM L-canavanine having only 16.4 ± 1.9% of cells in $G_0$/$G_1$ phase.

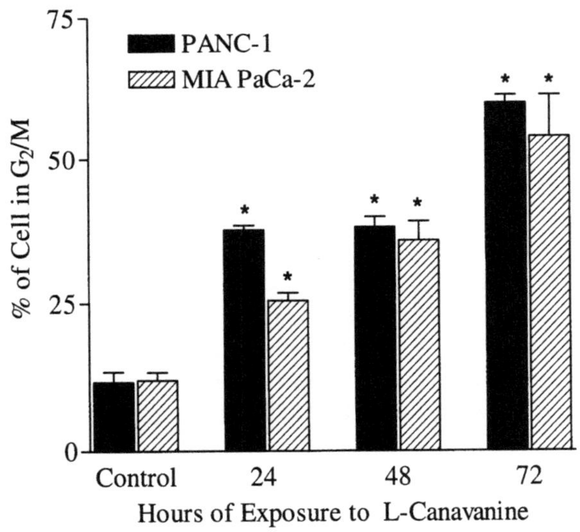


*Fig. 3.* The percentage of PANC-1 or MIA PaCa-2 cells in the $G_2$/M phase of the cell cycle after exposure to 3 mM L-canavanine. Cells were stained with propidium iodide and flow cytometric analysis was performed as described in 'Materials and methods'. A result was considered statistically significant (*) when $p \leq 0.05$.

To ensure that the effect of L-canavanine on the cell cycle of PANC-1 and MIA PaCa-2 cells was due to an intracellular event and not simply due to inhibition of L-arginine uptake into the cells, cells were grown in an L-arginine-depleted environment and the percentage of cells in each phase of the cell cycle was analyzed with propidium iodide and flow cytometry. Although limiting the amount of L-arginine in the media was cytotoxic to PANC-1 and MIA PaCa-2 cells, the percentage of cells in the $G_2$/M phase of the cell cycle, even with the cells completely deprived of L-arginine for 72 h, did not increase.

In order to assess the effect of high concentrations of other L-arginine analogs on cell cycle regulation, PANC-1 and MIA PaCa-2 cells were exposed to a series of L-arginine analogs. After 72 h of treatment in the presence of 3 mM concentrations of L-arginine, D-arginine, and L-homoarginine, there was no statistically significant increase in the percentage of cells in the $G_2$/M phase in either PANC-1 or MIA PaCa-2 cells. Similarly, neither the isosteric analog L-indospicine nor the primary metabolic product, L-canaline, induced a $G_2$/M block in either PANC-1 or MIA PaCa-2 cells (Fig. 4).

*Radiosensitization*

The dose-effect relationships of L-canavanine and radiation were examined using the median-effect plot to determine their potency ($D_m$), shape ($m$), and conformity ($r$), in both PANC-1 and MIA PaCa-2 cells. The $m$ and $D_m$ values for each treatment modality were used in the median-effect equation to calculate synergistic or antagonistic interactions. The $r$ values, indicating goodness of fit, all exceeded 0.90, indicating

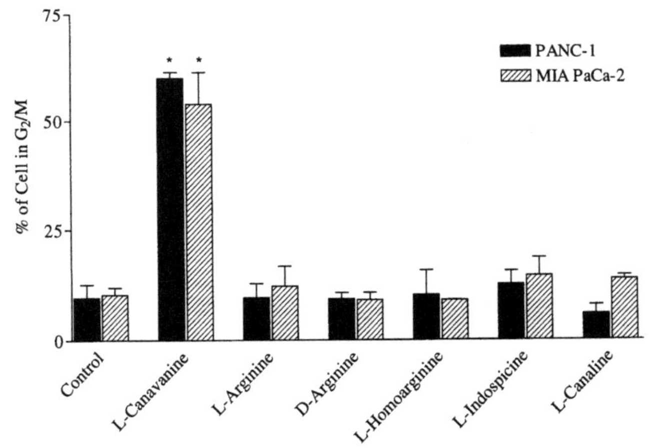


*Fig. 4.* The percentage of PANC-1 or MIA PaCa-2 cells in the $G_2$/M phase of the cell cycle after 72 h of exposure to 3 mM of one of a series of L-arginine analogs. Cells were stained with propidium iodide and flow cytometric analysis was performed as described in Materials and Methods. A result was considered statistically significant (*) when $p \leq 0.05$.

that median-effect analysis was valid for these data. CI values were calculated using $\alpha = 0$.

PANC-1 and MIA PaCa-2 cells were treated with L-canavanine at five different ratios of L-canavanine (mM) to radiation (Gy). The data from the evaluation of the L-canavanine to radiation ratio 1:1 and 1:5 are not shown because a median-effect analysis was not valid for the data. At the lowest dose in these treatment regimens, the cells received 4 Gy of radiation and 4 mM of L-canavanine (1:1) or 4 Gy of radiation and 0.75 mM (1:5). The number of cells killed at these ratios was so great that it precluded an accurate median-effect analysis.

When PANC-1 and MIA PaCa-2 cells were exposed to L-canavanine for 72 hours prior to irradiation, L-canavanine was found to be synergistic with radiation. While the combination index value is most accurate at the median-effect ($ED_{50}$) [29], L-canavanine was synergistic with radiation at all values between the $ED_{25}$ and the $ED_{75}$. L-Canavanine was synergistic with ionizing radiation at all ratios tested, with the 1:10 ratio of L-canavanine to radiation, having the greatest degree of synergy. At the $ED_{50}$, and a ratio of 1:10, both PANC-1 and MIA PaCa-2 cells that had been exposed to L-canavanine prior to radiation showed synergy with CI values of 0.530 and 0.455, respectively. Similarly, at a ratio of 1:20 synergy was observed. In PANC-1 cells the CI value at the $ED_{50}$ was 0.612; in MIA PaCa-2 cells the CI was 0.526 at the $ED_{50}$. Even at the lowest ratio analyzed (1:40 L-canavanine to radiation), there was still moderate synergy between the treatments. At a ratio 1:40, the $ED_{50}$ CI values were 0.692 (PANC-1) and 0.770 (MIA PaCa-2).

## Discussion

Exposure to L-canavanine induced significant cell cycle arrest, redistributing the cells in the $G_2$/M phase of the cell cycle in both PANC-1 and MIA PaCa-2 cells. The results observed in these experiments are in contrast with recently published data regarding the effect of L-canavanine on the cell cycle of A549 non-small cell lung cancer cells [30]. When A549 cells were exposed to 3 mM of L-canavanine for 72 h, a significant $G_1$ arrest was observed and no $G_2$/M arrest was induced. It is important to note that A549 cells harbor a wild-type p53 gene [31]. The tumor suppressor gene p53 codes for the p53 protein, that acts as a $G_1$ cell cycle checkpoint in response to cell stress. When the above authors repeated the experiment with the human cervical cancer cell line HeLa, and the human bladder cancer cell line HTB9 (p53 inactivated cell lines), no $G_1$ arrest was observed [30]. Mutations in the p53 gene occur in 50–70% of pancreatic ductal cancers [32]. Both PANC-1 and MIA PaCa-2 cells have mutated p53, and therefore, a defective $G_1$ checkpoint [33,34]. This difference in the functionality of the p53 gene may explain why the results of the current work differ from those previously reported for the A549 cell line.

It is known that $G_2$/M cell cycle arrest can also be induced as a result of nutrient deprivation [35]. Since L-arginine and L-canavanine are structurally similar, L-canavanine can compete with L-arginine for uptake into cells mediated by the system $y^+$ transporter [36]. Consequently, it was postulated that at the high concentrations of L-canavanine used in these experiments, L-arginine uptake into the cell might be reduced, and that the capacity of L-canavanine to cause cells to accumulate in the $G_2$/M phase of the cell cycle was solely a result of depletion of L-arginine reserves in the cell, and not due to L-canavanine incorporation into protein. To evaluate this potential mechanism of cell cycle dysregulation, cells were grown for 24, 48, and 72 h in L-arginine-reduced media at several different concentrations of L-arginine. Although depriving the cells of L-arginine was cytotoxic to the cells, it did not cause the cells to accumulate in the $G_2$/M phase of the cell cycle. This is in sharp contrast to the redistribution of cells in the cell cycle that occurs after exposure to L-canavanine. This data indicated that L-canavanine's ability to redistribute cells in the $G_2$/M phase of the cell cycle was most likely not a result of L-arginine depletion within the cells, but rather a result of a drug-related intracellular mechanism.

L-Canavanine had an $IC_{50}$ value of approximately 1 mM after 72 h of treatment. The greatest effect on the redistribution of cells in the cell cycle was seen at a concentration 3 mM, or three times the $IC_{50}$. In order to ensure that such high concentrations of drug were not affecting the integrity of the cellular membrane or the ionic strength of the media, and thereby inducing the cell cycle arrest, the cells were treated with identical concentrations of several other L-arginine analogs.

L-Arginine and D-arginine did not have any affect on the cell cycle progression when compared to control cells. Since the guanidino group of L-arginine and D-arginine have a $pK_a$ value that is over 5 units higher than that of L-canavanine, they are both much more basic than L-canavanine. This suggests that it is unlikely that the capacity of L-canavanine to redistribute cells in the cell cycle is due to a change in the basicity or ionic strength of the media solution. The non-cytotoxic L-arginine analog L-homoarginine, which can, like L-canavanine, act as a substrate for arginyl-tRNA synthetase and can be incorporated into protein in place of L-arginine [37], also lacked the capacity to alter cell cycle progression. The inability of this analog to induce a $G_2$/M block suggests that L-canavanine's effect on the cell cycle is not solely due to the fact that it is incorporated into protein by arginyl-tRNA synthetase. Rather, these results point to the importance of the oxyguanidino group of L-canavanine and the production of structurally aberrant proteins. The guanidino group of L-homoarginine will be predominantly protonated at physiological pH, and when incorporated into protein in place of

L-arginine, may, like L-arginine, be able to form the crucial ionic interactions required to maintain the structural and conformational integrity of proteins.

The inability of L-canaline, a toxic L-canavanine metabolite, to alter cell cycle progression is significant, because it supports recent data indicating that the cytotoxic properties of L-canavanine in pancreatic cancer cells are independent of L-canaline's cytotoxicity [38]. L-canaline has an $IC_{50}$ value of approximately 400 µM when exposed to PANC-1 and MIA PaCa-2 cells. Even at L-canaline concentrations of 3 mM (over 12 times the $IC_{50}$ concentration) no accumulation of cells in the $G_2$/M phase was observed in either PANC-1 or MIA PaCa-2 cells. As evidenced by their distinctly different effects on the cell cycle, it is likely that L-canavanine's mechanism of toxicity is quite different from that of L-canaline.

The effectiveness of radiation is cell cycle-dependent, with the $G_2$ and M phases being the most radiosensitive phases of the cell cycle [39,40]. Since exposure to L-canavanine for 72 h redistributed the cells in the $G_2$/M phase, the antiproliferative effectiveness of the combination of L-canavanine and radiation was evaluated in PANC-1 and MIA PaCa-2 cells. Using median-effect analysis and combination index plots to quantify the degree of synergy, the data from these studies indicate that L-canavanine synergistically enhances the effectiveness of radiation when the cells were exposed to L-canavanine prior to irradiation.

To evaluate whether the capacity of L-canavanine to interact synergistically with ionizing radiation was schedule-dependent, two additional sets of experiments were conducted (data not shown). L-Canavanine was exposed to PANC-1 and MIA PaCa-2 cells both before and after irradiation, and also after irradiation. Prior to the experiment, it was postulated that exposure to L-canavanine after the radiation insult might also enhance the effects of radiation, but via a different mechanism. After a significant insult, such as radiation, there is an upregulation of DNA repair proteins to repair the damage caused by the radiation [41]. It was thought that if L-canavanine were present, it might be incorporated into newly synthesized proteins, rendering them ineffective. However, when the cells were exposed to L-canavanine for 72 h before and after irradiation, or for 72 h after being irradiated, the effect of the two agents was only additive, or just slightly synergistic, suggesting that L-canavanine does not significantly alter the DNA repair processes in cells exposed to ionizing radiation. Consequently, it is likely that the observed radiosensitive effect of L-canavanine can be at least partially attributed to its ability to cause cells to accumulate in the radiosensitive $G_2$/M phase of the cell cycle, thereby maximizing the capacity of ionizing radiation to irreparably damage DNA.

The potential for L-canavanine to be an effective radiosensitization agent may be enhanced by its capacity to preferentially localize in the pancreas [9]. This unique property of L-canavanine could maximize its effectiveness as a radiosensitization agent, since it will preferentially concentrate in the tissue receiving ionizing radiation. Although the present studies are small-scale studies, they afford preliminary data that provide the basis for further preclinical studies, and strongly suggest that L-canavanine might hold promise as a radiosensitization agent for patients with pancreatic cancer.

## Acknowledgements

The authors gratefully acknowledge the American Foundation for Pharmaceutical Education Fellowships for fellowship support of AKB.

## References

1. Greenlee RT, Hill-Harmon MB, Murray T, Thun M: Cancer statistics, 2001. CA Cancer J Clin 51: 15–39, 2001
2. Warshaw AL, Fernandez-del Castillo C: Pancreatic carcinoma. N Engl J Med 326: 455–465, 1992
3. Rosenthal GA: The biological effects and mode of action of L-canavanine, a structural analogue of L-arginine. Q Rev Biol 52: 155–178, 1977
4. Thomas DA, Rosenthal GA, Gold DV, Dickey K: Growth inhibition of a rat colon tumor by L-canavanine. Cancer Res 46: 2898–2903, 1986
5. Green MH, Brooks TL, Mendelsohn J, Howell SB: Antitumor activity of L-canavanine against L1210 murine leukemia. Cancer Res 40: 535–537, 1980
6. Naha PM: An experimental model for selective inhibition of proliferating cells by chemotherapeutic agents. Cell Biol Int Rep 4: 829–839, 1980
7. Naha PM, Silcock JM, Fellows L: Reappraisal of L-canavanine as an anti-tumour agent. Cell Biol Int Rep 4: 155–166, 1980
8. Crooks PA, Rosenthal GA: Use of L-canavanine as a therapeutic agent for the treatment of pancreatic cancer. U.S. Patent 5,552,440. September 3, 1996
9. Thomas DA, Rosenthal GA: Toxicity and pharmacokinetics of the nonprotein amino acid L-canavanine in the rat. Toxicol Appl Pharmacol 91: 395–405, 1987
10. Rosenthal GA, Dahlman DL: L-Canavanine and protein synthesis in the tobacco hornworm *Manduca sexta*. Proc Natl Acad Sci USA 83: 14–18, 1986
11. Rosenthal GA, Lambert J, Hoffmann D: Canavanine incorporation into the antibacterial proteins of the fly, *Phormia terranovae* (Diptera), and its effect on biological activity. J Biol Chem 264: 9768–9771, 1989
12. Rosenthal GA, Reichhart JM, Hoffmann JA: L-Canavanine incorporation into vitellogenin and macromolecular conformation. J Biol Chem 264: 13693–13696, 1989
13. Mattei E, Damasi D, Mileo AM, Delpino A, Ferrini U: Stress response, survival and enhancement of heat sensitivity in a human melanoma cell line treated with L-canavanine. Anticancer Res 12: 757–762, 1992
14. Laszlo A, Li GC: Effect of amino acid analogs on the development of thermotolerance and on thermotolerant cells. J Cell Physiol 154: 419–432, 1993

15. Ackerman WW, Cox DC, Dinka S: Control of histones and DNA synthesis with canavanine, puromycin, and polio virus. Biochem Biophys Res Commun 19: 745–750, 1965
16. Hare JD: Reversible inhibition of DNA synthesis by the arginine analog canavanine in hamster and mouse cells *in vitro*. Exptl Cell Res 58: 170–174, 1969
17. Schactele CF, Rogers P: Canavanine death in *Escherichia coli*. J Mol Biol 34: 474–489, 1965
18. Amorino GP, Freeman ML, Choy H: Enhancement of radiation effects *in vitro* by the estrogen metabolite 2-methoxyestradiol. Radiat Res 153: 384–391, 2000
19. Buolamwini JK: Cell cycle molecular targets in novel anticancer drug discovery. Curr Pharm Des 6: 379–392, 2000
20. Choy H, Rodriguez FF, Koester S, Hilsenbeck S, Von Hoff DD: Investigation of taxol as a potential radiation sensitizer. Cancer 71: 3774–3778, 1993
21. Bodner WR, Hilaris BS, Mastoras DA: Radiation therapy in pancreatic cancer: Current practice and future trends. J Clin Gastroenterol 30: 230–233, 2000
22. Green MH, Ward JF: Enhancement of human tumor cell killing by L-canavanine in combination with γ-radiation. Cancer Res 40: 4180–4182, 1983
23. Rosenthal GA: Preparation and colorimetric analysis of L-canavanine. Anal Biochem 77: 147–151, 1977
24. Rosenthal GA: The preparation and colorimetric analysis of L-canaline. Anal Biochem 51: 354–361, 1973
25. Bence AK, Crooks PA: Synthesis of L-indospicine. Syn Commun 32: 2075–2082, 2002
26. Cory AH, Owen TC, Barltrop JA, Cory JG: Use of an aqueous soluble tetrazolium/formazan assay for cell growth assays in culture. Cancer Commun 3: 207–212, 1991
27. Chou TC: Derivation and properties of Michaelis-Menten type and Hill type equations for reference ligands. J Theor Biol 59: 253–276, 1976
28. Chou TC, Talalay P: Quantitative analysis of dose-effect relationships: The combined effects of multiple drugs or enzyme inhibitors. Adv Enzyme Regul 22: 27–55, 1984
29. Merlin JL: Concepts of synergism and antagonism. Anticancer Res 14: 2315–2319, 1994
30. Ding Y, Matsukawa Y, Ohtani Fujita N, Kato D, Dao S, Fujii T, Naito Y, Yoshikawa T, Sakai T, Rosenthal GA: Growth inhibition of A549 human lung adenocarcinoma cells by L-canavanine is associated with p21/WAF1 induction. Jpn J Cancer Res 90: 69–74, 1999
31. Gu Y, Turck CW, Morgan DO: Inhibition of CDK2 activity *in vivo* by an associated 20K regulatory subunit. Nature 366: 707–710, 1993
32. Redston MS, Caldas C, Seymour AB, Hruban RH, da Costa L, Yeo CJ, Kern SE: p53 mutations in pancreatic carcinoma and evidence of common involvement of homocopolymer tracts in DNA microdeletions. Cancer Res 54: 3025–3033, 1994
33. Bouvet M, Bold RJ, Lee J, Evans DB, Abbruzzese JL, Chiao PJ, McConkey DJ, Chandra J, Chada S, Fang B, Roth JA: Adenovirus-mediated wild-type p53 tumor suppressor gene therapy induces apoptosis and suppresses growth of human pancreatic cancer. Ann Surg Oncol 5: 681–688, 1998
34. Mogaki M, Hirota M, Chaney WG, Pour PM: Comparison of p53 protein expression and cellular localization in human and hamster pancreatic cancer cell lines. Carcinogenesis 14: 2589–2594, 1993
35. Wang SR, Su HL, Chiu CC, Huang MH, Tsai CY, Tsai JJ, Yu CL: Alteration of intracellular DNA and RNA patterns by liver arginase studied with flow cytometry. Chung Hua I Hsueh Tsa Chih (Taipei) 50: 267–272, 1992
36. Schmidlin A, Wiesinger H: Transport of L-arginine in cultured glial cells. Glia 11: 262–268, 1994
37. Rosenthal GA, Harper L: L-Homoarginine studies provide insight into the antimetabolic properties of L-canavanine. Insect Biochem Mol Biol 26: 389–394, 1996
38. Swaffar DS, Ang CY: Growth inhibitory effect of L-canavanine against MIA PaCa-2 pancreatic cancer cells is not due to conversion to its toxic metabolite canaline. Anticancer Drugs 10: 113–118, 1999
39. Terasima T, Tolmach LJ: X-ray sensitivity and DNA synthesis in synchronous populations of HeLa Cells. Science 140: 490–492, 1963
40. Sinclair WK, Morton RA: X-ray sensitivity during the cell generation cycle of cultured Chinese hamster cells. Radiat Res 29: 450–474, 1966
41. Brown KD, Lataxes TA, Shangary S, Mannino JL, Giardina JF, Chen J, Baskaran R: Ionizing radiation exposure results in up-regulation of Ku70 via a p53/ataxia-telangiectasia-mutated protein-dependent mechanism. J Biol Chem 275: 6651–6656, 2000

*Molecular and Cellular Biochemistry* **244:** 45–48, 2003.


# The clinical syndrome of creatine transporter deficiency

Ton J. deGrauw,[1] Kim M. Cecil,[2] Anna W. Byars,[1] Gajja S. Salomons,[3] William S. Ball[2] and Cornelis Jakobs[3]

*Divisions of [1]Neurology; [2]Radiology, Childrens Hospital Medical Center, Cincinnati, OH, USA; [3]Metabolic Units, VU Medical Center, Amsterdam, The Netherlands*

## Abstract

To describe the clinical, spectroscopic and neuropsychological features of the first family diagnosed with a defect in the creatine transporter.

Proton Magnetic Resonance Spectroscopy (MRS) indicated an absence of creatine and phosphocreatine in the brain of a male patient characterized by developmental delay, mild epilepsy and severe expressive language impairment. Subsequent genetic testing revealed a defect in the X-linked creatine transporter (SLC6A8/CT1), with a hemizygous mutation in the patient and a heterozygous mutation for the female carriers.

Magnetic resonance imaging and spectroscopy examinations were performed on a 1.5T clinical MR Scanner. Neuropsychological examinations were performed on the index patient and maternal relatives.

Preliminary spectroscopy results indicate the disorder prevents transport of creatine and phosphocreatine in the brain of the affected male. However, the skeletal muscle demonstrates the presence of creatine and phosphocreatine which correlates clinically with normal structure and function. Female carriers demonstrated impairments in confrontational naming and verbal memory assessments.

This new neurological syndrome is associated with developmental delay, mild epilepsy, severe language impairment. MR Spectroscopy is a non-invasive method for obtaining a preliminary diagnosis of this disorder. Muscle creatine uptake may be normal in this disorder. (Mol Cell Biochem **244**: 45–48, 2003)

*Key words*: creatine, brain, metabolism, magnetic resonance

## Introduction

One of the first inborn errors of metabolism recognized with proton magnetic resonance spectroscopy (MRS) was a defect in creatine metabolism. A MRS examination revealed absent creatine and phosphocreatine which was eventually explained as a defect of guanidino acetate methyl transferase (GAMT, EC 2.1.1.2) [1]. This defect of the last step in creatine synthesis presents with central nervous system (CNS) symptoms such as developmental delay, seizures and movement disorders. A deficiency of the other enzyme necessary for creatine synthesis (L-arginine:glycine amidinotransferase, AGAT; EC 2.1.3.1) was recently reported [2] and this disorder was also indicated by proton MRS. In early 2000, we recognized a defect of the X-linked creatine transporter (SLC6A8/CT1) in a 7 year old boy presenting with developmental delay, expressive dysphasia and partial epilepsy [3]. This report describes additional findings of the index patient and assessments of the female relatives.

## Case report

Our patient has been previously described [3, 4]. Briefly, he has mental retardation with severe expressive speech language impairment, having no active speech at age 7 years.

*Address for offprints*: T.J. deGrauw, Division of Neurology, Childrens Hospital Medical Center, 3333 Burnet Avenue, Cincinnati, OH 45229, USA (E-mail: t.degrauw@chmcc.org)

His partial epilepsy is easily controlled with carbamazepine. Macrocrania at age 6 prompted a Magnetic Resonance Imaging (MRI) with proton MRS examination. MR imaging of the brain was normal including size of the ventricles; proton MRS showed absence of the creatine/phosphocreatine peak. His head circumference has remained stable at the 95th percentile for over 1 year. On examination he has normal muscle tone, muscle bulk and strength. Cardiac and echocardiogram examinations are normal. Neuropsychological evaluation showed good gross motor skills. He did not speak during the evaluation. Given his very low level of function, the Mental Scale of the Bayley Scales of Infant Development, 2nd edition [5], was administered. On this measure, his raw score was 92, which indicates a developmental age of 13 months. On the Vineland Scales of Adaptive Behavior [6], he obtained standard scores (mean = 100, S.D. = 15) of 32 on Communication, less than 20 on Daily Living Skills, and 49 on Socialization. These scores place him in the moderately to severely impaired range of adaptive functioning.

He was treated with creatine monohydrate to a maximum dose of 7.5 g three times/day (800 mg/kg/day). This dose was maintained for 3 months without any clinical improvement or spectroscopic evidence of creatine restoration (Fig. 1). No adverse effects were noted. While on creatine treatment, Cerebro-Spinal Fluid (CSF) was obtained showing normal creatine concentration (62 μmol/l, N: 38.8 ± 4.8 μmol/l). This indicates that a different uptake mechanism was likely present through the choroid plexus. In addition, proton and phosphorus MR spectroscopy of the left thigh demonstrated the presence of creatine and/or phosphocreatine in muscle (Figs 2a and 2b).

He is the only child of healthy parents. His mother and maternal grandmother have a history of severe learning disability. His maternal aunt was a below average student. All three are confirmed genetic carriers of the disorder [4]. Proton MRS of the mother and maternal aunt demonstrate slightly reduced creatine and phosphocreatine levels in the parietal white matter (Figs 3a and 3b). MRS performed in other predominantly gray matter structures indicated essentially normal creatine and phosphocreatine levels. The maternal uncle has severe mental retardation. He is not available for studies at the present.

The patient's mother, age 27 years, had a Full Scale IQ of 76 on the Wechsler Abbreviated Scale of Intelligence (WASI, [7]). Academic achievement scores on the Wide Range Achievement Test, 3rd edition (WRAT-3, [8]) were in the borderline to low average range (Reading: 88, Arithmetic: 77). A confrontational naming deficit and a severe verbal memory deficit were noted. Her spoken language, however, was fluent and appropriate. Visual-constructional ability and fine motor skills were intact.

The maternal aunt, age 30 years, had a Full Scale IQ of 88 on the WASI. Academic achievement scores on the WRAT-3 were in the average range (Reading: 103, Arithmetic: 108). Similar to her sister, her spoken language was fluent and appropriate but a mild confrontational naming weakness and a very mild verbal memory deficit were noted.

The maternal grandmother, age 64 years, had a Full Scale IQ of 79 on the WASI. Academic achievement scores on the WRAT-3 were in the low average range (Reading: 88, Arithmetic: 84). Her spoken language was fluent but somewhat odd and occasionally tangential. Similar to her daughters, a

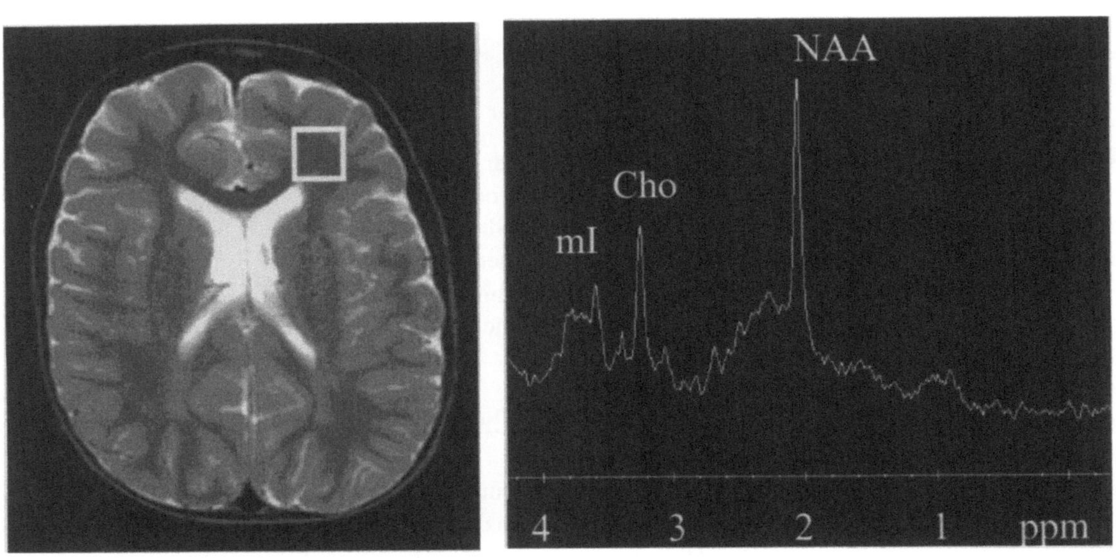


*Fig. 1.* MR Imaging and Spectroscopy of index patient following creatine supplementations. (a) T2 weighted image and (b) Short echo (TE 35 msec, TR 2000 msec) single voxel PRESS volume (8 cc).

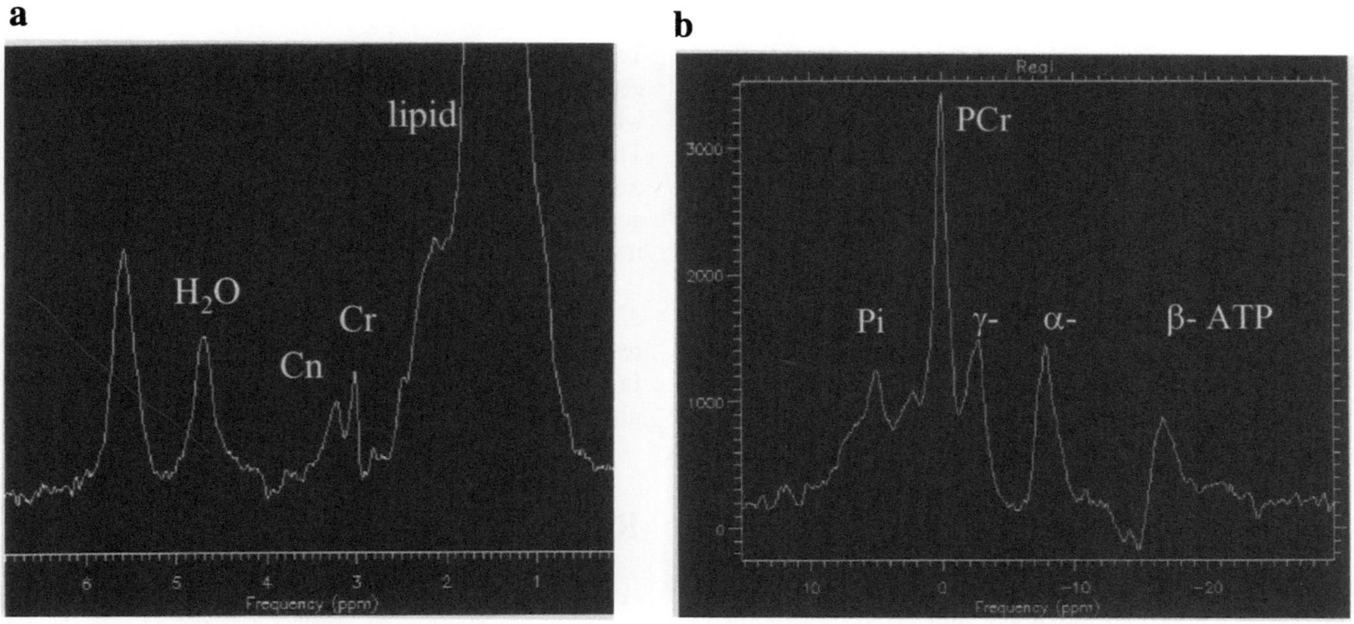


*Fig. 2.* (a) Proton MRS of thigh muscles of index patient at 7 years. Creatine and phosphocreatine resonance appears at 3.0 ppm. Other resonances include lipids (0.3–2.7 ppm), carnitine (3.2 ppm) and residual water (4.7 and 5.6 ppm). (b) Limited Phosphorus MRS of thigh muscles of patient WP at 7 years. Phosphocreatine resonance appears at 0.0 ppm. Other resonances include adenosine triphosphate (γ at −2.5, α at −8.2 and β at −17 ppm). The patient awoke during the procedure, and reference compound was not obtained.

mild confrontational naming deficit was noted. Verbal memory was not assessed due to time constraints. In contrast to her daughters' findings, visual-constructional ability and fine motor skills were severely impaired.

## Discussion

The discovery of a defect in the X-linked creatine transporter (SLC6A8/CT1) in a 6 year old male using proton MRS with

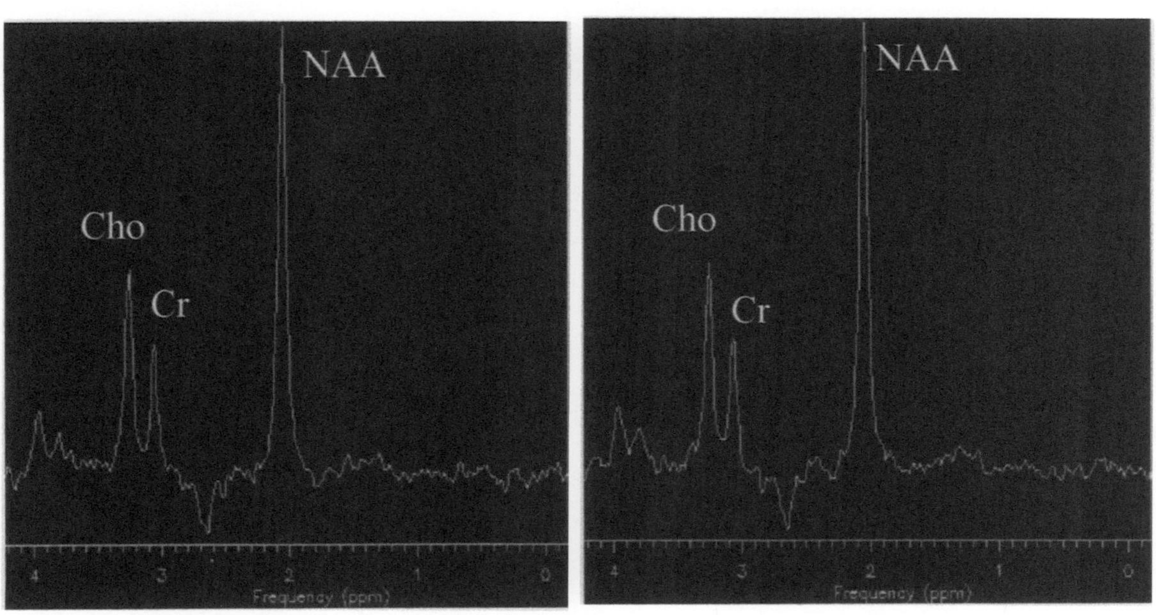


*Fig. 3.* Long echo (TE 288 msec, TR 2000 msec) single voxel MRS within parietal white matter of the (a) mother and (b) maternal aunt of the index patient.

no significant imaging findings provided a new approach towards diagnosing mental retardation syndromes. In this patient, brain involvement appears to be the most significant feature. Unfortunately, the preliminary attempts at creatine restoration using oral supplementation failed to demonstrate any improvement. The uptake of creatine into the CSF remains unexplained.

Our patient does not have clinical evidence of myopathy, which was initially not expected, since muscle contains the highest concentration of creatine and creatine depletion in animals causes severe muscle weakness. It is very surprising that creatine is present in the muscles of our patient. As far as we know, two creatine transporter genes are present in humans. SLC6A8/CT1 is located at Xq28 and is expressed in most tissues, including the CNS, muscle and heart [9]. A second transporter was found at 16p11.2 (SLC6A8/CT2), which is expressed in testis only. Of course, it is possible that additional transporters (both specific or non-specific) exist that carry creatine into muscle and other organs. On the other hands, diffusion processes from the blood stream, independent of the creatine transporter, may provide the necessary creatine requirements in muscle. Further studies may elucidate this. In GAMT deficient patients, the main feature of the disease is also the brain involvement, although some children suffering from GAMT deficiency have developmental motor delay and muscle weakness. The low creatine levels arising from the absence of creatine biosynthesis in those patients may indeed affect muscle. Reduced creatine levels (8.8 mmol/kg; normals ca 19–21 mmol/kg) have been found by proton and phosphorus MRS of the muscle in a GAMT deficient patient [10]. It seems clear that the congenital creatine deficiency syndromes affect the CNS more than any other organ in the human body without any good explanation [3, 4].

Our patient has a diagnosis of mental retardation with specific developmental speech language impairment (SLI). Children with specific SLI sometimes have abnormalities on MRI [11]. However, it is not clear how these imaging abnormalities relate to the SLI. Also, MR spectroscopy abnormalities have been found in patients with dyslexia [12], but these abnormalities did not involve the creatine concentrations levels.

With regard to the neuropsychological findings, the patient is so globally impaired that meaningful patterns of performance cannot be identified. However, in the female carriers of the genetic defect, there is the interesting finding of confrontational naming deficit in all three who were assessed. The remainder of their neuropsychological profiles vary, however. Two of the female relatives demonstrated intact visual-perceptual and visual-constructional ability and fine motor skills whereas the third was significantly impaired in these areas. These findings suggest that it will be useful to assess cognition and language in female carriers in order to fully characterize the effect of the gene responsible for the creatine transporter deficiency.

We continue to study this family to fully characterize this disorder. Hopefully, this new disorder will encourage further research into the dynamics of creatine transport into tissues. In addition, the significance of creatine metabolism for cognitive function should be explored.

## References


1. Stockler S, Hozbach U, Hanefeld F: Creatine deficiency in the brain: A new, treatable inborn error of metabolism. Pedatr Res 36: 509–413, 1994
2. Item CB, Stockler-Ipsiroglu S, Stromberger C, Muhl A, Alessandri MG, Bianchi MC, Tosetti M, Fornai F, Cioni G: Arginine: Glycine amidinotransferase (AGAT) deficiency: The third inborn error of creatine metabolism. J Inherit Metab Dis 24: 118, 2001
3. Cecil KM, Salomons GS, Ball WS Jr, Wong B, Chuck G, Verhoeven NM, Jakobs C, DeGrauw TJ: Irreversible brain creatine deficiency with elevated serum and urine creatine: A creatine transporter defect? Ann Neurol 49: 401–404, 2001
4. Salomons GS, van Dooren SJ, Verhoeven NM, Ecil KM, Ball WS, DeGrauw TJ, Jakobs C: X-linked creatine-transporter gene (SLC6A8) defect: A new creatine-deficiency syndrome. Am J Hum Genet 68: 1497–1500, 2001
5. Bayley N: In: Bayley Scales of Infant Development. The Psychological Corporation, San Antonio, 1993
6. Sparrow SS, Balla DA, Cicchetti DV: In: Vineland Adaptive Behavior Scales. American Guidance Service, Circle Pines, MN, 1984
7. Wechsler Abbreviated Scale of Intelligence. The Psychological Corporation, San Antonio, 1999
8. Wilkinson G: In: Wide Range Achievement Test, Revision Three. Wide Range Inc., Wilmington, DE, 1993
9. Snow RJ, Murphy RM: Creatine and the creatine transporter: A review. Mol Cell Biochem 224: 169–181, 2001
10. Ensenauer R: Presence of muscle creatine in a patient with guanidinoacetate methyltransferase (GAMT) deficiency. J Inherit Metab Dis 23: 212, 2000
11. Trauner D, Wulfeck B, Tallal P, Hesselink J: Neurological and MRI profiles of children with developmental language impairment. Dev Med Child Neurol 42: 470–475, 2000
12. Rae C, Lee MA, Dixon RM, Blamire AM, Thompson CH, Styles P, Talcott J, Richardson AJ, Stein JF: Metabolic abnormalities in developmental dyslexia detected by $^{1}H$ magnetic resonance spectroscopy. Lancet 351: 1849–1852, 1998

*Molecular and Cellular Biochemistry* **244:** 49–55, 2003.


# Creatine supplementation in health and disease: What is the evidence for long-term efficacy?

Wim Derave, Bert O. Eijnde and Peter Hespel
*Exercise Physiology and Biomechanics Laboratory, Department of Kinesiology, Faculty of Physical Education and Physiotherapy, Catholic University of Leuven, Leuven, Belgium*

## Abstract

Creatine supplementation is an established ergogenic aid in sports and is now claimed to have therapeutical applications in a variety of diseases. The available literature mainly covers the short-term (one to several weeks) effects of creatine supplementation on skeletal muscle function in health and disease, which is of little help to evaluate the long-term (two or more months) potential of creatine as a drug in chronic disorders, such as neurodegenerative diseases or muscular dystrophies. Recent findings in healthy humans indicate that the beneficial effect on muscle function and muscle total creatine content may disappear when creatine is continuously ingested for more than two or three months. The mechanism for this habituation to chronic creatine exposure is poorly understood. The primary purpose of the present review article is to critically evaluate the available evidence for long-term efficacy of creatine administration and to hypothesize about ways to optimize creatine administration regimens. (Mol Cell Biochem **244**: 49–55, 2003)

*Key words*: creatine supplementation, skeletal muscle, energy metabolism

## Introduction

Creatine has become a popular 'ergogenic' supplement in sports. It has been intensively studied which exercise disciplines do and which do not benefit from supplementation-induced creatine accumulation in muscle [1]. The use of creatine in sports has prompted interest in creatine as a potential agent to be used in clinical therapeutical conditions. The number of diseases for which creatine is hypothesized to be possibly effective, is rapidly growing. For some diseases, the expectations are based on positive evidence obtained in animal models, like Parkinson's disease [2], Huntington's disease [3] and amyotrophic lateral sclerosis [4]. However, there is also some preliminary positive evidence from small scale clinical trials, notably in gyrate atrophy [5], McArdle's disease [6], various forms of muscular dystrophy [7], genetic deficiencies of L-arginine:glycine amidinotransferase (AGAT) [8] or guanidinoacetate methyltransferase (GAMT) [9] and congestive heart failure [10]. In addition, some 'claims' have an anecdotal background or are based on pure hypotheses, such as in aging [11] and herpes [12].

In athletic practice, periods of creatine supplementation in conjunction with high intensity training, often are interspersed with periods of lower training workload without creatine supplementation. Such intervals obviously can not be defined in chronic and progressive neurological and muscular diseases. Therefore, when advised or examined as a clinical substance, creatine as a rule is administered on a continuous and long-term base. As an example, in a number of ongoing clinical trials exploring the potential benefit of long-term creatine supplementation in patients with Huntington's disease, amyotrophic lateral sclerosis, muscular dystrophies, or aging, creatine is supplemented daily at a rate of 3–10 g/day for several months up to 1 year. However, whether such creatine intake regimens are valid is entirely unclear. In fact, knowledge of the long-term effects of creatine supplementation, in relation to the creatine supplementation protocol used, is very poor. There is some evidence in the literature to suggest that long-term creatine supplementation can cause some tissue adaptations that eventually might render the body insensitive to creatine intake (as will be discussed below in 'Long-term creatine supplementation'). If

*Address for offprints*: P. Hespel, Exercise Physiology and Biomechanics Laboratory, Faculty of Physical Education and Physiotherapy, Tervuursevest 101, B-3001 Leuven, Belgium (E-mail: peter. hespel@flok.kuleuven.ac.be)

such mechanism proves to be true, this may conceivably compromise the likeliness of success in clinical trials evaluating the therapeutical benefit of creatine. The primary aim of the current paper is to compare literature data with regard to long-term creatine supplementation with knowledge on short-term effects. We try to critically evaluate the validity of extrapolating data from short-term studies in healthy subjects, to long-term effects in patient populations. Literature results on healthy as well as ill subjects are discussed. Furthermore, we hypothesize about strategies of how to overcome possible habituation to creatine in the long term, and we try to raise some pivotal issues for future research.

## Short-term creatine supplementation (< 2 months)

### *Studies in healthy humans*

The effects of short-term creatine supplementation (< 2 months) on human skeletal muscle are well documented. In their original paper in 1992, Harris *et al.* [13] showed that nearly every individual responds to several days of creatine supplementation by increasing muscle total creatine content, resulting from both increased phosphocreatine and free creatine. They also demonstrated that the increase of muscle creatine content resulting form creatine intake was highly dependent on the initial creatine content. Thus, short-term creatine supplementation in any person eventually resulted in a muscle creatine content in the range of 140–150 mmol/(kg d.w.). However, the greater increases occurred in the subjects with low creatine contents, whilst only a small, if any effect was seen in individuals with high initial muscle creatine content.

In a follow-up study by Greenhaff *et al.* [14], where in it was shown that creatine supplementation can facilitate post-exercise muscle phosphocreatine resynthesis in some but not all individuals, a clear-cut distinction was made between responders and non-responders to creatine supplementation. Despite the elegance of this study, the validity of a black-and-white presentation of sensitivity to creatine supplementation can be questioned. Given that total creatine content in human muscle is normally distributed [15], with an average value of ~ 125 mmol/(kg d.w.), it can be predicted that a large number of individuals is located in the 'gray zone' between responders and non-responders, as defined by Greenhaff *et al.* [14]. Several laboratories have also shown that subjects with the highest increase in total creatine (TCr) content (defined as the sum of phosphocreatine and free creatine) also exhibit the highest gain in muscle function [16–18]. Thus, taking the magnitude of rise in TCr content in muscle as a measure of the effectiveness of creatine supplementation with regard to enhancing muscle function seems appropriate. This is compatible with the tight coupling between metabolic and contractile characteristics in muscle cells.

Apart from the well-described effect of creatine on the contractile properties of muscle, creatine loading has recently also been shown to promote glycogen accumulation in human muscle [19–22]. It is interesting to consider whether also with regard to this metabolic effect of creatine supplementation, the increase in TCr is an appropriate criterion to distinguish between responders and non-responders. In a recent study we could not find a relationship between the increase in TCr content and the increase in glycogen content upon 5 weeks of creatine supplementation [19]. A likely explanation for the stimulatory effect of creatine on glycogen accumulation is the increase in the glucose transporter, GLUT4, expression [19]. Interestingly, in Fig. 1, we show the increase in GLUT4 expression is positively related to the increase in TCr content in 8 subjects who received 15 g creatine per day during 3 weeks of rehabilitation training following 2 weeks of cast immobilization of a leg. This could suggest that the change in TCr content is a good estimate of responsiveness to short-term creatine supplementation with regard to various metabolic adaptations in the muscle. Still, creatine supplementation does not only affect muscle, but also other tissues. Horn *et al.* have shown that creatine supplementation in rats leads to an accumulation of creatine in liver and kidney, but not in heart, brain and skeletal muscle [23]. Thus, evaluation of the efficacy of creatine supplementation to enhance muscle contractile and metabolic functions based on measurement of muscle TCr content is probably valid, but should not be extrapolated to other tissues and organs.

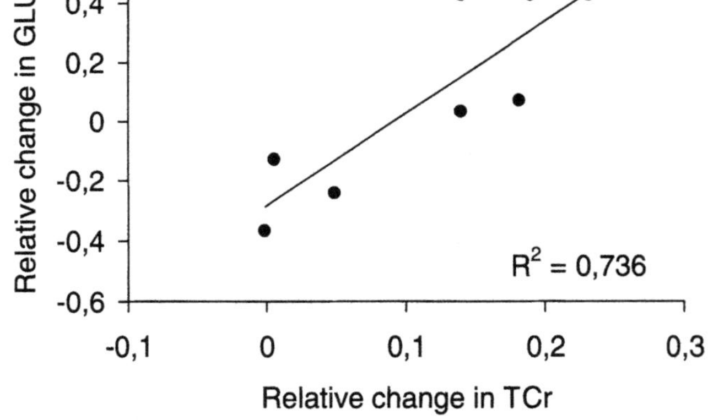


*Fig. 1.* The change in total creatine content of the right vastus lateralis muscle during three weeks of creatine supplementation, compared with the change muscle GLUT4 content during the same period. The change is calculated between the value obtained following 2 weeks of cast immobilization and the value obtained after 3 weeks of heavy rehabilitation training. Data are from Eijnde *et al.* [19].

*Clinical studies*

Similar to what is reported in healthy individuals, there is some evidence to suggest that short-term creatine supplementation can improve muscle strength and high-intensity exercise performance in patients with various neuromuscular disorders. In 1997, Tarnopolsky *et al.* investigated the potential of short-term creatine supplementation (3 weeks, 10 g/day down to 2 g/day) on muscle strength in seven patients with mitochondrial cytopathies, and found the intervention to improve handgrip strength by ~ 20%. Soon after, the same research group confirmed that 11 days of creatine supplementation (10 g/day for 5 days and 5 g/day thereafter) in 81 patients with various neuromuscular disorders, including dystrophies and other myopathies, increased maximal isometric and isokinetic strength of the knee extensors [24]. This finding of improved muscle performance has been obtained in other placebo-controlled studies, involving nine patients with McArdle's disease [6] and 36 patients with muscular dystrophies [25], who were supplemented with creatine for 5 and 8 weeks, respectively. Conversely, Klopstock *et al.* did not find a beneficial effect of short-term creatine supplementation (4 weeks, 20 g/day) on muscle performance in a placebo-controlled crossover trial in patients with mitochondrial diseases [26]. Furthermore, preliminary results from an uncontrolled study in patients with amyotrophic lateral sclerosis reported mild improvement in muscle force following one week of creatine supplementation (20 g/day).

The above observations tentatively indicate that short-term creatine supplementation can be a useful symptomatic therapy in some neuromuscular disease conditions. However, do these findings prove that creatine is an adequate long-term intervention strategy impacting on the pathogenesis of the aforementioned diseases? The answer is 'no'. It is not so much of a surprise that the short-term effect of creatine supplementation on muscle function is acutely present in patients as it is in healthy humans, but it does not address its therapeutical value in the long term. First, we need to know whether muscle creatine content can be maintained at an elevated level for longer periods in healthy subjects and in various groups of patients. Second, it must be investigated whether creatine supplementation for several months or years can delay or optimally-prevent the disease progression in neurological and muscular disorders. We address these questions in the following paragraphs.

## Long-term creatine supplementation (> 2 months)

*Studies on healthy humans*

Compared to the large number of studies that have examined short-term creatine supplementation in humans, there is a shortage of data on supplementation periods beyond 2 months (see Fig. 2). One of the first studies on the ability of creatine supplementation to keep muscle creatine elevated for several months was performed by Vandenberghe *et al.* [27]. Although only phosphocreatine was measured by NMR spectroscopy, this study was the first to clearly show that phosphocreatine can remain elevated compared to placebo throughout the first 10 weeks of supplementation, whereas a continuation of supplementation for an additional 10 weeks shows a slight decline in muscle phosphocreatine content [27]. As shown in Fig. 3, a more recent study from our laboratory has shown that the initial rise in TCr content in human vastus lateralis after 2 and 5 weeks of creatine supplementation has disappeared after 12 weeks of continued supplementation [28]. A similar findings has been reported by Volek *et al.*, as the elevation of TCr content of the human vastus lateralis was 22% ($p < 0.05$) after the one week loading phase (25 g/day) and was reduced to 10% (n.s.) following 12 weeks of creatine supplementation (5 g/day) [29]. Thus, effect sizes after 12 weeks are markedly smaller than after 1–5 weeks, despite continued creatine supplementation (Fig. 2). Thus, it seems that at a time point between 5 and 12 weeks, continuous creatine supplementation starts to fail keeping up the TCr in human muscle. Steenge [30] has shown that following 8 weeks of creatine supplementation (3 g/day) in women, muscle TCr is at the elevated level where it was after the initial 5 days of creatine loading (20 g/day). Thus, in this study habituation to continuous creatine supplementation did not take place during an 8 week period. Accordingly, these data are helpful

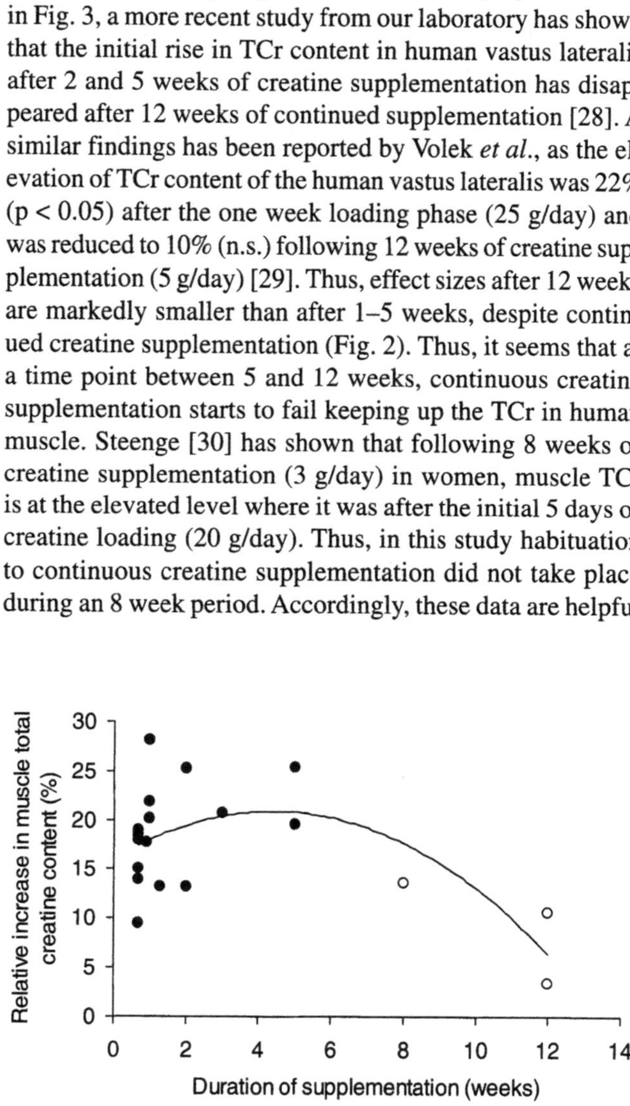


*Fig. 2.* Literature data reporting the increment in total creatine (TCr) content in human skeletal muscle in response to various durations of oral creatine supplementation. All studies were performed on healthy young humans and supplementation consisted of a loading phase of 5–14 days (15–20 g/day) and in some cases a subsequent period on maintenance dose (2–5 g/day). Filled circles indicate significant increments in TCr and open circles indicate non-significant effects. Data are from references [14, 16, 17, 21, 28, 29, 44–49] and additional unpublished observations from our laboratory.

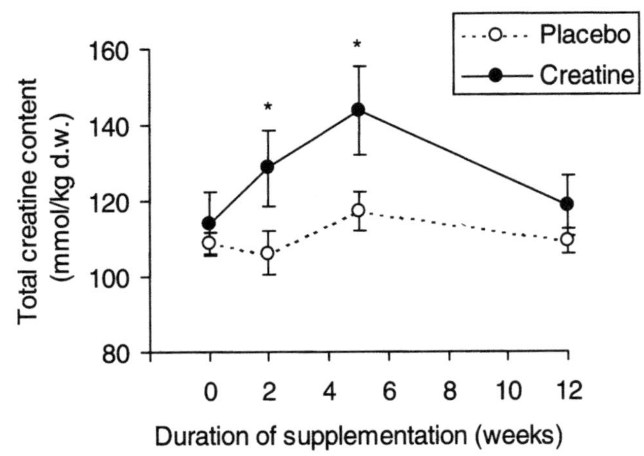


*Fig. 3.* Effect of 12 weeks of oral creatine supplementation on the total creatine content (sum of phosphocreatine and free creatine) of the right vastus lateralis. During the initial 2 weeks, creatine monohydrate or maltodextrine (placebo) was ingested at 20 g/day and the subject's right leg was cast-immobilized. The creatine dose was subsequently reduced to 15 g/day (weeks 2–5) and to 5 g/day (weeks 5–12) and during the ten weeks of rehabilitation. Data are means ± S.E. and are extracted from reference [28]. $*p < 0.05$ significant difference compared to the baseline value.

in defining the onset of failure of creatine supplementation, which, with the limited amount of presently available data, is presumed to lie between 2 and 3 months.

Recently, we evaluated the effects of creatine supplementation for 6 months (5 g/day) in males 55–75 years old (unpublished observations). Throughout this period the subjects participated in a resistance training program. Muscle biopsies were taken from the vastus lateralis for determination of total creatine content at baseline and after 3 and 6 months of supplementation. There was no increase in muscle TCr after 6 months. We can not define whether this failure of creatine supplementation to increase muscle creatine content was due to either the high age of the subjects, or to the long duration of the supplementation.

At present, it is unknown why during long-term creatine supplementation muscle TCr, after an initial increase, tends to return to baseline despite continued creatine intake. Interestingly, Guerrero-Ontiveros and Wallimann have shown that three months of high-dose creatine supplementation in rats results in a marked downregulation in the expression of creatine transporter isoforms in skeletal muscle [31]. An alternative mechanism for creatine-induced inhibition of creatine transport is proposed by Loike *et al.* [32]. They have observed that inhibition of protein synthesis by cycloheximide can prevent downregulation of creatine transport in L6 cells, suggesting that prolonged creatine exposure *in vitro* induces the synthesis of a protein that inhibits creatine transporter activity and/or localization [32]. Either mechanism provides an interesting hypothesis as to how muscle may control intracellular creatine homeostasis. However, it remains to be explored in humans whether the reduction in muscle TCr content beyond two months of creatine supplementation is due to downregulation of the creatine transporter. In fact we even don't know whether muscle creatine transport or the creatine transporter expression is a limiting factor in the creatine content in muscle cells *in vivo*. Murphy et al. have recently compared the TCr and creatine transporter content of various hindlimb muscles of the rat [33]. They surprisingly found the highest TCr content in the muscle type (white gastrocnemius) with the lowest creatine transporter expression. It therefore remains to be established what the exact role of creatine transporter is in muscle and if its expression is related to habituation in long-term creatine supplementation.

### *Clinical studies*

The efficacy of long-term creatine therapy in various pathological conditions is hardly studied and poorly understood at present. In this respect, it is important to distinguish between (1) diseases in which the pathogenesis is causally linked with a deficiency of creatine metabolism, (2) diseases in which creatine deficiency is a secondary symptom which still contributes to part of the pathogenesis, and (3) diseases without manifestation of creatine deficiency. In the first two categories, long-term creatine therapy may be able to slow or prevent the progression of the disease (category 1) or an aspect of the disease (category 2). In the third category, it is presently unsure whether creatine supplementation can provide a long-term therapeutic help.

*Category 1.* Diseases in which deficiency of creatine stores is the primary pathogenic mechanism are related to a genetic error causing deficiency of a protein involved in biosynthesis or transport of creatine, such as AGAT, GAMT and creatine transporter. Stockler *et al.* [9] have reported a case of an infant with GAMT deficiency and consequently severe creatine deficiency in body and brain. A 25-month treatment with creatine supplementation resulted in substantial clinical improvements and normalization of creatine stores. Similarly, the recently discovered mentally retarded infants with AGAT deficiency and absence of creatine in the brain, showed normalization of brain creatine content and functional improvement following creatine supplementation [8, 34]. A third disease is the recently found inborn deficiency in creatine transporter [35]. Although this disease is also characterized by absence of creatine in the brain, creatine substitution is of no help [36], since not the availability but rather the distribution of the creatine pool within the body is impaired. It can be concluded that even though no data of long-term clinical trials are available yet, AGAT and GAMT deficiency are diseases that may be partly or even completely cured by means of long-term creatine substitution therapy.

*Category 2.* Some diseases are characterized by severe creatine deficiency and muscle atrophy, although it is not the primary pathogenic factor. Gyrate atrophy is a syndrome of severe deterioration of vision, ultimately leading to blindness. It is caused by deficiency of ornithine-aminotransferase, resulting in elevated plasma ornithine concentrations, which inhibits creatine biosynthesis. Skeletal muscles of patients with gyrate atrophy exhibit reduced creatine content [37] and atrophied type II-fibers [38], symptoms that can be reversed by long-term (studied up to 15 years) oral creatine supplementation [5, 38]. However, the creatine therapy did not slow the visual deterioration of the patients [38]. Gyrate atrophy belongs to a category of diseases that can only be partly cured by long-term creatine supplementation, supposedly because creatine depletion is not the sole key pathogenic factor of the disease. Several other diseases, such as mitochondrial cytopathies, also belong to this category and future studies will need to shown if creatine can provide a therapeutical benefit in the long term.

*Category 3.* A potential therapeutic benefit of creatine is also anticipated in some diseases that are not typically characterized by depletion of the body creatine pool. In these disorders, creatine is not regarded as a substitution of a deficiency, but is intended to increase the survival of degenerating neurons or muscle cells. In amyotrophic lateral sclerosis and Huntington's disease, an increased cerebral energy reserve by extra creatine might be able to slow the progression of the neurodegeneration. Animal studies have shown that a 2% creatine supplementation works neuroprotective [3, 4]. So far, only one study on patients with ALS has been published by Mazzini *et al.* [39] and they did not detect a slowing of the deterioration of muscle function after 6 months of creatine supplementation (3 g/day). A possible explanation for this discrepancy between human and mouse studies, may be the duration of continuous supplementation in humans. Since ALS is not characterized by considerable creatine deficiency, the expected elevation of neuronal and muscular creatine content is probably only temporary and may wear off as habituation to creatine supplements takes place after several months, similarly to what occurs in healthy subjects (see above). However, results of large-scale placebo-controlled clinical trials are pending and will soon provide us the answer.

## Improving the response to long-term creatine supplementation

In one of the previous sections we discussed the difficulty of keeping muscle TCr content elevated for periods longer than 2 months in healthy humans. The search for ways to avoid this habituation to creatine supplementation is much like a 'trial and error', and should therefore be conducted on healthy humans. Here, we would like to propose two hypothetical strategies, which still await experimental proof. The first strategy is to try to maximize cellular creatine uptake by combining chronic creatine supplementation with exercise and/or with carbohydrate ingestion. The second strategy is to try to avoid the presumed downregulation of the creatine synthesis and transport system by intermittent creatine supplementation.

The former strategy, i.e. to promote creatine uptake, has been shown to be effective in short-term supplementation. Harris *et al.* [13] have compared the effects of creatine supplementation with or without a daily 60-min submaximal cycling exercise bout. The mean muscle TCr concentration increased by 37% when creatine was combined with exercise, compared to 26% with creatine alone. Thus, submaximal exercise promotes muscle creatine accumulation, an effect that is limited to the exercising muscles, as later shown by Robinson *et al.* [21]. However, in several of the diseases for which creatine has a potential therapeutic benefit, patients are not able to exercise. Therefore, maximization of muscle creatine uptake by co-ingestion of carbohydrate may be more appropriate. Several consistent findings by the research group of Greenhaff have clearly shown that the insulin release in response to ingestion of relatively large amounts of carbohydrate or carbohydrate plus protein potentiate creatine accumulation in human muscle [40–43]. However, it is not known whether this is a valuable strategy in chronically maximizing creatine accumulation in humans. If the restoration of baseline TCr concentrations in long-term creatine supplementation is due to a downregulation of the creatine transporter, then co-ingestion of carbohydrates will probably speed up the downregulation and prove unable to prolong TCr elevation. Further research is warranted on this issue.

Possibly a more direct way to overcome the reduced effectiveness of long-term supplementation would be to avoid downregulation of the creatine synthesis and transport system. Although the latter is not definitely proven to take place in humans, it currently presents the most likely explanation. Ideally, one should find an intermittent creatine supplementation protocol which keeps TCr content in muscle elevated at all time but still allows periods without supplementation to stimulate the expression of enzymes involved in endogenous creatine synthesis and transport. The important factors in designing such a protocol are (1) the time it takes for Cr to leave the muscle following cessation of supplementation, i.e. the 'wash-out' period, and (2) the duration of creatine supplementation after which the presumed downregulation takes place.

As to the former, several studies have indicated that one month without supplementation is enough to return the muscle TCr concentration to baseline level [27, 44, 45]. Interestingly, Hultman *et al.* have shown that 21 days following cessation of creatine loading, muscle TCr was still elevated compared to baseline. Thus, during long-term discontinuous creatine supplementation, an interval without supplementation of less than 3 weeks will probably not cause a decrease in TCr.

The duration of creatine supplementation, after which the presumed downregulation takes place, is more difficult to define. As stated above (see 'Long-term creatine supplementation'), TCr probably decreases beyond 8 weeks of creatine supplementation, no matter if creatine ingestion is continued or not. Thus, waiting 8 weeks before including a non-supplementation period is probably too late to prevent a decline in TCr. Therefore, supplementation periods of 2–6 weeks, interspersed with periods of 2 weeks without supplementation is theoretically expected to be an effective protocol for long-term intermittent creatine supplementation. Nevertheless, more than one study will be necessary in the future to determine the effectiveness and the ideal protocol of discontinuous creatine supplementation.

## Conclusions

In today's literature, most but not all studies report a positive effect of short-term (less than 2 months) oral creatine supplementation on muscle function in both healthy and ill humans. On the other hand, this phenomenon could well be a temporary effect as several studies have now shown that continuous creatine intake for three months or more can lead to a habituation in healthy muscle. Over recent years, the application possibilities of creatine as a nutritional supplement have been expanded to the clinical field, but many pivotal questions on the long-term efficacy of creatine still remain unanswered.

## Acknowledgements

The authors gratefully acknowledge the assistance by Monique Ramaekers. Creatine research in our laboratory was supported by grants from the 'Onderzoeksraad K.U.-Leuven' (grant # OT94/31 and OT99/38), from the Flemish Fund for Scientific Research (FWO-Vlaanderen grant #'s G.0331.98 and G.0255.01). Creatine supplements were kindly provided by either Degussa AG (Trostberg, Germany), MJJ Power (Brussels, Belgium), the Avicena Group Inc. (Cambridge, USA) or Numico Research (Wageningen, The Netherlands). Wim Derave is a recipient of a post-doctoral fellowship from the Flemish Fund for Scientific Research (F.W.O.-Vlaanderen). Bert O. Eijnde obtained a research fellowship from the 'Onderzoeksraad K.U. Leuven' (grant #OT99/38).

## References

1. Hespel P, Op't Eijnde B, Derave W, Richter EA: Creatine supplementation: Exploring the role of the creatine kinase/phosphocreatine system in human muscle. Can J Appl Physiol 26(suppl): 79–103, 2001
2. Matthews RT, Ferrante RJ, Klivenyi P, Yang L, Klein AM, Mueller G, Kaddurah-Daouk R, Beal MF: Creatine and cyclocreatine attenuate MPTP neurotoxicity. Exp Neurol 157: 142–149, 1999
3. Matthews RT, Yang L, Jenkins BG, Ferrante RJ, Rosen BR, Kaddurah-Daouk R, Beal MF: Neuroprotective effect of creatine and cyclocreatine in animal models of Huntington's disease. J Neurosci 18: 156–163, 1998
4. Klivenyi P, Ferrante J, Matthews RT, Bogdanov MB, Klein AM, Andreassen OA, Mueller G, Wermer M, Kaddurah-Daouk R, Beal MF: Neuroprotective effects of creatine in a transgenic animal model of amyotrophic lateral sclerosis. Nat Med 5: 347–350, 1999
5. Heinanen K, Nanto-Salonen K, Komu M, Erkintalo M, Alanen A, Heinonen OJ, Pulkki K, Nikoskelainen E, Sipila I, Simell O: Creatine corrects muscle $^{31}$P spectrum in gyrate atrophy with hyperornithinaemia. Eur J Clin Invest 29: 1060–1065, 1999
6. Vorgerd M, Grehl T, Jager M, Muller K, Freitag G, Patzold T, Bruns N, Fabian K, Tegenthoff M, Mortier W, Luttmann A, Zange J, Malin JP: Creatine therapy in myophosphorylase deficiency (McArdle disease): A placebo-controlled crossover trial. Arch Neurol 57: 956–963, 2000
7. Felber S, Skladal D , Wyss M, Kremser C, Koller A, Sperl W: Oral creatine supplementation in Duchenne muscular dystrophy: A clinical and $^{31}$P magnetic resonance spectroscopy study. Neurol Res 22: 145–150, 2000
8. Item CB, Stockler-Ipsiroglu S, Stromberger C, Muhl A, Alessandri MG, Bianchi MC , Tosetti M, Fornai F, Cioni G: Arginine:glycine amidinotransferase deficiency: The third inborn error of creatine metabolism in humans . Am J Hum Genet 69: 1127–1133, 2001
9. Stockler S, Hanefeld F, Frahm J: Creatine replacement therapy in guanidinoacetate methyltransferase deficiency, a novel inborn error of metabolism. Lancet 348: 789–790, 1996
10. Gordon A, Hultman E, Kaijser L, Kristjansson S, Rolf CJ, Nyquist O, Sylven C: Creatine supplementation in chronic heart failure increases skeletal muscle creatine phosphate and muscle performance. Cardiovasc Res 30: 413–418, 1995
11. Tarnopolsky MA: Potential benefits of creatine monohydrate supplementation in the elderly. Curr Opin Clin Nutr Metab Care 3: 497–502, 2000
12. Ness SR, McCarty MF: Does supplemental creatine prevent herpes recurrences? Med Hypotheses 57: 310–312, 2001
13. Harris RC, Söderlund K, Hultman E: Elevation of creatine in resting and exercised muscle of normal subjects by creatine supplementation. Clin Sci 83: 367–374, 1992
14. Greenhaff PL, Bodin K, Soderlund K, Hultman E: Effect of oral creatine supplementation on skeletal muscle phosphocreatine resynthesis. Am J Physiol 266: E725–E730, 1994
15. Harris RC, Hultman E, Nordesjo LO: Glycogen, glycolytic intermediates and high-energy phosphates determined in biopsy samples of musculus quadriceps femoris of man at rest. Methods and variance of values. Scand J Clin Lab Invest 33: 109–120, 1974
16. Casey A, Constantin-Teodosiu D, Howell S, Hultman E, Greenhaff PL: Creatine ingestion favorably affects performance and muscle metabolism during maximal exercise in humans. Am J Physiol (Endo Metab) 271: E31–E37, 1996
17. Snow RJ, McKenna MJ, Selig SE, Kemp J, Stathis CG, Zhao S: Effect of creatine supplementation on sprint exercise performance and muscle metabolism. J Appl Physiol 84: 1667–1673, 1998
18. Vandenberghe K, Van Hecke P, Van Leemputte M, Vanstapel F, Hespel P: Phosphocreatine resynthesis is not affected by creatine loading. Med Sci Sports Exerc 31: 236–242, 1999

19. Eijnde BO, Urso B, Richter EA, Greenhaff PL, Hespel P: Effect of oral creatine supplementation on human muscle GLUT4 protein content after immobilization. Diabetes 50: 18–23, 2001
20. Nelson AG, Arnall DA, Kokkonen J, DAY AR, Evans J: Muscle glycogen supercompensation is enhanced by prior creatine supplementation. Med Sci Sports Exerc 33: 1096–1100, 2001
21. Robinson TM, Sewell DA, Hultman E, Greenhaff PL: Role of submaximal exercise in promoting creatine and glycogen accumulation in human skeletal muscle. J Appl Physiol 87: 598–604, 1999
22. Eijnde BO, Richter EA, Henquin JC, Kiens B, Hespel P: Effect of creatine supplementation on creatine and glycogen content in rat skeletal muscle. Acta Physiol Scand 171: 169–176, 2001
23. Horn M, Frantz S, Remkes H, Laser A, Urban B, Mettenleiter A, Schnackerz K, Neubauer S: Effects of chronic dietary creatine feeding on cardiac energy metabolism and on creatine content in heart, skeletal muscle, brain, liver and kidney. J Mol Cell Cardiol 30: 277–284, 1998
24. Tarnopolsky M, Martin J: Creatine monohydrate increases strength in patients with neuromuscular disease. Neurology 52: 854–857, 1999
25. Walter MC, Lochmuller H, Reilich P, Klopstock T, Huber R, Hartard M, Hennig M, Pongratz D, Muller-Felber W: Creatine monohydrate in muscular dystrophies: A double-blind, placebo- controlled clinical study. Neurology 54: 1848–1850, 2000
26. Klopstock T, Querner V, Schmidt F, Gekeler F, Walter M, Hartard M, Henning M, Gasser T, Pongratz D, Straube A, Dieterich M, Muller-Felber W: A placebo-controlled crossover trial of creatine in mitochondrial diseases. Neurology 55: 1748–1751, 2000
27. Vandenberghe K, Goris M, Van Hecke P, Van Leemputte M, Vangerven L, Hespel P: Long-term creatine intake is beneficial to muscle performance during resistance training. J Appl Physiol 83: 2055–2063, 1997
28. Hespel P, Op't Eijnde B, Van Leemputte M, Urso B, Greenhaff PL, Labarque V, Dymarkowski S, Van Hecke P, Richter EA: Oral creatine supplementation facilitates the rehabilitation of disuse atrophy and alters expression of muscle myogenic factors in humans. J Physiol (Lond) 536: 625–633, 2001
29. Volek JS, Duncan ND, Mazzetti SA, Staron RS, Putukian M, Gómez AL, Pearson DR, Fink WJ, Kraemer WJ: Performance and muscle fiber adaptations to creatine supplementation and heavy resistance training. Med Sci Sports Exerc 31: 1147–1156, 1999
30. Steenge GR: Factors affecting creatine accumulation in human skeletal muscle. PhD-thesis, University of Nottingham, 1999
31. Guerrero-Ontiveros ML, Wallimann T: Creatine supplementation in health and disease. Effects of chronic creatine ingestion *in vivo*: Down-regulation of the expression of creatine transporter isoforms in skeletal muscle. Mol Cell Biochem 184: 427–437, 1998
32. Loike JD, Zalutsky DL, Kaback E, Miranda AF, Silverstein SC: Extracellular creatine regulates creatine transport in rat and human muscle cells. Proc Natl Acad Sci USA 85: 807–811, 1988
33. Murphy R, McConell G, Cameron-Smith D, Watt K, Ackland L, Walzel B, Wallimann T, Snow R: Creatine transporter protein content, localization, and gene expression in rat skeletal muscle. Am J Physiol (Cell Physiol) 280: C415–C422, 2001
34. Bianchi MC, Tosetti M, Fornai F, Alessandri' MG, Cipriani P, De Vito G, Canapicchi R: Reversible brain creatine deficiency in two sisters with normal blood creatine level. Ann Neurol 47: 511–513, 2000
35. Salomons GS, van Dooren SJ, Verhoeven NM, Cecil KM, Ball WS, Degrauw TJ, Jakobs C: X-linked creatine-transporter gene (SLC6A8) defect: A new creatine-deficiency syndrome. Am J Hum Genet 68: 1497–1500, 2001
36. Cecil KM, Salomons GS, Ball WS Jr, Wong B, Chuck G, Verhoeven NM, Jakobs C, Degrauw TJ: Irreversible brain creatine deficiency with elevated serum and urine creatine: A creatine transporter defect? Ann Neurol 49: 401–404, 2001
37. Heinanen K, Nanto-Salonen K, Komu M, Erkintalo M, Heinonen OJ, Pulkki K, Valtonen M, Nikoskelainen E, Alanen A, Simell O: Muscle creatine phosphate in gyrate atrophy of the choroid and retina with hyperornithinaemia – clues to pathogenesis. Eur J Clin Invest 29: 426–431, 1999
38. Sipilä I, Rapola J, Simell O, Vannas A: Supplementary creatine as a treatment for gyrate atrophy of the choroid and retina. N Engl J Med 304: 867–870, 1981
39. Mazzini L, Balzarini C, Colombo R, Mora G, Pastore I, De Ambrogio R, Caligari M: Effects of creatine supplementation on exercise performance and muscular strength in amyotrophic lateral sclerosis: Preliminary results. J Neurol Sci 191: 139–144, 2001
40. Green AL, Simpson EJ, Littlewood JJ, MacDonald IA, Greenhaff PL: Carbohydrate ingestion augments creatine retention during creatine feeding in humans. Acta Physiol Scand 158: 195–202, 1996
41. Steenge GR, Simpson EJ, Greenhaff PL: Protein- and carbohydrate-induced augmentation of whole body creatine retention in humans. J Appl Physiol 89: 1165–1171, 2000
42. Steenge GR, Lambourne J, Casey A, MacDonald IA, Greenhaff PL: Stimulatory effect of insulin on creatine accumulation in human skeletal muscle. Am J Physiol 275: E974–E979, 1998
43. Green AL, Hultman E, MacDonald IA, Sewell DA, Greenhaff PL: Carbohydrate ingestion augments skeletal muscle creatine accumulation during creatine supplementation in humans. Am J Physiol (Endocrinol Metab ) 271: E821–E826, 1996
44. Hultman E, Soderlund K, Timmons JA, Cederblad G, Greenhaff PL: Muscle creatine loading in men. J Appl Physiol 81: 232–237, 1996
45. Febbraio MA, Flanagan TR, Snow RJ, Zhao S, Carey MF: Effect of creatine supplementation on intramuscular TCr, metabolism and performance during intermittent, supramaximal exercise in humans. Acta Physiol Scand 155: 387–395, 1995
46. Parise G, Mihic S, MacLennan D, Yarasheski KE, Tarnopolsky MA: Effects of acute creatine monohydrate supplementation on leucine kinetics and mixed-muscle protein synthesis. J Appl Physiol 91: 1041–1047, 2001
47. McKenna MJ, Morton J, Selig SE, Snow RJ: Creatine supplementation increases muscle total creatine but not maximal intermittent exercise performance. J Appl Physiol 87: 2244–2252, 1999
48. Balsom PD, Soderlund K, Sjodin B, Ekblom B: Skeletal muscle metabolism during short duration high-intensity exercise: Influence of creatine supplementation. Acta Physiol Scand 154: 303–310, 1995
49. Bellinger BM, Bold A, Wilson GR, Noakes TD, Myburgh KH: Oral creatine supplementation decreases plasma markers of adenine nucleotide degradation during a 1-h cycle test. Acta Physiol Scand 170: 217–224, 2000

*Molecular and Cellular Biochemistry* **244:** 57–62, 2003.


# A role for guanidino compounds in the brain

Midori Hiramatsu
*Tohoku University of Community Service and Science, Sakata, Yamagata, Japan*

## Abstract

Guanidino compounds of guanidinoethanesulfonic acid, guanidinoacetic acid, guanidinosuccinic acid, N-acetylarginine, β-guanidinopropionic acid, creatinine, γ-guanidinobutyric acid, arginine, guanidine, methylguanidine, homoarginine and α-guanidinoglutaric acid are present in the mammalian brain. These guanidino compounds except for arginine and guanidine induce seizures and convulsions in rat, rabbit and cat by intracisternal injection.

Hirudonine, audonine, α-keto-δ-guanidinovaleric acid, N,N′-dibenzoylguanidine and phenylethylguanidine are also convulsants. Levels of creatinine, guanidinoethanesulfonic acid, creatinine, guanidinoacetic acid and methylguanidine in animal brain were changed at pre- and during convulsions induced by pentylentetrazol, amygdala kindling, iron-induced epileptogenesis and so on. These convulsions are thought to be due to depressed functions of serotonergic neurons and accumulated free radicals.

Arginine is a substrate of nitric oxide production by nitric oxide synthase. α-Guanidinoglutaric acid is a generator of superoxide, hydroxyl radicals and nitric oxide, and induced C6 glial cell death. On the other hand, aminoguanidine is a free radical scavenger. Energy formation by creatine metabolism may inhibit apoptosis induced by pathogenesis. Free radical generation/ reaction and energy generation by guanidino compounds must be important key role in the brain. (Mol Cell Biochem **244**: 57–62, 2003)

*Key words*: Guanidino compounds, brain, convulsion, free radicals, energy generation

## Introduction

Guanidino compounds are present in the mammalian brain. I would like to review the role of guanidino compounds in the brain against convulsion, interaction of neurotransmitters and amino acids, reaction with free radicals and toxicity as a summary of 27 years of study.

## Guanidino compounds induce seizures and convulsions

Guanidino compounds induce seizures and convulsion. Jinnnai and Mori *et al.* [1] first published reports on guanidino compounds induced convulsions. The paper showed that the amount of guanidino compounds of arginine, γ-guanidinobutyric acid, glycocyamine (guanidinoacetic acid), taurocyamine (guanidinoethane sulfonic acid) and α-N-acetylarginine in rabbit brain were 0.1–0.2, 0.03–0.05, 0.01–0.03, 0.01–0.03 and –0.02 μM/g, respectively. The content of arginine is twice those of other guanidino compounds. These guanidino compounds are suggested to be synthesized from amino acids with transamidination of the amidino group from arginine. γ-Guanidinobutyric acid, guanidinoacetic acid and taurocyamine are synthesized from γ-aminobutyric acid, glycine and taurine. These guanidino compounds can induce tonic and clonic convulsions [1]. We found that the contents of taurocyamine, glycocyamine, γ-guanidinobutyric acid and arginine in the rat brain and the levels of these guanidino compounds are lower than those of liver [2].

In addition, there are methylguanidine, β-guanidinopropionic acid, guanidinosuccinic acid, α-guanidinoglutaric acid, N-acetylarginine, homoarginine and creatinine in the mammalian brain, and these guanidine compounds induce convulsions as well. Guanidinosuccinic acid and α-guanidinoglutaric acid are suggested to be synthesized from as-

*Address for offprints*: M. Hiramatsu, Tohoku University of Community Service and Science, 3-5-1 Iimoriyama, Sakata, Yamagata 998-8580, Japan (E-mail: midori@koeki-u.ac.jp)

partic acid and glutamic acid by transamidination with arginine. Aminoguanidine is an artificially synthesized compound and is a nitric oxide inhibitor (Table 1).

N,N′-Dibenzoylguanidine and phenylethylguanidine are also synthesized substances. N,N′-Dibenzoylguanidine induces convulsions in the mouse and rabbit by intraperitoneal injection . The pattern of convulsions in the rabbit shows that clonic movements of vibrissae and face appeared, and then tonic forelimb convulsions were induced while the head was jerked backward [3]. Phenylethylguanidine also induces seizures in cats by intraperitoneal injection [4]. Hirudonine, arcaine and audoine, which are transamidination substances of polyamines, induce seizures in rabbits [5]. α-Keto-δ-guanidinovaleric acid, of which was found in high amounts in the urine of argininemia, induced epileptic-like discharges in rabbit [6]. Intraventricular injection of α-guanidinoglutaric acid induced epileptic-like discharges in cats, rabbits and rats [7, 8]. We conclude from the above summary almost all guanidino compounds are convulsants.

## Endogenous levels of guanidino compounds after convulsion

During convulsions induced by pentylenetetrazol and electroconvulsive shock, creatinine levels in the mouse brain was elevated [9]. When pentylenetetrazol was intraperitoneally injected into rabbit, concentrations of guanidino compounds in cerebrospinal fluid were changed. That is the levels of creatinine and guanidinoacetic acid were increased after the convulsion. The levels returned to control level 2 h after injection. The arginine level gradually decreased 48 h after injection and it increased dramatically 3 days post-injection [10].

Injection of ferric chloride solution into the sensory motor cortex of rats induces epileptic-like discharges and the model is regarded as post-traumatic epilepsy [11]. In the model the levels of guanidinoacetic acid and methylguanidine in seven areas of rat brain started to increase 15 min after in-

*Table 1.*

| Guanidino compounds | | Amino acids | |
|---|---|---|---|
| Guanidine | $NH_2$<br>$NH = C - NH_2$ | | |
| Methylguanidine | $NH - CH_3$<br>$NH = C - NH_2$ | | |
| Guanidinoacetic acid | $NH$<br>$NH_2 - C - NH - CH_2 - COOH$ | $NH_2 - CH_2 - COOH$ | Glycine |
| β - Guanidinopropionic acid | $NH$<br>$NH_2 - C - NH - CH_2 - CH_2 - COOH$ | | |
| Guanidinoethanesulfonic acid | $NH$<br>$NH_2 - C - NH - CH_2 - CH_2 - SO_3H$ | $NH_2 - CH_2 - CH_2 - SO_3H$ | Taurine |
| Guanidinosuccinic acid | $NH \quad COOH$<br>$NH_2 - C - NH - CH - CH_2 - COOH$ | $COOH$<br>$NH_2 - CH - CH_2 - COOH$ | Aspartic acid |
| β- Guanidinobutyric acid | $NH$<br>$NH_2 - C - NH - CH_2 - CH_2 - CH_2 - COOH$ | $NH_2 - CH_2 - CH_2 - CH_2 - COOH$ | γ- Aminobutyric acid |
| γ - Guanidinoglutaric acid | $NH \quad COOH$<br>$NH_2 - C - NH - CH - CH_2 - CH_2 - COOH$ | $COOH$<br>$NH_2 - CH - CH_2 - CH_2 - COOH$ | Glutamic acid |
| Arginine | $NH \quad NH_2$<br>$NH_2 - C - NH - CH_2 - CH_2 - CH_2 - CH - COOH$ | | |
| *N* - Acethylarginine | $NH \quad NH - CO - CH_3$<br>$NH_2 - C - NH - CH_2 - CH_2 - CH_2 - CH - COOH$ | | |
| Homoarginine | $NH \quad NH_2$<br>$NH_2 - C - NH - CH_2 - CH_2 - CH_2 - CH - COOH$ | | |
| Creatinine | $NH - C = O$<br>$NH = C$<br>$N - CH_2$<br>$CH_3$ | | |
| Aminoguanidine | $NH - NH_2$<br>$NH = C - NH_2$ | | |

jection, and they decreased 60 min after injection. The levels rose again 2 months after injection, in the areas where the epileptogenetic focus was completely formed [12]. In the case of left amygdala kindling rat, levels of guanidinoacetic acid and methylguanidine in left and right amygdala were elevated 28 days after the last seizure [13].

In the three kinds of animal epileptic models, we found an elevation of guanidinoacetic acid and methylguanidine in the brain with epileptogenetic focus. The possible metabolic pathway of methylguanidine from guanidinoacetic acid through creatinine during seizures or convulsion was shown in Fig. 1. In the mechanism for induction of convulsions or seizures, methylguanidine and guanidinoacetic acid may play a role in the brain.

## Guanidino compounds affect neurotransmitters and amino acids

When N,N′-dibenzoylguanidine induced myoclonus and tonic convulsions, 5-hydroxytryptamine (serotonin) levels in the mouse brain was elevated [3]. Intravenous injection of α-guanidinoglutaric acid into the rat induces epileptic-like discharges as well. In this model 5-hydroxytryptamine levels in the right and left cortex was examined, and we found that the level in the cortex was elevated 3 min after injection, and it was decreased 10 min after injection. It recovered to control levels 30 min after injection [8]. In addition, we observed that an elevation of 5-hydroxytryptamine levels in the rabbit brain during convulsions induced by phenylethylguanidine [4]. These results suggest that 5-hydroxytryptamine may be involved in the mechanism for seizures and convulsion.

Synaptosomal membrane fluidity affects neuronal function. Guanidinoethane sulfonic acid decreased membrane fluidity and taurine increased it [14]. The administration of guanidinoethane sulfonic acid lowered taurine level in brain, liver and muscle of mouse [15]. As the chemical structure of guanidinoethane sulfonic acid is close to taurine, they may act competitively.

## Guanidino compounds react with free radicals

Recently there are some reports on free radicals associated with apoptosis. It is suggested that guanidino compounds on apoptosis function during α-guanidinoglutaric acid in a water solution was found to generate a lot of hydroxyl and car-

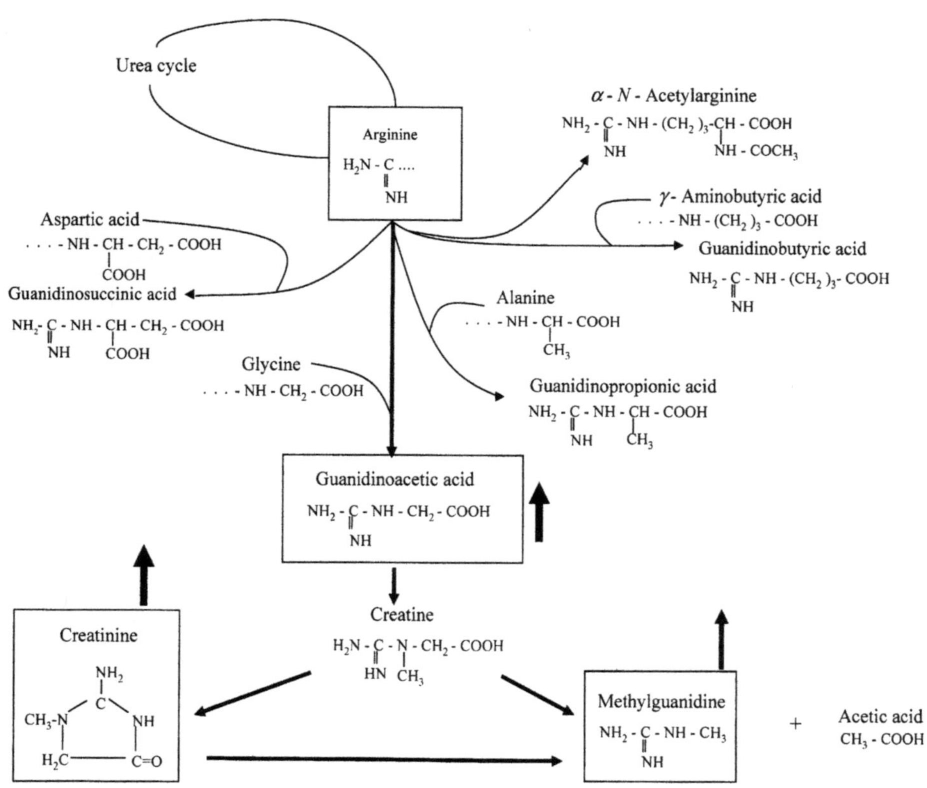


*Fig. 1.* The possible pathway of guanidine compounds in the seizure mechanism.

bon centered radicals as observed by electron spin resonance technique with spin tarp, and the generation of both radicals is dose dependent manner [16]. Methylguanidines are synthesized from creatinine with hydroxyl radical [17]. Superoxide and hydroxyl radicals decreased synaptosomal membrane fluidity in the rat cortex, and α-guanidinoglutaric acid and methylguanidine also lowered membrane fluidity [14]. The decrease in membrane fluidity by α-guanidinoglutaric acid is thought to be due to its free radical generation.

α-Guanidinoglutaric acid levels were found to be elevated dramatically in the epileptogenetic focus of the cat 24 h after application of cobalt into the sensory motor cortex [18]. So, the mechanism for induction of cobalt-induced epileptogenetic focus is thought to be due to free radical generation by α-guanidinoglutaric acid. However, we observed an elevation of extracellular hydroxyl radical in the hippocampus after intraventricular injection of α-guanidinoglutaric acid into rat using microdialysis method with salichilate [19]. In addition, we found a dramatic elevation of extracellular glutamate level in the hippocampus 60 min after injection, and the elevation was maintained for an additional hour. These phenomena suggest that the induction of seizures by α-guanidinoglutaric acid is due to neuronal cell death induced by glutamate.

The mechanism for generating free radicals is thought to be as follows, guanidino compounds have guanidium ion of conjugate base, which is easily donates an electron, generating a superoxide forming a hydroxyl radical. These free radicals may damage neuronal membranes and induce apoptosis or necrosis.

In addition, we observed the generation of nitric oxide from α-guanidinoglutaric acid in the rat brain homogenate with the Fenton reagent, which generates hydroxyl radicals [20]. It is possible to induce formation of the peroxynitrate ion, which is a toxic via induction of neuronal cell death.

## Toxicity of guanidino compounds

The fact that glutamate induces neuronal cell death, is thought to be due to peroxinitrate ion formed with superoxide and nitric oxide generated by NO synthetase by stimulation of calcium influx after stimulation of the N-methyl-D-aspartate (NMDA) receptor by glutamate. Glutamate induces C6 glial cell death, and α-guanidinoglutaric acid also induces C6 glial cell death. C6 cell death by α-guanidinoglutaric acid is 2.5 times higher than glutamate [20].

Guanidino compounds affected development of the chicken egg [21]. Arginine, creatinine, a-guanidinoglutaric acid and methylguanidine in 16 kinds of guanidino compounds inhibited the first developmental stage of 0–7 days in the chick.

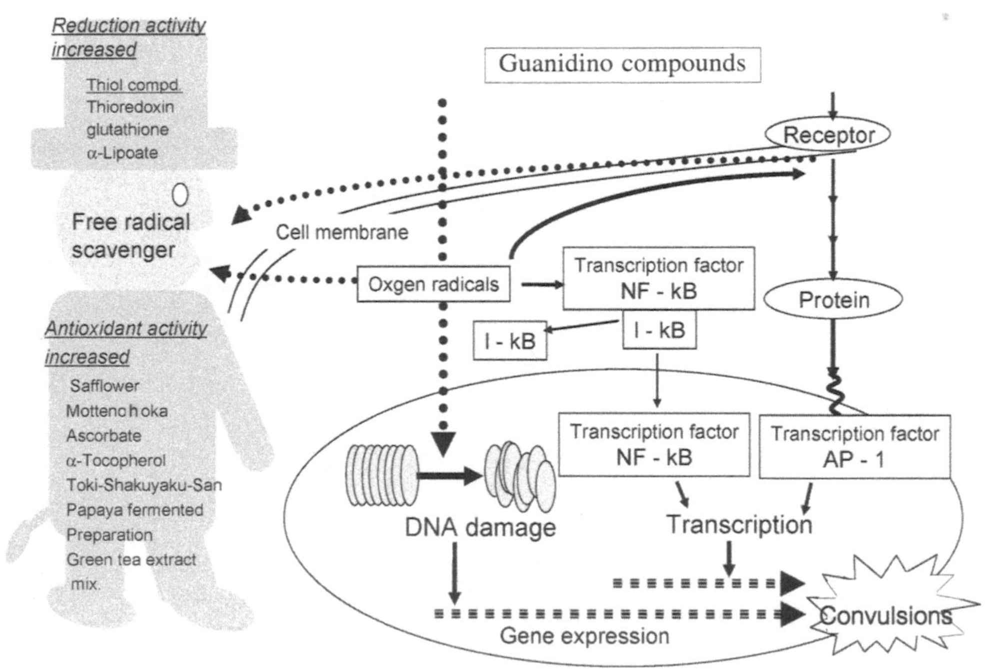


*Fig. 2.* Oxidative stress and redox regulation in seizures.

Except for arginine, we observed a phenomena in that creatinine, α-guanidinoglutaric acid and methylguanidine were toxic for chick development and we found an elevation of these guanidine compounds in the animal brain during seizures and convulsion. They might therefore be toxic for neurons.

## Prospect

Recently we found an elevation of 8-hydroxy-2′-deoxyguanosine in the brain which was formed during iron-induced epileptogenesis [22]. However, no affect of α-guanidinoglutaric acid on 8-hydroxy-2′-deoxyguanosine was found in the rat brain. The mechanism for seizures induced by α-guanidinoglutaric acid is not likely to involve DNA oxidation by free radicals.

Recently modulation of energy metabolism by creatine suggested an inhibition of apoptosis induced by glutamate [23, 24]. The generation of free radicals in the mitochondria by low activities of the electron transfer complexes is shown that to affect mitochondrial DNA in Parkinson's disease and Alzheimer diseases. Guanidino compounds can be a donor of electrons from the amidino group. From these phenomena, the study of the electron transport system in the mitochondria, apoptosis associated with transcription factors and signal transduction by guanidino compounds should be examined (Fig. 2).

## Conclusions

The role of guanidino compounds in the brain is as follows: (1) Membrane fluidity and substance competition between guanidino compounds and amino acids, for example, guanidinoethane sulfonic acid (taurocyamin) and taurine can increase or decrease fluidity. Guanidinosuccinic acid and aspartate acts on NMDA receptor competitively. (2) Generation of free radicals, and their reaction and an affection on signal transduction. (3) Energy production by creatine metabolism may be important for protection of neuronal cell death and apoptosis.

## References

1. Jinnai D, Mori A, Mukawa J, Ohkusu H, Hosotani M, Mizuno A: Biochemical and physiological studies on guanidine compounds induced convulsion. Choong Jpn J Brain Physiol 106: 3668–3673, 1969
2. Mori A, Hiramatsu M, Takahashi K, Kohsaka M: Guanidino compounds in rat organs. Comp Biochem Physiol 51B: 143–144, 1975
3. Nakae I: Synthesis of N,N′-dibenzoylguanidine and its convulsive action. Neurosciences 7: 205–217, 1981
4. Shirakabe Y: Synthesis of phenylethylguanidine and its effects on neurotransmitters of mouse brain and electroencephalogram of cat. Neurosciences 8: 161–170, 1982
5. Mori A, Hiramatsu M, Numoto A, Robin Y: Action convulsante des derives diguanidiques, l'hirudonine l'arcaine et l'audonine. CR Soc Biol 175: 755–760, 1981
6. Marescau B, Hiramatsu M, Mori A: α-keto-γ-guanidinovaleric acid induced electroencephalographic epileptiform discharges in rabbits. Neurochem Pathol 1: 203–209, 1983
7. Mori A, Watanabe Y, Shindo S, Akagi M, Hiramatsu M: α-Guanidinoglutaric acid and epilepsy. In: A. Lowenthal, A. Mori, B. Marescau (eds). Urea Cycle Diseases. Plenum Press, New York, 1983, pp 465–470
8. Shiraga H, Hiramatsu M, Mori A: Convulsive activity of α-guanidinoglutaric acid and the possible involvement of 5-hydroxytryptamine in the α-guanidinoglutaric acid induced seizure mechanism. J Neurochem 47: 1832–1836, 1986
9. Hiramatsu C: Guanidino compounds in mouse brain II. Guanidino compound levels in brain in relation to convulsions. Okayama Igakkai Zasshi 92: 427–433, 1980 ( in Japanese)
10. Hiramatsu M, Edamatsu R, Fujikawa N, Shirasu A, Yamamoto Y, Suzuki S, Mori A: Analyses of guanidino compounds in CSF collected by implanted catheter into cisterna magna; changes in guanidino compounds in CSF of rabbits during convulsion. Brain Res 455: 38–42, 1988
11. Willmore LJ, Sypert GW, Munson JB: Chronic focal epileptiform discharges induced by injection of iron into rat and cat cortex. Science 200: 1501–1503, 1978
12. Fukushima M: Guanidino compounds in iron-induced epileptogenic foci of rats. Okayama Igakkai Zasshi 99: 787–806, 1987 ( in Japanese)
13. Hisayasu Y, Morimoto K, Otsuki S: Increase of methylguanidine and guanidinoacetic acid in the brain of amygdala-kindled rats. Epilepsia 32: 761–766, 1986
14. Hiramatsu M, Ohba S, Edamatsu R, Kadowaki D, Mori A: Effect of guanidino compounds on membrane fluidity of rat synaptosomes. In: P.P. Deyn, B. Marescau, V. Stalon, L.A. Qureshi (eds). Guanidino Compounds in Biology and Medicine. John Libbey & Co. Ltd., 1992, pp 387–393
15. Hiramatsu M, Niiya-Nishihara H, Mori A: Effect of taurocyamine on taurine and other amino acids in the brain, liver and muscle of mice. Neurosciences 8: 289–294, 1982
16. Hiramatsu M, Edamatsu R, Kohno M, Mori A: Scavenging of free radicals by Sho-saiko-to-go-keishi-kashakuyaku-to. In: E. Hosoya, Y. Yamamura (eds). Recent Advances in the Pharmacology of KAMPO Medicines. Excepter Medica, 1988, pp 120–127
17. Hiramatsu M, Edamatsu R, Kohno M, Mori A: The reactivity of guanidino compounds with hydroxyl radicals. In: A. Mori, B.D. Cohen, H. Koide (eds). Guanidines. Plenum Press, New York, 1989, pp 97–105
18. Mori A, Akagi M, Katayama Y, Watanabe Y: α-Guanidinoglutaric acid in cobalt-induced epileptogenic cerebral cortex of cats. J Neurochem 35: 603–605, 1980
19. Hiramatsu M, Komatsu M, Kido T: The relationship between reactive oxygen species and the α-guanidinoglutaric acid induced seizure mechanism. In: A. Mori, M. Ishida, J.F. Clark (eds). Guanidino Compounds: 5. Blackwell Science Asia Pty Ltd., 1999, pp 193–198
20. Hiramatsu M, Komatsu M: The detection of nitric oxide from α-guanidinoglutaric acid by hydroxyl radicals using an electron spin resonance method with spin trap. In: P.P. Deyn, B. Marescau, I.A. Qureshi, A. Mori (eds). Guanidino Compounds. John Libbey & Co. Ltd., 1997, pp 31–36

21. Seki S, Yuyama N, Hiramatsu M: Effect of guanidino compounds on hen egg development. In: A. Lowenthal, A. Mori, B. Marescau (eds). Urea Cycle Diseases. Plenum Press, New York, 1983, pp 465–470
22. Komatsu M, Hiramatsu M, Willmore LJ: Zonisamide reduces the increase in 8-hydroxy-2′-deoxyguanosine levels in the brain formed during iron-induced epileptogenesis in rats. Epilepsia 41: 1091–1094, 2000
23. Holtzman D, Kekelize T: Guanidino compounds in studies of brain ATP metabolism. In: A. Mori , M. Ishida, J.F. Clark (eds). Guanidino Compounds: 5. Blackwell Science Asia Pty Ltd., London, 1999 pp 3–15
24. Matthew RT, Kaddurah-Daouk R, Beal M: The neuroprotective effect of creatine and its analogs in animal models of neurodegenerative diseases. In: A. Mori, M. Ishida, J.F. Clark (eds). Guanidino Compounds: 5. Blackwell Science Asia Pty Ltd., London, 1999, pp 105–115

*Molecular and Cellular Biochemistry* **244:** 63–67, 2003.


# Electron paramagnetic resonance imaging of nitric oxide organ distribution in lipopolysuccaride treated mice

Aki Hirayama,[1] Sohji Nagase,[1] Atsushi Ueda,[1,2] Keigyou Yoh,[1] Takaaki Oteki,[1] Mami Obara,[1] Kenji Takada,[1] Yukari Shimozawa,[1] Kazumasa Aoyagi[3] and Akio Koyama[1]

[1]*Institute of Clinical Medicine, University of Tsukuba;* [2]*Institute for Life Support Technology, Yamagata Public Corporation for the Development of Industry;* [3]*Tsukuba College of Technology*

## Abstract

The recent development of electron paramagnetic resonance (EPR) permits its application for *in vivo* studies of nitric oxide (NO). In this study, we tried to obtain 3D EPR images of endogenous NO in the abdominal organs of lipopolysuccaride (LPS) treated mice. Male ICR mice, each weighing about 30 g, received 10 mg/kg of LPS intraperitoneally. Six hours later, a spin trapping reagent comprised of iron and an N-dithiocarboxy sarcosine complex ($Fe(DTCS)_2$, Fe 200 mM, DTCS/Fe = 3) were injected subcutaneously. Two hours after this treatment, the mice were fixed in a plastic holder and set in the EPR system, equipped with a loop-gap resonator and a 1 GHz microwave. NO was detected as an NO-$Fe(DTCS)_2$ complex, which had a characteristic 3-line EPR spectrum. NO-$Fe(DTCS)_2$ complexes in organ homogenates were also measured using a conventional X-band EPR system. NO-$Fe(DTCS)_2$ spectra were obtained in the upper abdominal area of LPS treated mice at 8 h after the LPS injection. 3D EPR tiled and stereoscopic images of the NO distribution in the hepatic and renal areas were obtained at the same time. The NO-$Fe(DTCS)_2$ distribution in abdominal organs was confirmed in each organ homogenate using conventional X-band EPR. This is the first known EPR image of NO in live mice kidneys. (Mol Cell Biochem **244**: 63–67, 2003)

*Key words*: nitric oxide, electron paramagnetic resonance, EPR imaging, lipopolysuccaride

## Introduction

Free radical species including nitric oxide (NO) are characterized by their high reactivity and short life-time *in vivo* [1]. These chemical features make their direct measurement difficult. However, these characteristics also generate a need for their *in vivo*, real time direct detection. Despite instability and the low concentration of NO, several methods for direct detection have been developed. Electron paramagnetic resonance (EPR) is a method which meets this requirement. NO is a paramagnetic substance having an unpaired electron in one of its orbits. Usually an NO-EPR study is performed using spin trapping reagents, which turn unstable radicals to relatively stable radical species. Conventionally, EPR measurements are done with an X-band EPR system, which uses microwaves of a frequency of around 9 GHz. Such an X-band EPR system is not applicable to *in vivo* measurements, because of the loss of dielectric in water. Recent development of a low frequency EPR (L-band EPR) have resolved this problem. In combination with the improvement of both spin trapping and L-band EPR techniques, the *in vivo* NO measurement and imaging by EPR is now possible.

Recently several studies were reported which were successful in detecting and imaging endogenous NO using EPR [2–6]. The landmark study in this area was done by Yoshimura *et al.* in 1997 [3]. They clearly imaged a 2D projection of

*Address for offprints*: A. Hirayama, Institute of Clinical Medicine, University of Tsukuba, 1-1-1 Amakubo, Tsukuba-City, Japan 305-8545 (E-mail: aki-hira@md.tsukuba.ac.jp)

endogenous NO in the liver of LPS treated mice. After this, 2 reports showed EPR images of exogenously given NO. Fujii imaged NO from the vasodilating agent ISDN using a $^{15}N$ isotope [4] and Yokoyama imaged NO-Fe$(DTCS)_2$ distribution [5]. Jackson *et al.* reported a round shaped EPR image from the liver area of LPS-treated mice [6].

Despite these excellent works, it is still difficult to image detailed NO organ distributions without using special reagents or isotopes. In this study, we attempt to obtain EPR images which distinguish minute organ distributions of NO. Among the EPR spin trapping reagents, Fe$(DTCS)_2$ has a relatively long half life [7] compared to other such reagents. This chemical feature allows us enough scanning time and leads to obtain high quality EPR signals. Using this, we attempt to obtain 3D NO EPR images which indicate with low invasivity the detailed abdominal organ distribution of NO.

## Materials and methods

### *Animal*

Male ICR mice, each weighing about 30 g, were employed. Mice were fed with standard chow *ad libitum* and allowed free access to water.

### *Experimental procedure*

The mice received 10 mg/kg of LPS intraperitoneally. Six hours later, a complex of $FeCl_3$ and *N*-(dithiocarboxy) sarcosine (DTCS, 600 mM, DTCS:Fe = 3:1) was injected subcutaneously. Two more hours after this treatment (8 h after the LPS injection), the mice were anesthetized with Pentobarbital 0.05 mg/kg, fixed in a plastic holder and set in the EPR system, which described below. EPR imaging studies were performed, continuously following which the mice were sacrificed and *ex vivo* studies were done using a process described below. Control mice received the same treatment using the same volume of saline instead of LPS.

### *The* in vivo *EPR imaging system and the EPR measuring condition*

An *in vivo* EPR system made by JEOL (Tokyo, Japan), equipped with a loop-gap resonator (31 mm diameter) and a 1 GHz microwave unit was used (Fig. 1). EPR images were constructed using the center peak of the NO-Fe$(DTCS)_2$ spectrum. EPR conditions for this were: magnetic field: 37.0 ± 4.3 mT, microwave power: 0.25 mW, time constant: 0.03 sec, field gradient: 1.0 mT/cm, changing direction: 30 degree steps (provides 6 spectra for each projection) and accumulation of the signal: 5 times. 3D cross-sectional images were obtained using ESR-CT ver. 1.136 (JEOL), and 3D stereoscopic images were reconstructed with IRIS explorer software (IRIS Co., UK) using the cross-sectional data.

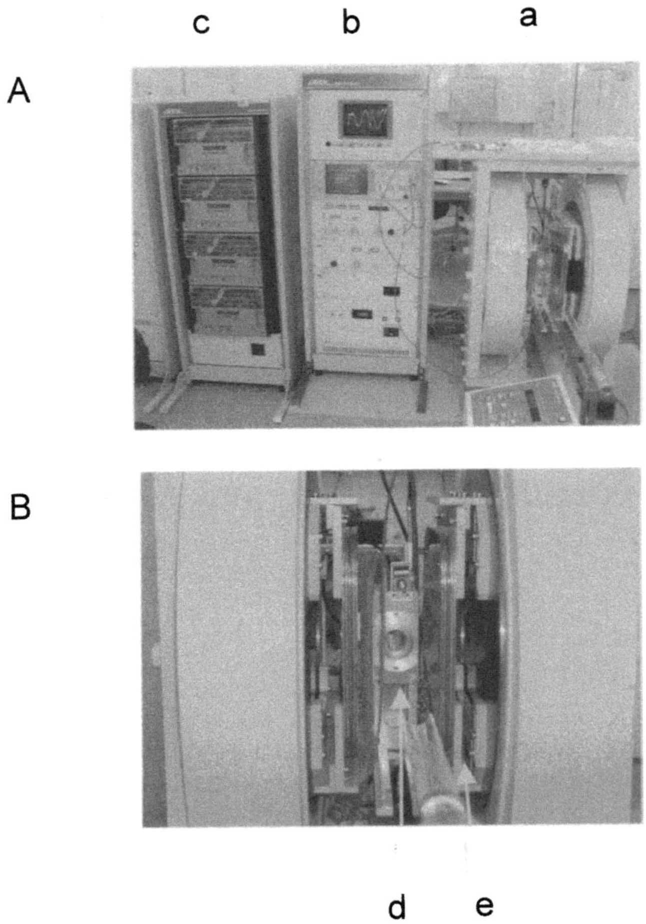


*Fig. 1. In vivo* ESR system. (A) The system consists of a resonator and magnet unit (a), microwave controlling units (b), amplifiers (c) and a power supply unit (not shown). (B) A close up view of the resonator unit. A 31 mm diameter loop-gap resonator (d) and the gradient coils (e) are shown in this photograph.

### Ex vivo *NO detection*

To confirm the signal localization, NO-Fe$(DTCS)_2$ signals from the organ homogenates were measured using a conventional X-band EPR. Soon after the imaging studies, the mice were sacrificed and the livers, kidneys and spleens were removed. The removed organs were mixed with 4 times the sample weight of 1.15% potassium chloride solution and homogenized on ice. NO-Fe(DTCS)$_2$ in the homogenates from these organs was measured using the TR-25 X-band ESR equipment (JEOL) and a flat quarz EPR cell (Labotec). The X-band EPR measurement conditions for these experiments were: frequency of microwave: 9.420 GHz, magnetic field: 337 ± 5 mT, modulation width: 0.1 mT, time constant: 0.1 sec, sweep time: 4 min and accumulation of the signal: 5 times.

*Reagents*

DTCS-Na was purchased from Dojindo Co. (Kumamoto, Japan). $FeCl_3$ was obtained from Wako Chemical Co. (Tokyo, Japan). All other chemicals were purchased from Sigma-Aldrich Japan (Tokyo, Japan).

## Results

In vivo *detection of NO*

Figure 2A shows a typical X-band spectrum of the NO-Fe(DTCS)$_2$ complex solution *in vitro*. This NO-Fe(DTCS)$_2$ complex showed a 3-line sharp signal. Figure 2B shows a typical *in vivo* L-band spectrum of the NO-Fe(DTCS)$_2$ complex at 6 h after the LPS injection. This spectrum was obtained by a 5 times accumulation of each 30 sec scan. NO-Fe(DTCS)$_2$ spectra were detected by an *in vivo* L-band EPR at 6–8 h after the LPS injection (data not shown). The spectra *in vivo* have a low S/N due to high noise levels and show signal broadening. The mice which received DTCS and saline instead of LPS showed no EPR signal.

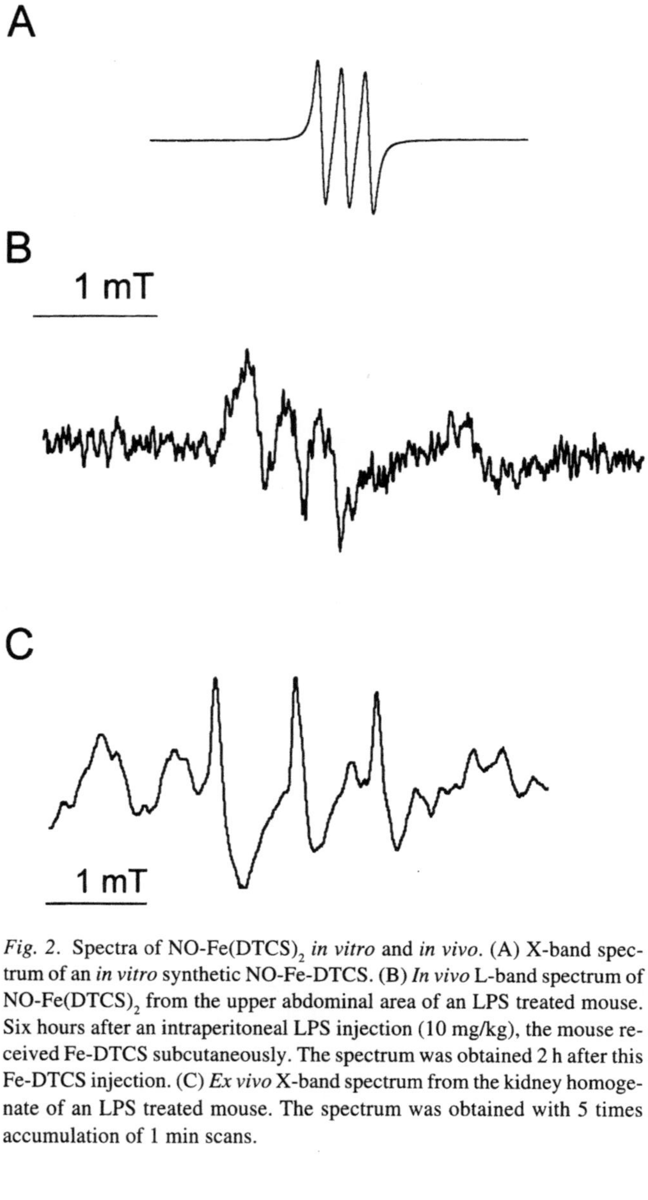


*Fig. 2.* Spectra of NO-Fe(DTCS)$_2$ *in vitro* and *in vivo*. (A) X-band spectrum of an *in vitro* synthetic NO-Fe-DTCS. (B) *In vivo* L-band spectrum of NO-Fe(DTCS)$_2$ from the upper abdominal area of an LPS treated mouse. Six hours after an intraperitoneal LPS injection (10 mg/kg), the mouse received Fe-DTCS subcutaneously. The spectrum was obtained 2 h after this Fe-DTCS injection. (C) *Ex vivo* X-band spectrum from the kidney homogenate of an LPS treated mouse. The spectrum was obtained with 5 times accumulation of 1 min scans.

*EPR images*

The center peak of the 3-line NO-Fe(DTCS)$_2$ signal was extracted and used for image construction. A 3D tiled image of the axial section of the abdominal lesion at 6 h after the LPS injection is shown in Fig. 3A. The red areas indicate high signal intensities and blue areas indicate low signal intensities. There is a red area in the dorsal and frontal view, which is consistent with the liver. There are also 2 small red areas under the liver, which are consistent with the kidneys. Enlarged images of the upper abdominal area are shown in Figs 3B and 3C. Figure 3B is expressed by a 16 point colour, and Fig. 3C is made by 256 points on a gray scale. Strong signal areas were noticed which were consistent with the liver. In this image, the renal area is recognized to have a crescent shape. The signal intensity of the renal area is less than that of the hepatic area. A 3D stereoscopic view of the NO abdominal distribution is shown in Figs 4A and 4B. The organ distribution of NO was confirmed with these 2 images.

Ex vivo *detection*

The organ images obtained in this study represent high signal areas but do not exactly show the organ shapes, because EPR detects only unpaired electrons. This, then, confirms NO-Fe(DTCS)$_2$ in the organ homogenates using conventional X-band EPR. After the imaging study, the abdominal organs were resected and the existence of an NO-Fe(DTCS)$_2$ complex was confirmed with conventional X-band EPR. EPR

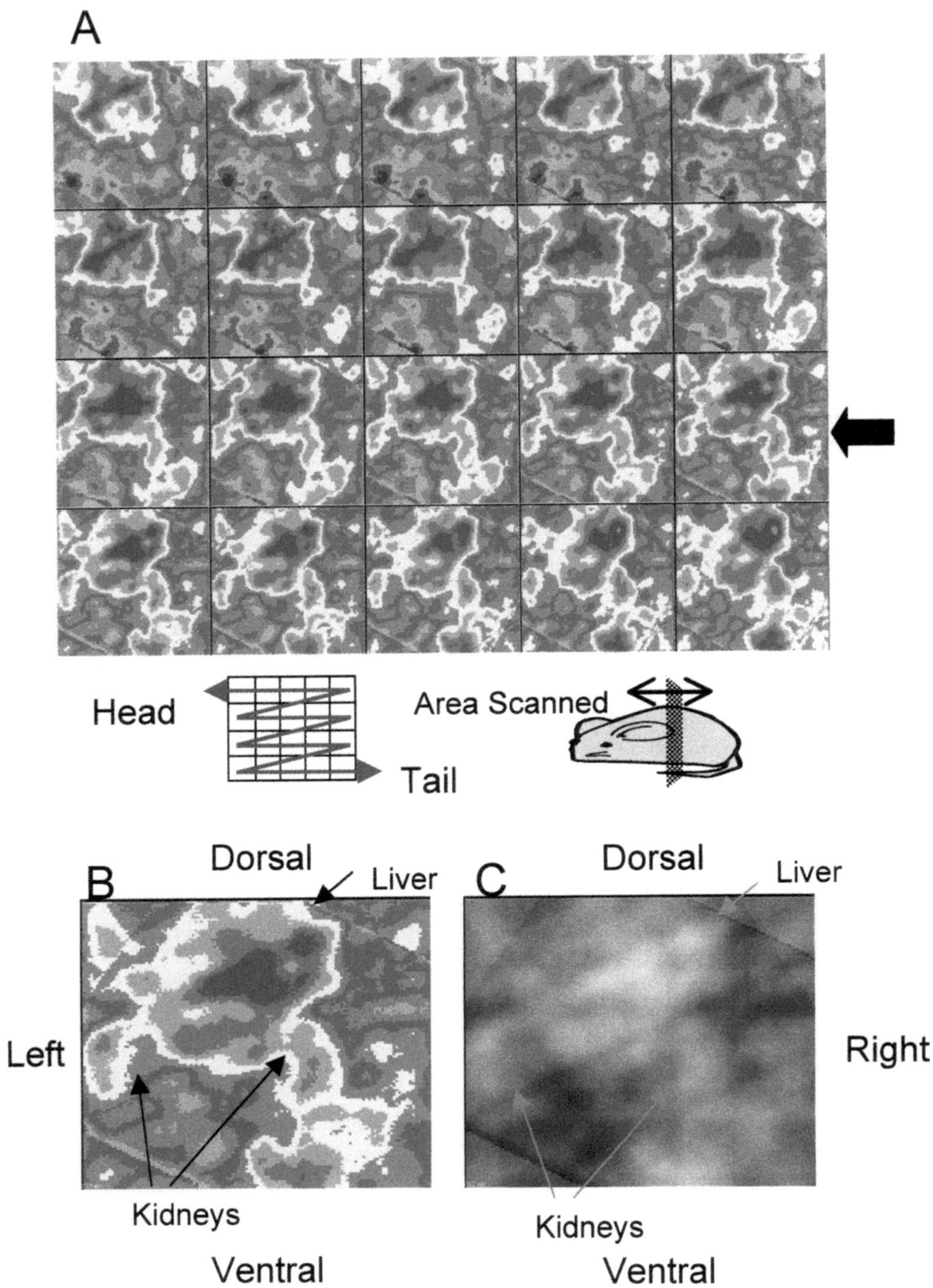


*Fig. 3.* (A) 3D tiled image of the abdominal area from an LPS treated mouse (axial section). High signal intensity is indicated as red and low intensity as blue. A high signal is recognized in a domain equivalent to the liver. There are also 2 high intensity areas below the liver domain which are consistent with the kidneys. An image indicated with an arrow is enlarged in Figs 3B and 3C. (B, C) The enlarged images of Fig. 3A expressed as a 16 point colour scale (B) and 256 point gray scale (C). In this image, the renal areas are recognized as crescent shapes. The signal intensity of the renal area is less than that of the liver area. A high intensity area on the lower right is artifact.

signals were confirmed in the homogenates of the liver and the kidneys. The NO-Fe(DTCS)$_2$ spectrum detected in the kidney homogenate is shown in Fig. 2C. No NO complexes were detected in the homogenates from the spleen and the small intestines.

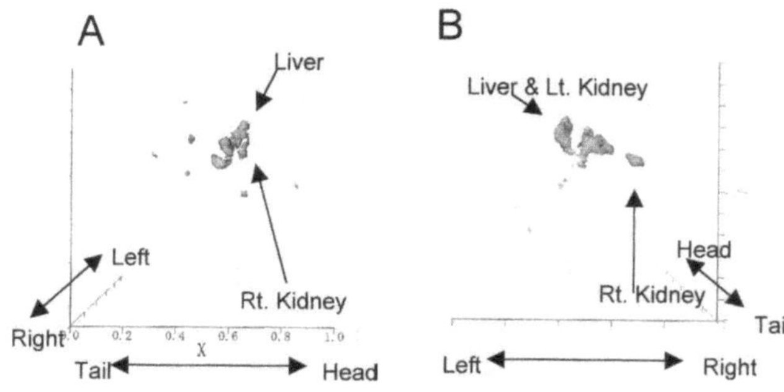


*Fig. 4.* A 3D stereoscopic view of the NO-Fe(DTCS)$_2$ distribution in the abdominal area: longitudinal view (A) and axial view (B). The left kidney was shaded by the liver in this view.

## Discussion

LPS induced endogenous NO was captured by an Fe-DTCS complex and EPR images of the kidneys and liver were successfully obtained. As far as we can tell, this is the first EPR image of NO in the renal area of live mice. LPS induces iNOS in various organs including the liver, lungs, kidneys and brain. Fujii and Berliner first detected NO by *in vivo* EPR in the upper abdominal area of LPS treated mice using N-methyl-D-glucamine dithiocarbamate (MGD) as a spin trap [8]. The major organ to produce NO by LPS treatment is the liver, but iNOS induction is also observed in the kidneys. In this study, we imaged an NO-Fe(DTCS)$_2$ image in the bilateral renal areas. In the kidney, LPS exposure results in an increase of iNOS expression mainly in glomeruli and inner medulla [9] [10]. NO production from iNOS in the kidney becomes maximal at 6–8 h after LPS injection [8]. On the other hand, NO trapped in other organs may accumulate in the kidney via circulation and filtration. LPS induced endotoxemia also affects renal hemodynamics and reduces renal cortical blood flow [11]. Therefore, the accumulated NO-DTCS may play a relatively minor role in our EPR image of the renal area. Further studies are required to clarify this point. The other major organ to produce NO in sepsis is the lung. EPR detection, however, of NO-Fe(DTCS)$_2$ in the chest area is difficult because of low signal to noise ratio due to respiratory movement.

Though NO was successfully imaged in this study, there are still some problems to be solved for the development and expansion of EPR imaging. The improvement of L-band EPR and spin trapping reagents are two major factors to be developed *in vivo* EPR. The sensitivity of the L-band EPR system is enough to detect NO spin adducts but there is still a high noise level and artifacts appear in the images. The NO-Fe(DTCS)$_2$ complex has enough stability to construct 3D images which show the EPR signal for more than 30 min, but the required reaction time between Fe-DTCS and NO is still long. Moreover, the Fe-DTCS complex has a high toxicity [12], and to detect the EPR signal we still need a high concentration of DTCS. Improvements in these areas will allow more detailed *in vivo* imaging of NO and other radical species using EPR.

## Acknowledgements

We thank Dr. B.D. Cohen for his great assistance in our research. We also thank Drs. H. Yokoyama, K. Oikawa, H. Ohya and H. Kamada, Institute for Life Support Technology, Yamagata Public Corporation for Development of Industry, for their wide range support in our EPR study. This study was supported by the Grant-in-Aid for Scientific Research from the Japanese Society for the Promotion of Science, Japan (13770595) and the University of Tsukuba Research Project.

## References

1. Ignarro LJ: Biosynthesis and metabolism of endothelium-derived nitric oxide. Annu Rev Pharmacol Toxicol 30: 535–560, 1990
2. Kuppusamy P, Ohnishi ST, Numagami Y, Ohnishi T, Zweier JL: Three-dimensional imaging of nitric oxide production in the rat brain subjected to ischemia-hypoxia. J Cereb Blood Flow Metab 15: 899–903, 1995
3. Yoshimura T, Yokoyama H, Fujii S, Takayama F, Oikawa K, Kamada H: *In vivo* EPR detection and imaging of endogenous nitric oxide in lipopolysaccharide-treated mice. Nat Biotechnol 14: 992–994, 1996
4. Fujii S, Suzuki Y, Yoshimura T, Kamada H: *In vivo* three-dimensional EPR imaging of nitric oxide production from isosorbide dinitrate in mice. Am J Physiol 274: G857–862, 1998
5. Yokoyama H, Fujii S, Yoshimura T, Ohya NH, Kamada H: *In vivo* ESR-CT imaging of the liver in mice receiving subcutaneous injection of nitric oxide-bound iron complex. Magn Reson Imaging 15: 249–253, 1997
6. Jackson SK, Madhani M, Thomas M, Timmins GS, James PE: Applications of *in vivo* electron paramagnetic resonance (EPR) spectroscopy: Measurements of pO(2) and NO in endotoxin shock. Toxicol Lett 120: 253–257, 2001
7. Pou S, Tsai P, Porasuphatana S, Halpern HJ, Chandramouli GV, Barth ED, Rosen GM: Spin trapping of nitric oxide by ferro-chelates: Kinetic and *in vivo* pharmacokinetic studies. Biochim Biophys Acta 1427: 216–226, 1999
8. Fujii H, Koscielniak J, Berliner LJ: Determination and characterization of nitric oxide generation in mice by *in vivo* L-band EPR spectroscopy. Magn Reson Med 38: 565–568, 1997
9. Morrissey JJ, McCracken R, Kaneto H, Vehaskari M, Montani D, Klahr S: Location of an inducible nitric oxide synthase mRNA in the normal kidney. Kidney Int 45: 998–1005, 1994
10. Ahn KY, Mohaupt MG, Madsen KM, Kone BC: *In situ* hybridization localization of mRNA encoding inducible nitric oxide synthase in rat kidney. Am J Physiol 267: F748–F757, 1994
11. Millar CG, Thiemermann C: Intrarenal haemodynamics and renal dysfunction in endotoxaemia: Effects of nitric oxide synthase inhibition. Br J Pharmacol 121: 1824–1830, 1997
12. Fujii S, Yoshimura T, Kamada H: Nitric oxide trapping efficiencies of water-soluble iron(III) complexes with dithiocarbamate derivatives. Chem Lett 9: 785–786, 1996

*Molecular and Cellular Biochemistry* **244:** 69–76, 2003.


# Restricted neuronal expression of ubiquitous mitochondrial creatine kinase: Changing patterns in development and with increased activity

J. Boero,[2] W. Qin,[2] J. Cheng,[2] T.A. Woolsey,[3] A.W. Strauss[1] and Z. Khuchua[1]

[1]*Department of Pediatrics of Vanderbilt University Medical Center, Nashville, TN;* [2]*Department of Pediatrics;* [3]*Department of Neurology and Neurological Surgery, Washington University School of Medicine, St. Louis, MO, USA*

## Abstract

Whereas ATP consumption increases with neural activity and is buffered by phosphocreatine (PCr), it is not known whether PCr synthesis by ubiquitous mitochondrial creatine kinase (uMtCK) supports energy metabolism in all neurons. To explore the possibility that uMtCK expression in neurons is modulated by activity and during development, we used immunocytochemistry to detect uMtCK-containing mitochondria. In the adult brain, subsets of neurons including layer Va pyramidal cells, most thalamic nuclei, cerebellar Purkinje cells, olfactory mitral cells and hippocampal interneurons strongly express uMtCK. uMtCK is transiently expressed by a larger group of neurons at birth. Neurons in all cortical layers express uMtCK at birth (P0), but uMtCK is restricted to layer Va by P12. uMtCK is detected in cerebellar Purkinje cells at birth, but localization to dendrites is only observed after P5 and is maximal on P14. Hippocampal CA1 and CA3 pyramidal neurons contain uMtCK-positive mitochondria at birth, but this pattern becomes progressively restricted to interneurons. Seizures induced uMtCK expression in cortical layers II–III and CA1 pyramidal neurons. In the cortex, but not in CA1, blockade of seizures prevented the induction of uMtCK. These findings support the concept that uMtCK expression in neurons is (1) developmentally regulated in post-natal life, (2) constitutively restricted in the adult brain, and (3) regulated by activity in the cortex and hippocampus. This implies that mitochondrial synthesis of PCr is restricted to those neurons that express uMtCK and may contribute to protect these cells during periods of increased energy demands. (Mol Cell Biochem **244**: 69–76, 2003)

*Key words*: brain, cerebellum, cortex, creatine kinase, hippocampus, interneuron, mitochondria, olfactory, seizures, somatosensory

## Introduction

Energy transduction from mitochondria to sites of ATP utilization is critical to very active cells. In neurons, cytoplasmic ATP is consumed by $Na^+$-$K^+$ ATPase as the membrane potential is restored. Phosphocreatine (PCr), in turn, buffers acute changes in ATP by donating its phosphoryl group to ADP, as catalyzed by cytosolic brain creatine kinase (B-CK). Thus, PCr concentration diminishes with increased brain activity, but ATP remains constant [1]. PCr is required during synaptic transmission [2], and variations of PCr synthesis are a sensitive marker of neuronal metabolism.

Pathological processes that affect energy metabolism in neurons are characterized by depletion of PCr. For example, inhibition of mitochondrial succinate dehydrogenase by 3-nitro-propionic acid (3-NPA) [3] depletes brain PCr and ATP, an effect reversed by creatine [4]. Furthermore, PCr delays neuronal damage and anoxic depolarization in hippocampal slices [5]; and a transient decrease in brain PCr during early post-natal development increases susceptibility to seizures [6], showing that PCr synthesis is required for neuronal development and function.

Mitochondrial creatine kinases (MtCK) synthesize PCr from ATP in the intermembranous space [7, 8]. MtCKs are

*Address for offprints*: Z. Khuchua, Department of Pediatrics, Vanderbilt University Medical Center, D2220 MCN, 1161 21st Avenue South, Nashville, TN 37232, USA (E-mail: zaza.khuchua@vanderbilt.edu)

encoded by two nuclear genes [9–12]. Sarcomeric MtCK (sMtCK) is expressed in skeletal muscle fibers and cardiac cells [13]. In the brain, kidney, intestinal epithelial cells, smooth muscle fibers and sperm cells, ubiquitous MtCK (uMtCK) catalyzes the same reaction [14, 15, 16]. Outside the mitochondrion, two cytoplasmic CKs that form homo or heterodimers restore ATP. Muscle CK (MCK) is expressed in the heart and skeletal muscle, but brain CK (BCK) is present in neurons, heart, kidney and gastrointestinal and urogenital systems [17].

The metabolic specialization of neurons in the brain is proposed as analogous to slow (highly oxidative) and fast (glycolytic) twich muscle fibers [18]. In the cortex and hippocampus, cytochrome c oxidase and lactate dehydrogenase activities are reciprocal, with higher oxidative capacity localized in dendritic tips spines and glomeruli. Cytochrome c oxidase activity in the rodent somatosensory cortex [19] and the primate visual cortex [20, 21] demarcates more metabolically active zones. However, uMtCK protein expression in Golgi type I neurons suggests metabolic heterogeneity among neurons [22–24]. In this study, we hypothesized that because neuronal populations mature at different rates, post-natal expression of uMtCK would vary regionally during development and that increased energy demands could modify uMtCK distribution and expression. We report that seizures induce uMtCK expression in cortical and hippocampal neurons and that uMtCK becomes restricted to a subpopulation of neurons during post-natal brain development. These findings imply that PCr synthesis by uMtCK is limited to a subset of neurons and that changes in energy demand regulate uMtCK expression.

## Materials and methods

### *uMtCK recombinant protein*

The coding sequence of human uMtCK cDNA was amplified by PCR using a λ clone provided by Dr. B. Perryman (University of Colorado, Denver, CO, USA) as template. A *Nde* I site that creates a new initiator methionine codon was generated at the 5′ end of the sequence encoding the mature uMtCK protein to remove the codons of the transit peptide. A *Xho* I site was created immediately after the translation stop codon by PCR using specific primers. The *Nde* I-*Xho* I PCR fragment containing the complete sequence of mature human uMtCK cDNA was subcloned into a pET21 vector in frame with a C-terminal $His_6$-tag (Novagen Inc., Madison, WI, USA). uMtCK protein was expressed by transformed *E. coli* BL21(DE3)pLys cells and purified by affinity chromatography through a $Ni^{2+}$ column [25]. Standard preparations yielded 15–20 mg of pure protein per L of culture with 140 IU/mg protein of CK activity (reverse reaction). Native agarose gel electrophoresis confirmed that all CK activity corresponded to uMtCK (data not shown).

### *Rabbit antiserum and Western blotting*

An adult New Zealand rabbit was immunized by intradermal injection with 1 mg of recombinant uMtCK protein in phosphate-buffered saline (PBS). To avoid cross-reactivity with cytosolic CK, diluted uMtCK antiserum (1:500) was incubated for 12 h at 4°C with nitrocellulose membranes previously adsorbed with mouse brain cytosolic fraction and stored at –20°C. The specificity of the antiserum was tested by Western blotting. Brain extracts were prepared in 10 mM Tris-HCl, pH 7.9, buffer with a Polytron homogenizer (setting 10) for 10 sec; stored at 4°C for 1 h; and centrifuged at 16,000 rpm for 30 min. The supernatants were stored at –70°C until use. Western blots were performed using a standard apparatus (Pharmacia Biotech Inc., Piscataway, NJ, USA). Nitrocellulose membranes were incubated with adsorbed primary antibody (1:10,000) in Phosphate-buffered saline (PBS) and 1% normal goat serum for 1 h at room temperature. Biotin-conjugated goat anti-rabbit antibodies (1:1000) were used to detect uMtCK bands as directed by the manufacturer (Vector Laboratories Inc., Burlingame, CA, USA).

### *Immunocytochemistry*

Male Swiss mice (12–14 weeks old) were anesthetized with sodium pentobarbital (40 mg/kg i.p.), perfused through the heart with saline followed by Bouin's solution (0.9% picric acid, 0.9% formaldehyde, 5% acetic acid) (Sigma Diagnostics, St. Louis, MO, USA), and stored in Bouin's at 4°C for 16 h. Pups were obtained from timed-pregnant female Swiss mice and sacrificed at different times after birth (postnatal day 0, P0). After dehydration and paraffin embedding, 8 µm coronal and sagittal sections were mounted in Superfrost plus® (Fisher Scientific, Pittsburgh, PA, USA) microscope slides. Indirect immunocytochemistry was performed as described by the manufacturer (Vector Laboratories) with 1:1000 dilution of rabbit anti-uMtCK serum. Peroxidase activity in the sections was detected by incubation in a 3,3′ diaminobenzidine solution with nickel chloride and mounted in Cytoseal 60 (Stephens Scientific, Riverdale, NJ, USA). Immunofluorescence detection of uMtCK protein was performed using 1:200 dilution of cy3-labeled donkey serum specific for rabbit IgG (Jackson Immunochemical Laboratories Inc., West Grove, PA, USA). Controls were performed with pre-immune rabbit serum (1:50 dilution). Photographs were taken using Elite

II film (400 ASA; Kodak, Rochester, NY, USA) with epifluorescence light in a Nikon microscope. In addition, images were digitized with a Zeiss LSM 310 confocal microscope, and a 3-dimensional representation of the intracellular distribution of uMtCK protein in single neurons was obtained with the VoxelView® program (Vital Images, Fairfield, IA, USA) from stacks of 12–14 images scanned every 0.4–0.45 μm.

*Induction of tonic-clonic seizures*

Adult male Swiss mice (12–14 weeks old) were injected with kainic acid (KA) (Sigma, St. Louis, MO, USA) (35 mg/kg), as described previously [26]. With this dose, the mortality was less than 10%, although seizures started about 30 min post-injection and continued for 2–3 h. Within minutes of the injection, decreased locomotion was followed by rigid posture (stiff tails) and automatisms (repetitive head bobbing and circling). Clonic forelimb movements preceded generalized tonic-clonic seizures in most of the animals. Forty eight h after KA injections, brain sections were obtained as described before and stained for uMtCK protein. An age matched control group received an injection of saline i.p. at the same time as the experimental group and were kept in identical housing conditions. The sections were counterstained with 1% thionin, and the number of uMtCK positive cells was counted in 10 fields at 1000× in the somatosensory and motor cortex layers II–III and V–VI, the subiculum, and hippocampal CA1 and CA3 pyramidal layers. In each field, the total number of cells and the number of uMtCK positive cells were counted. A second group of mice was treated similarly except for an injection of clonazepam (0.5 mg/kg) 10 min before, KA injection [27].

The Institutional Animal Care and Use Committees of the Washington University School of Medicine and Vanderbilt University Medical Center approved all aspects of animal care and experiments performed in this study.

# Results

*Specificity of rabbit anti-uMtCK antibody*

Western blots showed that purified uMtCK antiserum specifically reacts with uMtCK recombinant protein (lane R in Fig. 1A) and with a 42 kDA band in brain extracts (lane B in Fig. 1A). Immunostaining for uMtCK in the adult mouse brain showed that expression is restricted to a subpopulation of neurons. Anti-uMtCK antibodies stained subcellular organelles with the size, shape and localization appropriate for mitochondria (Fig. 1B). This perinuclear punctate pattern extends into dendrites, axons and terminals (Figs 2–6). uMtCK is not detected in glial or perivascular cells.

*Transient post-natal expression of uMtCK in the cortex and hippocampus*

At birth (P0), uMtCK is found in the somata and apical dendrites of cortical infragranular neurons and cells of the cortical plate (CP) (Fig. 2). Subplate cells are also very densely stained (not shown). However, as development progresses, uMtCK expression initially decreases by P2 in infragranular neurons and is followed by less staining in supragranular neurons by P5. At P7, few immunoreactive cells are detected in layer Va. uMtCK expression becomes restricted to pyramidal layer Va neurons by post-natal days 12–14 and persists in the adult brain.

In the hippocampus, all pyramidal neurons in CA1 and CA3 express uMtCK at birth, but uMtCK progressively decreases from P1 to adulthood (Fig. 3, left-sided panels). This is similar to the distribution of cytochrome c oxidase-rich mitochondria preferentially localized in dendritic terminals of CA1 pyramidal neurons [28]. In striking contrast, interneurons in the *stratum oriens* become darkly stained during development, consistent with their high mitochondrial content [29] and onset of spontaneous bursting discharges [30]. Although uMtCK is not detectable in granule cells of the dentate gyrus, interneurons in the polymorphic layer strongly express uMtCK at the same time as in CA1 and CA3. Other brain areas examined (striatum, thalamus, olfactory bulb, and brain stem nuclei) show no significant changes from their patterns at birth. These results demonstrate that uMtCK is expressed at birth and early in postnatal life by a larger number of neurons in those areas that continue developing postnatally. In several of these neurons, uMtCK is expressed transiently with a more restricted pattern of expression established by the second week of life.

*uMtCK expression in Purkinje cells parallels dendritic growth and branching*

Purkinje cells in the cerebellum strongly express uMtCK at birth (Fig. 3, right-sided panels). In these neurons, uMtCK-containing mitochondria are polarized between the apical dendrite and the nucleus where they accumulate until P6, when the enzyme is detected for the first time in primary and secondary branches of the apical dendrite. Between P7 and P14, uMtCK is detected more distally in dendrites, including branching points, with maximal expression at P14. Immunoreactive mossy fibers in the granular cell layer are detected after P5, but granular cells do not express uMtCK at any time during the development of the folium.

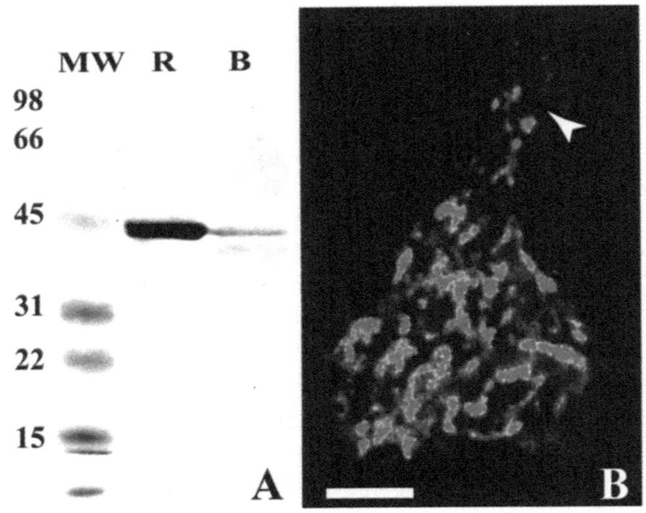


*Fig. 1.* Anti-uMtCK antibodies specifically recognize uMtCK protein in neuronal mitochondria. (Panel A) Purified anti-uMtCK rabbit antibodies react with recombinant uMtCK, lane R, and a 42 kDa protein in mouse brain mitochondrial extract, lane B, by Western blot. (Panel B) Three-dimensional representation of uMtCK protein distribution in an adult Purkinje neuron stained with anti-uMtCK antibodies. Stained mitochondria (red organelles) surround and obscure the nucleus and fill the dendrite proximally (arrow head). The image was generated with the VoxelView program using 12 consecutive images scanned at 0.42 μm intervals in the Z axis. Bar = 5 μm.

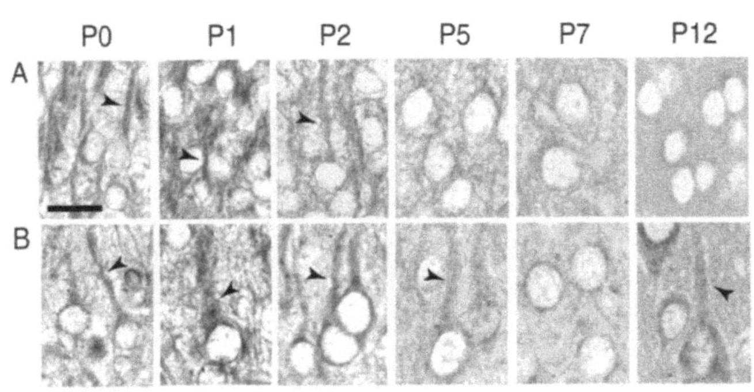


*Fig. 2.* Post-natal expression of uMtCK in the mouse neocortex at different ages. (Panel A) Cortical plate and supragranular layers. Strong uMtCK expression (dark granular precipitate) is detected in the somata and apical dendrites (arrow heads) of neurons in the cortical plate at P0 and P1. From days P2 to P5, uMtCK expression decreases and becomes undetectable by P7. uMtCK expression in supragranular layers is not restored in the adult mouse. (Panel B) Infragranular layers. Similar to supragranular layers, uMtCK is detected in the somata and apical dendrites (arrow heads) at birth but, after a transient decrease in expression (P7), uMtCK becomes constitutively expressed in layer Va pyramidal neurons by the end of the second week (P12). This pattern persists in the adult cortex. Bar = 100 μm applies to all pictures.

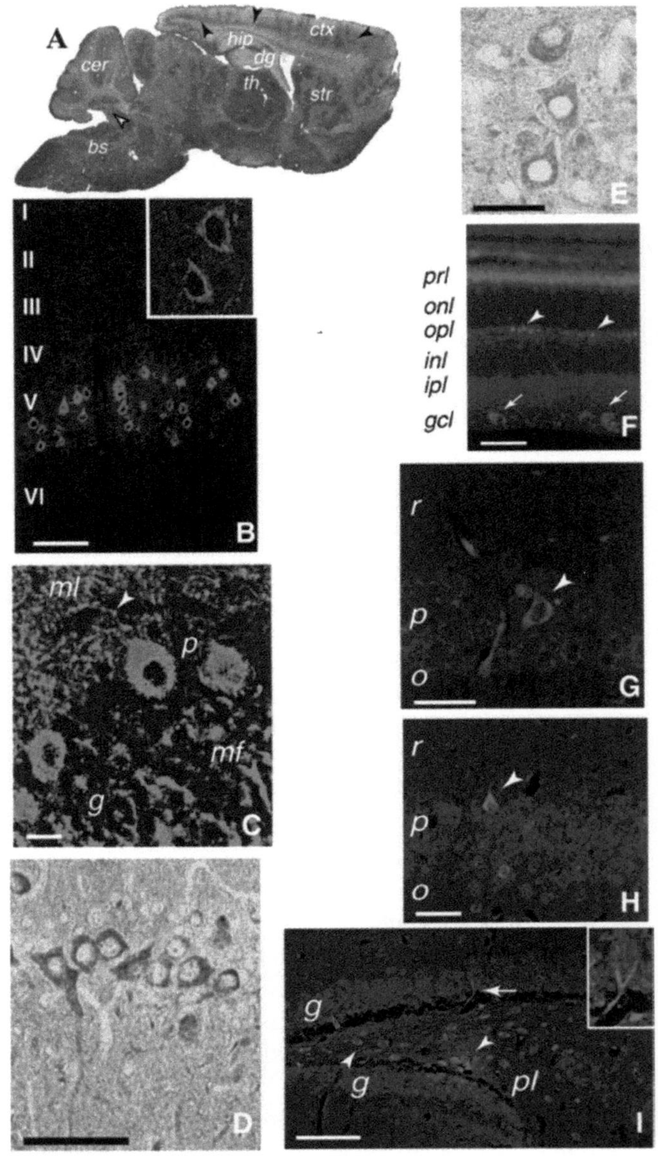


*Fig. 4.* Restricted uMtCK expression in the adult mouse. (Panel A) Distribution of uMtCK immunoreactivity (low power magnification) in the adult mouse brain. (Panel B) There is high uMtCK expression in layer Va pyramidal neurons of the somatosensory cortex (red fluorescense staining). The inset shows perikaryal punctate staining in two neurons. (Panel C) uMtCK expression (green fluorescence) is observed in Purkinje neurons (*p*) and mossy fibers (*mf*) ending in rosettes. The arrow head points to the apical dendrite. Note the absence of uMtCK in granule (*g*) cells. *ml* = molecular layer. (Panel D) Mitral cells of the olfactory bulb with stained somata and dendrites (dark brown 3,3′-diaminobenzidine precipitate) projecting to the glomeruli. uMtCK staining is absent in granular cells. (Panel E) Strong uMtCK expression in large neurons of the Gigantocellular Nucleus. (Panel F) Retina: uMtCK staining (red fluorescense) is observed in the mitochondria-rich inner segment of photoreceptors (*prl*), nerve endings in the outer plexiform layer (*opl*) (arrow heads), and occasional large neurons in the ganglion cell layer (*gcl*). *onl* = outer nuclear layer, *inl* = inner nuclear layer, *ipl* = inner plexiform layer. (Panels G–I) Arrow heads point to uMtCK-expressing interneurons in CA1, CA3 and the dentate gyrus, respectively. *p* = pyramidal layer, *r* = *stratum radiatum*, *o* = *stratum oriens*, *g* = granular cell layer, *pl* = polymorphic layer. Insets: stained interneurons in *pl* and axon terminal branching around granular cells in the dentate gyrus. Bar: A, 1 mm; B, 10 μm; C, 50 μm; D, 20 μm; F–H, 500 μm.

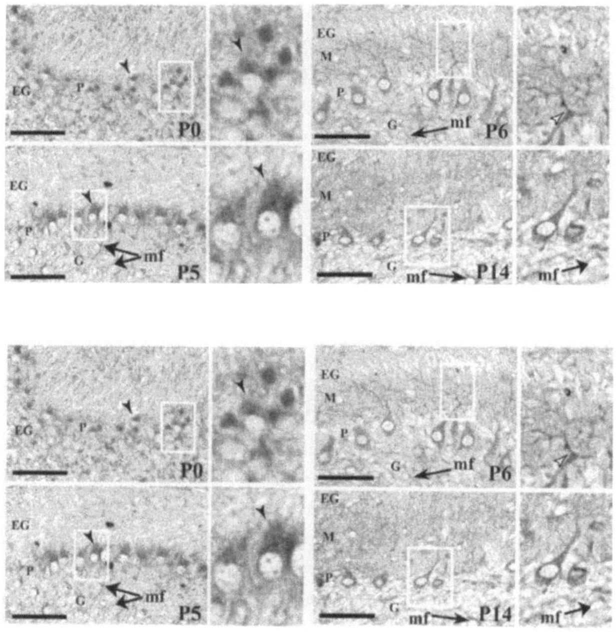


*Fig. 3.* uMtCK expression in the developing hippocampus and cerebellum. (Panel A) *Hippocampus.* Strong immunoreactive CA1 pyramidal dendrites (shown as dark granular precipitate) are observed on P1 and P5. However, uMtCK expression progressively decreases on P12–14 to adult levels. Insets show uMtCK distribution in apical dendrites. O = *stratum oriens*; P = pyramidal cell layer; R = *stratum radiatum*. (Panel B). *Cerebellum.* uMtCK-positive mitochondria accumulate between the perikaryon and the apical dendrite of Purkinje cells from P0 to P5 (black arrow head). On P6, uMtCK immunoreactivity is localized in the dendrite including the branching points (white arrow head). Maximal uMtCK expression is observed by P14 and then progressively declines to the levels observed in the adult mouse. Immunoreactive mossy fibers (mf) (arrows) are observed on P5, P6 and P14. EG = external granular layer; G = granular cell layer; M = molecular layer; P = Purkinje cell layer. Bar = 500 μm applies to all pictures.

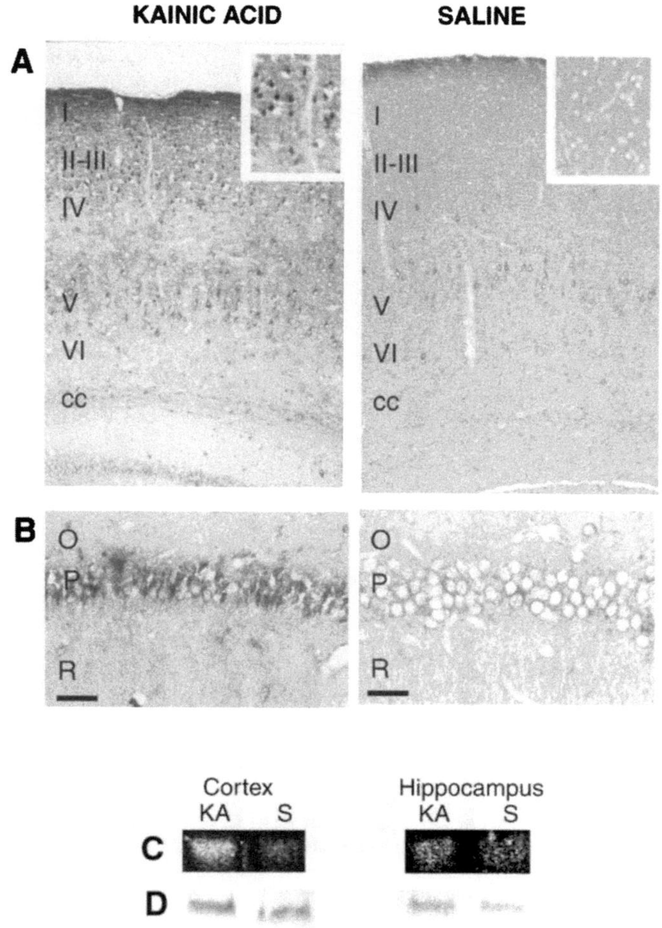


*Fig. 5.* Seizures induce uMtCK expression in the adult cortex and hippocampus. Panel A shows increased uMtCK staining (dark precipitates) in cortical supragranular (layers II–III) neurons 48 h after a single injection of KA in adult mice. The inset shows darkly stained layer III neurons. Control mice injected with saline do not express uMtCK enzyme in supragranular neurons. The inset shows control layer III neurons. Bar = 1 mm. (Panel B) CA1 region of a KA-treated mouse shows increased uMtCK expression in the pyramidal cell layer (P) when compared to saline control. Bar = 500 μm. (Panel C) Native gel assays show increased uMtCK activity in tissue extracts of the cortex and hippocampus of KA-treated mice (lanes KA) compared to saline (lanes S) controls. (Panel D) Western blot shows a moderate increase in uMtCK after seizures in the cortex and hippocampus. R = stratum radiatum, O = stratum oriens.

### *Restricted uMtCK expression in the adult mouse*

uMtCK expression in the adult mouse brain is restricted to a subpopulation of neurons. In the cortex, uMtCK staining in large layer Va pyramidal neurons forms a punctuate pattern localized in the periphery of the soma and at the base of the apical dendrite (Fig. 4B). Occasional large and heavily stained neurons are noted in layer IV. As in the neocortex, uMtCK expression in the pyriform and entorhinal cortices is limited to layer III neurons. In the thalamus, with the exception of the reticular nucleus, neurons in all other nuclei express uMtCK in varying intensities. Adult cerebellar Purkinje neurons and interneurons in the molecular layer (basket and stellate cells) also intensely express uMtCK (Fig. 4C). Although granule neurons do not contain uMtCK, heavily stained mossy fibers enter the granule cell layer and terminate in the 'rosettes'; uMtCK is also highly expressed in neurons of the deep cerebellar nuclei (not shown). In the olfactory bulb (Fig. 4D), the staining is prominent in the somata and apical dendrites of mitral cells as well as in the glomeruli and is similar to that observed in the rat [22]. Both periglomerular and granular neurons do not express uMtCK. In addition, several groups of large neurons located in the brainstem, such as those of the gigantocellular nucleus (Fig. 4E), intensely express uMtCK.

Selected groups of neurons express uMtCK in the adult retina (Fig. 4F). In striking contrast to the strong staining of the inner segment of photoreceptors, uMtCK protein is only detected in a fraction of synaptic terminals in the inner plexiform layer and occasional large ganglion cells. As in the brain,

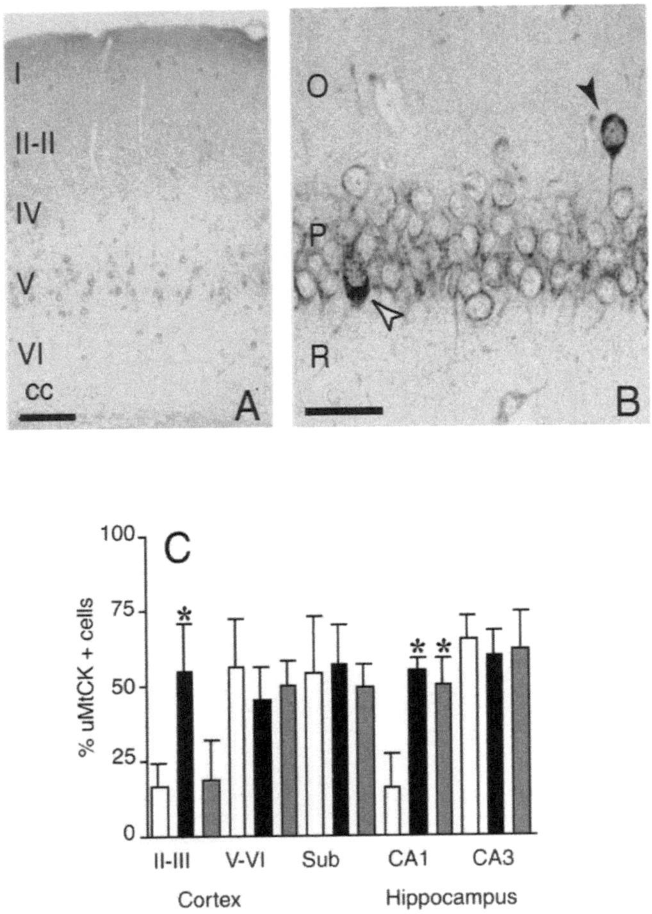


*Fig. 6.* Clonazepam blocks seizures and expression of uMtCK in the cortex but not in the hippocampus. (Panel A) Somatosensory cortex of clonazepam-pretreated mouse showing low uMtCK expression (dark precipitates) in supragranular neurons 48 h after KA injection. Bar = 1 mm. (Panel B) Strong uMtCK expression in CA1 neurons after administration of clonazepam and KA. Bkack and white arrowheads point to interneurons. Bar = 500 μm. (Panel C) Counts of uMtCK-positive cells in cortical layers II–III and V–VI, hippocampal CA1 and CA3, and the subiculum of saline (open bars), kainic acid (black bars), and clonazepam + kainic acid (grey bars)-treated mice. N = 5 mice per group. Average counts for the total number of cells and uMtCK-positive cells from 10 fields at a 1000× magnification were obtained for each area. Means ± S.D. were compared by the Student's independent *t*-test and significant differences accepted for $p < 0.05$.

uMtCK expression in the retina is restricted to a subset of neurons and their nerve endings.

Cell-type specificity of uMtCK expression is evident in the hippocampal formation (Figs 4G and 4H). Interneurons in CA1 and CA3 stratum oriens react intensely with uMtCK antibodies, while immunoreactivity in CA1 pyramidal neurons is low. In contrast, a small subset of large CA2-3 pyramidal neurons adjacent to CA1 cells strongly express uMtCK. In the dentate gyrus, uMtCK expression is very high in large pyramidal interneurons located in the granule cell layer and in medium-size cells in the polymorphic layer (Fig. 4I). Thus, in the adult mouse brain, uMtCK is expressed by a select population of neurons, some of which integrate local circuit output (large layer V pyramidal neurons, Purkinje cells, entorhinal layer III neurons) or are very active in regulating the activity of other neurons (CA1, CA3 and dentate gyrus interneurons).

*Neural activity induces uMtCK expression in the adult cortex and hippocampus*

The above results raised the question as whether uMtCK expression might be regulated by neural activity. To explore that hypothesis, we induced tonic-clonic seizures in adult mice by injections with kainic acid (KA). Layer II and III neurons of the somatosensory and motor cortices significantly increased uMtCK expression when studied 48 h after the seizures (Fig. 5A). In the hippocampus, uMtCK expression was also increased in CA1 but not in CA3 or the subiculum (Fig. 5B). Increased uMtCK activity was also detected by assays in native gels and Western blots (Figs 5C and 5D). To determine whether the change in uMtCK expression is secondary to increased neuronal activity or a direct effect of KA, a second group of animals was pre-treated with clonazepam, a benzodiazepine agonist that effectively blocks KA-mediated seizures in mice [27]. As shown in Fig. 6A and 6B), pre-treatment with clonazepam prevented KA-induced expression in cortical layers II and III neurons but not in CA1. These results were confirmed (Fig. 6C) by direct counting of uMtCK positive cells in several brain regions and show that uMtCK is induced by neuronal activity in cortical layers II–III of the adult mouse. The increase of uMtCK expression in CA1 pyramidal neurons was not prevented by clonazepam administration.

## Discussion

The main finding of this study is that restricted subsets of neurons in the mouse brain express uMtCK. The enzyme is highly expressed in regions that undergo post-natal development, but expression becomes more restricted in adult animals. uMtCK expression is also modified by changes in neuronal activity, as in CA1 pyramidal cells and cortical layer II–III neurons with seizures. Together, these results strongly suggest that changes in energy demands induce uMtCK expression. uMtCK-containing mitochondria were not only identified in the neuronal perikaryal cytoplasm but also in the proximal dendrites, axons and terminals. This distribution fits with higher PCr levels in the gray matter [31, 32] and a greater use of PCr during periods of increased neuronal activity [1].

Several types of cells that strongly express uMtCK, including layer V pyramidal cells [33–35], Purkinje neurons, and hippocampal interneurons [30] fire burst action potentials.

This suggests that high uMtCK activity is required for the efficient synthesis of PCr from mitochondrial ATP in these neurons, and it is likely that a large PCr reserve in these neurons is necessary for their normal function.

The selective increase of uMtCK expression in cortical supragranular neurons by seizures contrasts with the absence of expression in somatosensory layer IV [36], an area rich in cytochrome c oxidase activity [19]. Because the blockade of seizures by clonazepam abolished uMtCK induction in layer II–III after KA injections, we explain these changes as secondary to increased neuronal activity. However, uMtCK was not induced in layer IV even with the strong stimulus of seizures, suggesting a different metabolic adaptation. Interestingly, lactate dehydrogenase expression in the cortex is preferentially localized in layer V neurons but is very low in layer IV [18]. This pattern coincides with uMtCK staining not only in the neocortex, where several types of neurons in layer V are heavily labeled with 2-deoxy-glucose [37], but also in the deeper layers of the entorhynal and pyriform cortices, and might imply a metabolic specialization of the output neurons in these circuits. These findings argue for a functional and metabolic diversity of neurons, with uMtCK-expressing cells in the cortex having a higher glycolytic activity.

Early post-natal expression of uMtCK in the cortex was detected in a larger subset of neurons that in the adult mouse. Why might uMtCK expression be required by cortical plate and infragranular neurons at birth? Because most of uMtCK protein is localized in the apical dendrite, it is possible that PCr synthesis is crucial for dendritic growth and branching. Local PCr synthesis in dendrites would contribute to maintain ATP gradients and release and might promote neurite outgrowth [38, 39]. PCr might also promote mRNA 3′ UTR cleavage and polyadenylation [40] and, thus, contribute to regulate local protein synthesis in dendrites [41].

The transient decline of uMtCK expression in the cortex during the second week of postnatal life coincides *in vivo* with lower CK kinetic constants and PCr levels [42]. This increases neonatal susceptibility to hypoxia-induced seizures and is consistent with greater survival observed with creatine supplementation [6, 42]. A similar protective effect of creatine was observed in animal models of amyotrophic lateral sclerosis [43] and 1-methyl-4-phenyl-1,2,3,6-tetrahydropyridine (MPTP)-induced neurotoxicity [44]. Thus, the restricted expression of uMtCK in a subset of neurons might influence the cell-specificity of this therapeutic strategy in the brain. In addition, because uMtCK is expressed by a larger population of neurons at birth, creatine supplementation may prove useful to ameliorate energy deficits in the neonatal brain.

In summary, at birth, uMtCK is expressed by a large number of neurons in the brain; but expression becomes restricted to a small subset of cells early during post-natal development. In the cortex, cerebellum, and hippocampus of adult mice, uMtCK expression is restricted to neurons that have high energy requirements during bursts of activity. In CA1 and cortical supragranular neurons uMtCK expression is induced by increased activity and may help to protect these cells from energy deficits.

## Acknowledgements

We thank Jon Christensen for his excellent technical help. This work was supported by the National Institute of Health grants HL 52350, NS 17763, NS 28781, and the Spastic Paralysis Foundation of the Illinois-Eastern Iowa District of the Kiwanis International.

## References

1. Rango M, Castelli A, Scarlato G: Energetics of 3.5 sec neural activation in humans: A $^{31}$P MR spectroscopy study. Magnet Res Med 38: 878–883, 1997
2. Whittingham T, Lipton P: Cerebral synaptic transmission during anoxia is protected by creatine. J Neurochem 37: 1618–1621, 1981
3. Gould D, Gustine D: Basal ganglia degeneration, myelin alterations, and enzyme inhibition induced in mice by the plant toxin 3-nitropropanoic acid. Neuropathol Appl Neurobiol 8: 377–393, 1982
4. Matthews R, Yang L, Kenkins B, Ferrant R, Rosen B, Kaddurah-Daouk R, Beal M: Neuroprotective effects of creatine and cyclocreatine in animal models of Huntington's disease. J Neurosci 18: 156–163, 1998
5. Balestrino M, Rebaudo R, Lunardi G: Exogenous creatine delays anoxic depolarization and protects from hypoxic damage: Dose-effect relationship. Brain Res 816: 124–130, 1998
6. Holtzman D, Togliatti A, Khait I, Jensen F: Creatine increases survival and suppresses seizures in the hypoxic immature rat. Ped Res 44: 410–414, 1998
7. Jacobus W, Lehninger A: Creatine kinase of rat heart mitochondria. Coupling of creatine phosphorylation to electron transport. J Biol Chem 248: 4803–4810, 1973
8. Rojo M, Hovius R, Deme R, Nicolay K, Wallimann T: Mitochondrial creatine kinase mediates contact formation between mitochondrial membranes. J Biol Chem 266: 20290–20295, 1991
9. Haas R, Strauss A: Separate nuclear genes encode sarcomeric specific and ubiquitous human mitochondrial creatine kinase isoenzymes. J Biol Chem 265: 6921–6927, 1991
10. Payne R, Haas R, Strauss A: Structural characterization and tissue-specific expression of the mRNAs encoding isoenzymes from two rat mitochondrial creatine kinase genes. Biochim Biophys Acta 1089: 352–361, 1991
11. Steeghs K, Oerlemans F, Wieringa B: Mice deficient in ubiquitous mitochondrial creatine kinase are viable and fertile. Biochim Biophys Acta 1230: 130–138, 1995
12. Muhlebach S, Wirz T, Brandle U, Perriard J: Evolution of the creatine kinases. J Biol Chem 271: 11920–11929, 1996
13. Qin W, Khuchua Z, Klein A, Strauss A: Elements regulating cardiomyocyte expression of the human sarcomeric mitochondrial creatine kinase gene in transgenic mice. J Biol Chem 272: 25210–25216, 1997
14. Payne R, Friedman D, Grant J, Perryman M, Strauss A: CK isoenzymes are highly regulated during pregnancy in rat uterus and placenta. Am J Physiol 265: E624–E635, 1993

15. Payne R, Strauss A: Expression of mitochondrial creatine kinase genes. Mol Cell Biochem 133: 235–243, 1994
16. Payne R, Strauss A: Developmental expression of sarcomeric and ubiquitous mitochondrial creatine kinase is tissue-specific. Biochim Biophys Acta 1219: 33–39, 1994
17. Sistermans E, de Kok Y, Peters W, Ginsel L, Jap P, Wieringa B: Tissue- and cell-specific distribution of creatine kinase B: A new and highly specific monoclonal antibody for use in immunohistochemistry. Cell Tissue Res 280: 435–446, 1995
18. Collins R, Borowsky I: Metabolic architecture of brain: Oxidative and glycolytic systems. In: L. Na *et al.* (eds). Brain Work and Mental Activity. Alfred Benzon Symposium 31. Munksgaard, Copenhagen, 1991
19. Wong-Riley M, Welt C: Histochemical changes in cytochrome c oxidase of cortical barrels after vibrissal removal in neonatal and adult mice. Proc Natl Acad Sci USA 77: 2333–7233, 1980
20. Horton J, Hubel D: Regular patchy distribution of cytochrome c oxidase staining in primary visual cortex of macaque monkey. Nature 292: 762–764, 1981
21. Livingstone M, Hubel D: Thalamic inputs to cytochrome c oxidase-rich regions in monkey visual cortex. Proc Natl Acad Sci USA 79: 6098–6101, 1982
22. Friedman D, Roberts R: Compartmentation of brain-type creatine kinase and ubiquitous mitochondrial creatine kinase in neurons: Evidence for a creatine phosphate energy shuttle in the adult rat brain. J Comp Neurol 343: 500–511, 1994
23. Hemmer W, Zanolla E, Furter-Graves E, Eppenberger H, Walliman T: Creatine kinase isoenzymes in chicken cerebellum: Specific localization of brain-type creatine kinase in Bergman glial cells and muscle-type creatine kinase in Purkinje neurons. Eur J Neurosci 6: 538–549, 1994
24. Kaldis P, Hemmer W, Zanolla E, Holtzman D, Walliman T: 'Hot spots' of creatine kinase localization in brain: Cerebellum, hippocampus and choroid plexus. Dev Neurosci 18: 542–554, 1996
25. Khuchua Z, Qin W, Boero J, Cheng J, Payne R, Saks V, Strauss A: Octamer formation and coupling of cardiac sarcomeric mitochondrial creatine kinase are mediated by charge N-terminal residues. J Biol Chem 273: 22990–22996, 1998
26. Morrison R, Wenzel H, Kinoshita Y, Robbins C, Donehower L, Schwartzkroin P: Loss of p53 suppresor gene protects neurons from kainate-induced cell death. J Neurosci 16: 1337–1345, 1996
27. Kokate T, Cohen A, Karp E, Rogawski M: Neuroactive steroids protect against pilocarpine- and kainic acid-induced limbic seizures and status epilepticus in mice. Neuropharm 35: 1049–1056, 1996
28. Kageyama G, Wong-Riley G: Histochemical localization of cytochrome oxidase in the hippocampus: Correlation with specific neuronal types and afferent pathways. Neuroscience 7: 2337–2361, 1982
29. Schwartzkroin P, Kundel D: Morphology of identified interneurons in the CA1 region of guinea pig hippocampus. J Comp Neurol 232: 205–218, 1985
30. Schwartzkroin P, Mathers L: Physiological and morphological identification of a nonpyramidal hippocampal cell type. Brain Res 157: 1–10, 1978
31. Tsuji M, Mulkern R, Cook C, Meyers R, Holtzman D: Relative phosphocreatine and nucleoside triphosphate concentrations in cerebral g ray and white matter measured *in vivo* by $^{31}$P nuclear magnetic resonance. Brain Res 707: 146–154, 1996
32. Mason G, Chu W, Vaughan J, Ponder S, Twieg D, Adams D, Hetherington H: Evaluation of $^{31}$P metabolite differences in human cerebral gray and white matter. Magnet Res Med 39: 346–353, 1998
33. White E, Amitai Y, Gutnick M: A comparison of synapses onto the somata of intrinsically bursting and regular spiking neurons in layer V of rat SmI cortex. J Comp Neurol 342: 1–14, 1994
34. De la Peña E, Guijo-Barrientos E: Laminar localization, morphology, and physiological properties of pyramidal neurons that have the low-threshold calcium current in guinea-pig medial frontal cortex. J Neurosci 16: 5301–5311, 1996
35. Schwindt P, O'Brien J, Crill W: Quantitative analysis of firing properties of pyramidal neurons from layer 5 of rat somatosensory cortex. J Neurophysiol 77: 2484–2498, 1997
36. Woolsey T, Van der Loos H: The structural organization of layer IV in the somatosensory region (SI) of mouse cerebral cortex. The description of a cortical field composed of discrete cytoarchitectonic units. Brain Res 17: 205–242, 1970
37. McCasland J, Hibbard L, Rhoades R, Woolsey R: Activation of a widespread network of inhibitory neurons in barrel cortex. Somatosens Motor Res 14: 138–147, 1997
38. Tolkovsky A, Suidan H: Adenosine 5′-triphosphate synthesis and metabolism localized in neuritis of cultured sympathetic neurons. Neuroscience 23: 1133–1142, 1987
39. Skladchikova G, Ronn L, Berezin V, Bock E: Extracellular adenosine triphosphate affects neuronal cell adhesion molecule (NCAM)-mediated adhesion and neurite outgrowth. J Neurosci Res 57: 207–218, 1999
40. Hirose Y, Manley J: Creatine phosphate, not ATP, is required for 3′ end cleavage of mammalian pre-mRNA *in vitro*. J Biol Chem 272: 29636–29642, 1997
41. Crino P, Eberwine J: Molecular characterization of the dendrite growth cone: Regulated mRNA transport and local protein synthesis. Neuron 17: 1173–1187, 1996
42. Holtzman D, Meyers R, O'Gorman E, Khait I, Walliman T, Allred E, Jensen F: *In vivo* brain phosphocreatine and ATP regulation in mice fed a creatine analog. Am J Physiol 272: C1567–C1577, 1997
43. Klivenyi P, Ferrante R, Matthews R, Bogdanov M, Klein A, Andreassen O, Mueller G, Wermer M, Kaddurah-Daouk R, Beal M: Neuroprotective effects of creatine in transgenic animal model of amyotrophic lateral sclerosis. Nature Med 5: 347–350, 1999
44. Matthews R, Ferrante R, Klivenyi P, Yang L, Klein A, Mueller G, Kaddurah-Daouk R, Beal M: Creatine and cyclocreatine attenuate MPTP neurotoxicity. Exp Neurol 157: 142–149, 1999

*Molecular and Cellular Biochemistry* **244:** 77–81, 2003.


# Monocarboxylates and glucose utilization as energy substrates in rat brain slices under selective glial poisoning – a $^{31}$P NMR study

Takaaki Kitano,[1] Naoko Nisimaru,[2] Eriko Shibata,[2] Hideo Iwasaka,[1] Takayuki Noguchi[1] and Isao Yokoi[2]

[1]*Department of Anesthesiology;* [2]*Department of Physiology, Faculty of Medicine, Oita Medical University, Oita-gun, Oita, Japan*

## Abstract

We have investigated effects of various energy substrates including glucose, lactate and pyruvate on the recovery of the high energy phosphate levels after high-$K^+$ stimulation in rat brain slices by using $^{31}$P NMR. It was found that lactate, pyruvate and glucose almost equally supported the recovery of phosphocreatine (PCr) levels after high-$K^+$ stimulation (60 mM, 8 min) in artificial cerebrospinal fluid (ACSF). In iodoacetic acid (IAA) and fluorocitrate (FC)-pretreated slices, whereas *glucose* was unable to be utilized, the recovery of the PCr level after high-$K^+$ stimulation in ACSF containing *lactate* was completely abolished, the recovery of the PCr in ACSF containing *pyruvate* was unaffected. These results indicate that neurons themselves can utilize pyruvate as an exogenous energy substrate, but not lactate, without glial support. In intact brain, glucose may be metabolized to pyruvate in glial cells and then transported to neurons as an energy substrate. These suggest an astrocyte-neuron *pyruvate* shuttle mechanism of the brain energy metabolism *in vivo*.

We also investigated the effect of ischemic-preconditioning in FC-pretreated slices, which showed that the PCr levels recovered substantially in ACSF containing lactate after high-$K^+$ stimulation. This indicates that after the preconditioning, such as ischemia, neurons themselves acquired the ability to utilize *lactate* as an energy substrate. (Mol Cell Biochem **244**: 77–81, 2003)

*Key words*: NMR, phosphocreatine, creatine, pyruvate, brain

## Introduction

Glucose is considered to be the main substrate in the production of the energy for the central nervous system and functional activities in the brain are coupled to the rate of glucose utilization [1]. On the other hand, monocarboxylates such as pyruvate and lactate can also be used as substrates for energy metabolism in the brain [2–4]. Schurr *et al.* [5, 6] reported that lactate can support synaptic transmission of CA1 pyramidal cells in rat hippocampal slices. However, Cox and Bachelard [7] and Kanatani *et al.* [8] reported that replacing glucose in the perfusion medium with either pyruvate or lactate failed to maintain the evoked synaptic response and neural activity. It is still unclear whether monocarboxylates are utilized as energy substrates to maintain neural activity in brain.

Brain consists of heterogeneous cell populations of neurons and glial cells. Glial cells envelope capillaries with their end feet and form part of the blood brain barrier. Therefore, neurons are in contact with capillaries indirectly through glia [9]. Tsacopolous and Magistretti [10] have proposed an astrocyte-neuron lactate shuttle mechanism, in which glia cells feed neurons with lactate produced from glucose. Yoshioka *et al.* [4] have already shown that neurons themselves can utilize pyruvate for an energy substrate.

In order to establish the ability of neurons to metabolize the monocarboxylate such as lactate and pyruvate as energy

*Address for offprints*: T. Kitano, Department of Anesthesiology, Oita Medical University, Oita-gun, 879-5593 Oita, Japan (E-mail: tkitano@oita-med.ac.jp)

substrates, recoveries of high-energy phosphates in the intact and ischemic-preconditioned brain slices after high-$K^+$ stimulation were measured by using phosphorus nuclear magnetic resonance ($^{31}P$ NMR).

## Materials and methods

Male Wistar rats (6–10 weeks) were anesthetized with diethyl ether, and then decapitated. After the brains were quickly removed, the cerebral hemispheres were sectioned coronally into 400 μm-thick sections using a microslicer (DTK-3000, Dosaka EM). The sections were incubated in a standard ACSF, bubbled vigorously with a 95% $O_2$ plus 5% $CO_2$ gas mixture at 25°C for 1 h. Standard ACSF contained (mM): NaCl, 126; KCl, 5; $MgSO_4$, 1.3; $CaCl_2$, 2.4; $NaHCO_3$, 26; glucose, 10; HEPES, 5 at pH 7.4. In order to examine the effects of the energy substrates, 10 mM glucose was replaced with 10 mM lactate or pyruvate. The concentration of fluorocitrate (FC, Sigma), a selective glial toxin used, was 100 μM according to previous reports [4, 6, 11–13]. Iodoacetic acid (IAA), an irreversible glycolytic inhibitor, was applied for 12 min before stimulation, when glucose as the energy substrate was replaced with lactate or pyruvate. Ischemic-preconditioned slices were obtained from brain hemispheres, of which contra-lateral middle cerebral artery was occluded permanently with 5-0 nylon suture 48 h before the brain slices were prepared [14].

Brain slices for a single experiment were prepared from two to four brains. The glass tube and perfusion system used for the NMR measurements were the same as those in a previous report [4]. Stimulation of the brain slices was induced by changing the medium to a high-$K^+$ solution (60 mM $K^+$). $^{31}P$ NMR spectra were obtained using a Bruker AMX300wb spectrometer operating at 121 MHz (7.5 T) according to previous paper [4]. Free induction decays (FID) were obtained by 45° radio frequency pulses repeated at intervals of 3.8 sec. An accumulation of 64 FIDs were taken as a unit of observation. After Fourier-transformed and phase corrected, the baseline was corrected to eliminate the broad signal in the spectra. The 'saturation' factors were taken to be 1.0 for the peaks of interest [15]. There were mainly six peaks in the $^{31}P$ NMR spectra of the cerebral slices superfused with well-oxygenated standard ACSF in the resting condition (see Fig. 1A). The PCr peak served as the reference at 0 p.p.m. The PCr peaks were analyzed by the one-peak deconvolution method in the spectra arising from 64 FIDs. Therefore, time resolution in the changes in the PCr leves was 4 min.

Because each slice preparation had a different concentration of substances, changes in these metabolites were expressed as percent changes relative to control values. Control

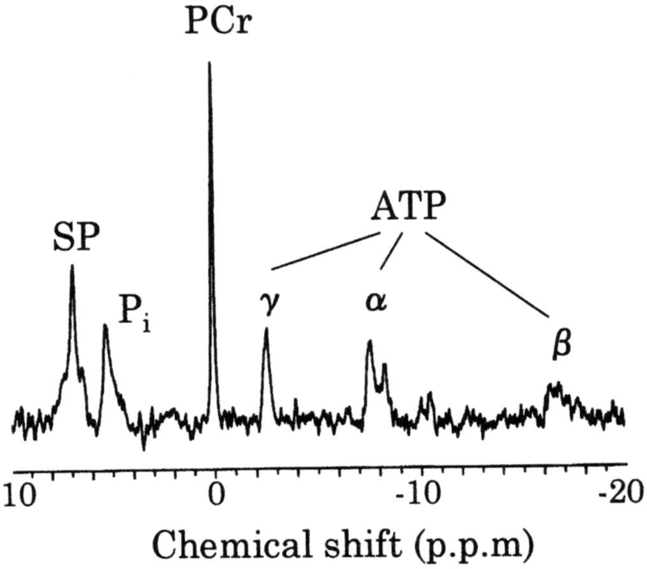


*Fig. 1.* $^{31}P$ NMR spectrum of rat brain slices under the oxygenated resting condition in ACSF containing glucose. The spectrum represents an accumulation of 256 FIDs. The peaks correspond to, from left, SP, $P_i$, PCr, and γ-, α-, and β-phosphates of ATP. Line broadening, 10 Hz.

values are the average of four measurements for each of the metabolites before high-$K^+$ stimulation.

Statistical analysis was performed by one-way ANOVA and Fisher's protected least significant difference multiple comparison analysis using Statview 4.5J (Abacus Concepts, Inc., USA). A p value of less than 0.05 was considered significant. All data were expressed as mean ± S.E.M.

## Results

Figure 1 shows a typical $^{31}P$ NMR spectra of normal brain slices superfused with well-oxygenated standard ACSF in the resting condition. Peaks of sugar phosphates (SP), inorganic phosphate ($P_i$), phosphocreatine (PCr), γ-, α-, and β-phosphates of ATP are resolved. During high-$K^+$ stimulation the peak amplitudes of $P_i$ increased and those of PCr decreased. When the slices were replaced to the standard ACSF after high-$K^+$ stimulation, the peak amplitudes of $P_i$ and PCr were recovered to their control levels within 1 h. The peak amplitudes of ATP and SP, however, remained almost unchanged during high-$K^+$ stimulation. In the brain slices pretreated with IAA, whereas glucose was unable to be utilized, the PCr levels decreased and then recovered to the control levels in ACSF containing 10 mM lactate or pyruvate following high-$K^+$ stimulation almost as well as in standard ACSF (Table 1).

Table 1.

| Time (min) from high K$^+$ | Glucose | Glucose+IAA | Glucose+FC | Pyruvate+IAA | Pyruvate+IAA+FC | Lactate+IAA | Lactate+IAA+FC | Lactate+IAA+FC+IPC |
|---|---|---|---|---|---|---|---|---|
| N | 7 | 5 | 4 | 2 | 4 | 6 | 4 | 5 |
| Control | 1.10 ± 0.08 | 1.28 ± 0.10 | 0.98 ± 0.14 | 1.14 ± 0.02 | 0.89 ± 0.11 | 1.10 ± 0.07 | 0.92 ± 0.15 | 1.09 ± 0.16 |
| High-K$^+$ | 0.50 ± 0.06 | 0.37 ± 0.08 | 0.21 ± 0.02 | 0.50 ± 0.16 | 0.34 ± 0.12 | 0.32 ± 0.07 | 0.31 ± 0.18 | 0.60 ± 0.16 |
| 8 | 0.78 ± 0.08 | 0.30 ± 0.14$^†$ | 0.42 ± 0.05$^†$ | 0.78 ± 0.19 | 0.40 ± 0.04$^†$ | 0.64 ± 0.12 | 0.22 ± 0.05$^†$ | 0.36 ± 0.04$^†$ |
| 16 | 0.96 ± 0.11 | 0.19 ± 0.03$^†$ | 0.80 ± 0.13 | 0.79 ± 0.21 | 0.78 ± 0.11 | 0.72 ± 0.08 | 0.27 ± 0.07$^†$ | 0.50 ± 0.05$^†$ |
| 24 | 1.00 ± 0.05 | 0.19 ± 0.03$^†$ | 1.05 ± 0.19 | 0.86 ± 0.01 | 1.00 ± 0.19 | 0.66 ± 0.07$^†$ | 0.26 ± 0.04$^†$ | 0.68 ± 0.12$^†$ |
| 32 | 0.94 ± 0.05 | 0.19 ± 0.03$^†$ | 1.12 ± 0.18 | 0.79 ± 0.07 | 0.94 ± 0.14 | 0.97 ± 0.15 | 0.41 ± 0.13$^†$ | 0.66 ± 0.10 |
| 40 | 1.01 ± 0.07 | 0.28 ± 0.07$^†$ | 1.07 ± 0.10 | 0.81 ± 0.17 | 0.72 ± 0.07$^†$ | 1.03 ± 0.09 | 0.23 ± 0.03$^†$ | 0.66 ± 0.09$^†$ |
| 48 | 0.91 ± 0.10 | 0.18 ± 0.03$^†$ | 1.32 ± 0.27$^†$ | 0.91 ± 0.01 | 0.72 ± 0.05 | 0.97 ± 0.07 | 0.39 ± 0.03$^†$ | 0.71 ± 0.16 |
| 56 | 0.93 ± 0.07 | 0.18 ± 0.04$^†$ | 0.94 ± 0.17 | 1.16 ± 0.18 | 0.77 ± 0.02 | 0.85 ± 0.16 | 0.28 ± 0.05$^†$ | 0.84 ± 0.20 |

IAA: Iodoacetic acid 0.2 mM for 12 min, FC: Fluorocitrate 100 μM for 2 h, IPC: Ischemic preconditioning, mean ± S.E., $^†p < 0.05$ vs. glucose. Stimulation of the brain slices was induced by changing the ACSF to a high-K$^+$ solution (60 mM).

PCr levels of brain slices pretreated with 100 μM FC during and after high-K$^+$ stimulation in ACSF containing glucose, lactate or pyruvate were summarized in Fig. 2. In ACSF containing glucose or pyruvate, the PCr levels decreased following high-K$^+$ stimulation and then recovered to the control level in about 30 min. In ACSF containing lactate, however, the PCr levels decreased following high-K$^+$ stimulation and were unable to recover at least for 1 h (Fig. 2 and Table 1).

Figure 3 compares changes in the PCr levels in ACSF containing lactate following high-K$^+$ stimulation among different conditions of brain slices (Table 1). In intact brain slices, the PCr levels in ACSF containing lactate were able to recover after high-K$^+$ stimulation, but in brain slices pretreated with FC the PCr levels in ACSF containing lactate were unable to recover after high-K$^+$ stimulation as shown Fig. 2. In ischemic-preconditioned slices obtained from the brain hemisphere, of which contralateral middle cerebral artery was occluded 48 h before, the PCr levels in ACSF containing lactate could be recovered after high-K$^+$ stimulation even in the slices pretreated with FC.

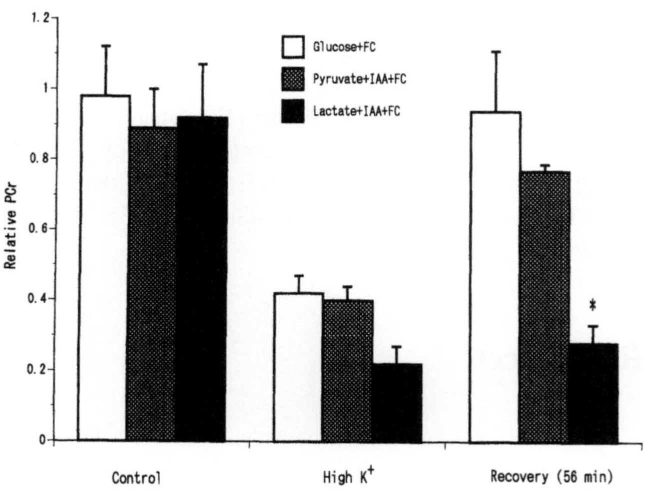


*Fig. 2.* Changes in the levels of PCr during and after high-K$^+$ stimulation in slices in ACSF containing different substrates and 100 μM FC. In the slices in standard ACSF, the PCr levels recovered after high-K$^+$ stimulation regardless of the presence of FC (white bars). In IAA-pretreated slices in ACSF containing lactate, however, the recovery of PCr level after high-K$^+$ stimulation was completely abolished (black bars). In IAA-pretreated slices in ACSF containing pyruvate, the PCr level recovered well after high-K$^+$ stimulation (hatched bars). $^*p < 0.05$ vs. glucose + FC.

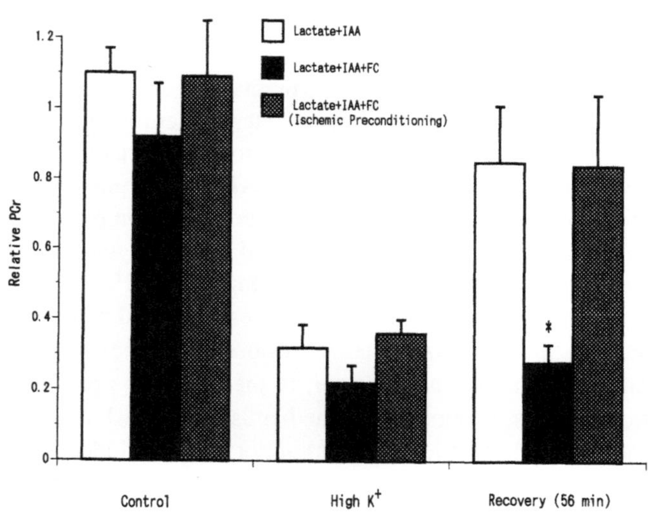


*Fig. 3.* Changes in the PCr levels in ACSF containing lactate during and after high-K$^+$ stimulation under various conditions of brain slices. In intact brain slices (white bars), the PCr levels were able to recover after high-K$^+$ stimulation, but in brain slices pretreated with FC (black bars) they were unable to recover. In ischemic-preconditioned slices (hatched bars) the PCr levels could recover after high-K$^+$ stimulation in lactate ACSF even in the slices pretreated with FC. $^*p < 0.05$ vs. lactate + IAA.

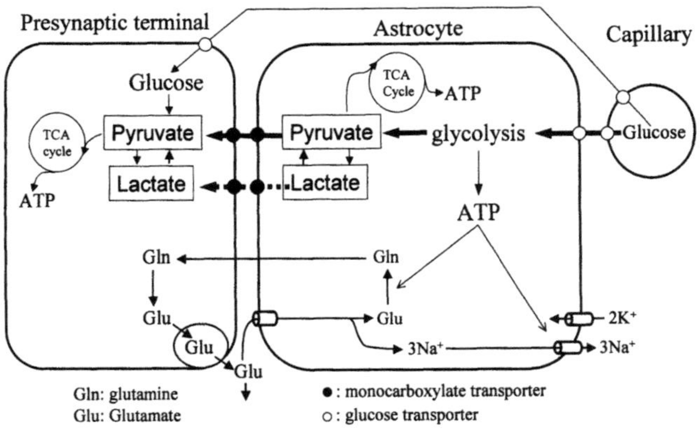


*Fig. 4.* Energy metabolism of the brain *in vivo*. This shows an astrocyte-neuron pyruvate shuttle mechanism. Under ischemic pre-stressed condition, even in the mature neuron, the ability to uptake and utilize lactate may be induced.

## Discussion

Yoshioka *et al.* [4] showed that neurons with glial support can utilize lactate and pyruvate as well as glucose, but those without glial support can not utilize lactate, as exogenous energy substrates in brain slices.

Brain consists of heterogeneous cell populations of neurons and glial cells. Glial cells, mainly astrocytes, wrap up intraparenchymal capillaries with their end-feet [9]. Thus, astrocytes play an important role in the distribution of energy substrates from the circulation to neurons *in vivo* [3, 10, 16]. Following neuronal activation and synaptic glutamate release in the brain, the increase of glutamate reuptake into astrocytes may occur, which accelerates glucose uptake from capillaries to astrocytes via activation of the $Na^+/K^+$-ATPase. Glucose is then processed glycolytically to pyruvate by astrocytes. The transport of pyruvate from astrocytes to neurons may be operated by monocarboxylate transporters. Pyruvate then enters the TCA cycle and generates ATP in neurons. These mechanisms may be taken as an astrocyte-neuron pyruvate shuttle as shown in Fig. 4, instead of an astrocyte-neuron lactate shuttle by Tsacopolous and Magestretti [10].

In this study, the neurons in the slices pretreated with IAA and FC can not utilize lactate to restore the PCr levels after high-$K^+$ stimulation. However, in the ischemic-preconditioned slices, obtained from the hemisphere of which contralateral side of middle cerebral artery was occluded 48 h before, the PCr levels were restored after high-$K^+$ stimulation in ACSF containing lactate, even though slices were pretreated with IAA and FC. It is well known that pretreatment of the brain with a sublethal stress induces resistance to subsequent stress. Previous studies have shown that pretreatment of the brain with a sublethal ischemia induces resistance to subsequent periods of ischemia that normally would give rise to neuronal damage, and the induction of such tolerance has been reported to be correlated with the selective synthesis of certain proteins, including heat-shock proteins [17, 18]. Belayev *et al.* [19] showed that bilateral ischemic tolerance of rat hippocampus and bilateral c-fos mRNA expression were induced only by prior unilateral transient focal ischemia. Furthermore, lactate is also known as the main metabolic fuel for the brain during the early neonatal period [20]. These might suggest that under stressed condition the mature neuron induces and expresses the ability to utilize lactate, which is induced in immature neurons originally (Fig. 4).

Gore and Mcllwain [21] showed that increased concentrations of potassium brought about changes which could be induced by electrical stimulation in slices of the cerebral cortex. Takei *et al.* [15] and Yoshioka *et al.* [4] have confirmed that high-$K^+$ induces a decrease in high energy phosphates in brain slices, also indicating the involvement of calcium in energy-requiring processes under high-$K^+$ stimulation. Machiyama *et al.* [22] suggested that high-$K^+$ stimulation might induce the release of the excitatory neurotransmitters and then activate energy metabolism. From these results, the high-$K^+$ stimulation may induce energy consumption in neurons due to the release of excitatory neurotransmitters by depolarization of neurons.

In summary, the normal neurons can utilize pyruvate without glial support as well as glucose, but not lactate, as exogenous energy substrates. The ability to utilize lactate in matured neuron can be induced after stress preconditioning such as ischemia.

## Acknowledgements

We thank Professor emeritus Kazuhiro Yamada for advice on the manuscript. This work was supported in part by a Grant-in-Aid for Scientific Research (13671594) from the Ministry of Education, Culture, Sports, Science and Technology in Japan.

## References

1. Sokoloff L: Relation between physiological function and energy metabolism in the central nervous system. J Neurochem 29: 13–26, 1977
2. Mcllwain H, Bachelard HS: Metabolic, ionic and electrical phenomena in separated cerebral tissue. In: Biochemistry and the Central Nervous System, 5th edn. Livingstone, Edinburgh, 1985, pp 54–83
3. Magistretti PJ: Brain energy metabolism. In: M.J. Zimond *et al.* (eds). Fundamental Neuroscience. Academic Press, San Diego, 1999, pp 389–413
4. Yoshioka K, Nisimaru N, Yanai S, Shimoda H, Yamada K: Characteristics of monocarboxylates as energy substrates other than glucose in

rat brain slices and the effect of selective glial poisoning – a 31P NMR study. Neurosci Res 36: 215–26, 2000
5. Schurr A, Dong WQ, Reid KH, West CA, Rigor BM: Lactic acidosis and recovery of neuronal function following cerebral hypoxia *in vitro*. Brain Res 438: 311–314, 1988
6. Schurr A, Payne RS, Miller JJ, Rigor BM: Glia are the main source of lactate utilized by neurons for recovery of function posthypoxia. Brain Res 774: 221–224, 1997
7. Cox DW, Bachelard HS: Attenuation of evoked field potentials from dentate granule cells by low glucose, pyruvate + malate, and sodium

*Molecular and Cellular Biochemistry* **244:** 83–88, 2003.


# Creatine supplementation during college football training does not increase the incidence of cramping or injury

Michael Greenwood,[1] Richard B. Kreider,[2,5] Charlie Melton,[2] Christopher Rasmussen,[5] Stacy Lancaster,[2] Edward Cantler,[2] Purvis Milnor[3] and Anthony Almada[4]

[1]*Human Performance Laboratory, Department of HPESS, Arkansas State University, Jonesboro, AR;* [2]*Exercise and Sport Nutrition Laboratory, Department of Human Movement Sciences and Education, The University of Memphis, Memphis, TN;* [3]*The Milnor Clinic, Memphis, TN;* [4]*MetaResponse Sciences, Inc., Laguna Niguel, CA;* [5]*Exercise and Sport Nutrition Laboratory, Department of Health, Human Performance, and Recreation, Waco, TX, USA*

## Abstract

The purpose of this study was to examine the effects of creatine supplementation on the incidence of injury observed during 3-years of NCAA Division IA college football training and competition. In an open label manner, athletes participating in the 1998–2000 football seasons elected to take creatine or non-creatine containing supplements following workouts/practices. Subjects who decided to take creatine were administered 15.75 g of creatine for 5 days followed by ingesting an average of 5 g/day thereafter administered in 5–10 g doses. Creatine intake was monitored and recorded by research assistants throughout the study and ranged between 34–56% of players during the course of the study. Subjects practiced or played in environmental conditions ranging from 8–40°C (mean 24.7 ± 9°C) and 19–98% relative humidity (49.3 ± 17%). Injuries treated by the athletic training staff were recorded and categorized as cramping, heat/dehydration, muscle tightness, muscle strains/pulls, non-contact joint injuries, contact injuries, and illness. The number of missed practices due to injury/illness was also recorded. Data are presented as the total number of treated injuries for creatine users/total injuries observed and percentage occurrence rate of injuries for creatine users for all seasons. The incidence of cramping (37/96, 39%), heat/dehydration (8/28, 36%), muscle tightness (18/42, 43%), muscle pulls/strains (25/51, 49%), non-contact joint injuries (44/132, 33%), contact injuries (39/104, 44%), illness (12/27, 44%), number of missed practices due to injury (19/41, 46%), players lost for the season (3/8, 38%), and total injuries/missed practices (205/529, 39%) were generally lower or proportional to the creatine use rate among players. Creatine supplementation does not appear to increase the incidence of injury or cramping in Division IA college football players. (Mol Cell Biochem **244**: 83–88, 2003)

*Key words*: exercise, nutrition, ergogenic aids, safety, sport injuries, athletic training

## Introduction

Creatine supplementation has been reported to increase strength, enhance work performed during repetitive sets of muscle contractions, improve repetitive sprint performance, and increase body mass/fat free mass [1–3]. Consequently, creatine has become one of the most popular nutritional supplements among athletes. The only reported side effect from creatine supplementation in the scientific/medical literature has been weight gain [2,4]. However, concerns have been raised over the medical safety of creatine supplementation among athletes even though extensive research has yielded

*Address for offprints*: R.B. Kreider, Exercise and Sport Nutrition Laboratory, Center for Exercise, Nutrition and Preventive Health Research, Department of Health, Human Performance and Recreation, Baylor University, P.O. Box 97313, Waco, TX 76798-7313, USA (E-mail: Richard_Kreider@baylor.edu)

no negative effects in the areas of endogenous creatine synthesis [5–7], renal function [8–15], muscle and liver efflux [10,13,16–18], blood volume and electrolyte status [19–23], blood pressure [24, 25], or general markers of medical safety [13, 14, 18, 25–36]. Additionally, there have been only anecdotal reports that creatine supplementation during intense training in hot/humid environments may predispose athletes to increase incidence of muscle cramping, dehydration, and/or musculoskeletal injuries such as muscle pulls/strains [37–39]. Therefore, the purpose of this study was to examine the effects of creatine supplementation on the incidence of cramping/injury observed during 3-years of NCAA Division IA college football training and competition.

## Materials and methods

### *Subjects*

Injury rates of approximately 130 Division IA National Collegiate Athletic Association (NCAA) college football players participating in the 1998–2000 football seasons at an area University were monitored during this study. Subjects who volunteered to participate in this safety study chose whether they wanted to take creatine or non-creatine containing supplements during training. This included approximately 85 athletes on the 1998 squad and 20–30 new athletes that joined the team in 1999 and 2000 to replace athletes who exhausted eligibility, were not invited to return, and/or no longer chose to play on the team. All subjects underwent pre-season medical examinations and were cleared to participate in football according to NCAA criteria. In addition, all athletes were provided medical supervision by the team athletic training staff and physicians throughout the course of this observation period. A subset of 116 athletes volunteered to donate blood and/or urine samples at various times during the initial 21 months of the study. Subjects were informed as to the experimental procedures and signed informed consent statements in adherence with the Internal Review Board for use of human subjects guidelines at The University of Memphis and the American College of Sports Medicine. Subjects were 19.6 ± 1 years (range 18–23 years), 184 ± 7 cm (range 161–200 cm), 102 ± 19 kg (range 70–146 kg) upon reporting to fall football camp during the 1998, 1999 or 2000 seasons.

### *Methods and procedures*

Injuries that were treated by the athletic training staff during the 1998, 1999 and 2000 college football seasons were monitored during this study. This was accomplished by having research assistants attend all training/practice sessions during this period and recording all injuries treated by the athletic training staff during practices and games. Research assistants confirmed the type/category and degree of injury to the athlete with the athletic training staff when the injury occurred and after practice/competition to ensure that the injury was accurately recorded. Additionally, a tabulation of missed practices was recorded to monitor the length of time an athlete was out of action due to their specific injury. Injuries were categorized as cramping, heat disorders (e.g. dehydration, heat syncope), muscle tightness, muscle pulls/strains, non-contact joint injuries (e.g. twisted ankles), contact injuries (e.g. injuries resulting from collisions), illness, and the number of missed practices due to injury. These injuries were considered significant injuries because they required medical attention during and/or following practices/games and they involved some limitation from participating in training sessions and/or games. Since football is a contact sport, minor injuries that did not limit the athletes' ability to participate in practices/games (e.g. bruises, general soreness, etc.) and/or require significant medical treatment were not recorded.

### *Supplementation protocols*

Subjects who volunteered to participate in the safety study chose whether they wanted to take creatine or non-creatine containing supplements during training. These subjects had reported no current use of creatine supplements. Subjects who chose to take creatine were administered in an open label manner 15.75 g/day of creatine monohydrate for 5 days and an average of 5 g/day thereafter in 5–10 g doses following supervised training sessions. Creatine was added to sports drinks or carbohydrate/protein drinks that were offered to all players following training sessions, practices, and games by research assistants working with the strength and conditioning staff while also witnessing the supplement ingestion. When subjects were not on campus, they were provided creatine to take on their own and self-reported compliance. If for some reason a subject a fell behind in taking creatine, subjects were administered up to 10 g/day in order to catch them up to an average of 5 g/day. All remaining athletes were asked whether they were taking creatine during training on their own when they started training with the team and when an injury was observed. During the course of the study, 45–56% of the athletes were documented as taking creatine. Here again, the percentage of creatine users fluctuated between 45–56% due to new athletes joining the team over the three-year period in order to replace athletes who exhausted eligibility and/or athletes who no longer chose to play on the team.

### *Training*

Training consisted of summer resistance training/conditioning drills (1–2 h/day, 4 days/week), fall football camp (3–6 h/day,

6-days/week), practicing/competing during football season (2–4 h/day, 6 days/week) off-season resistance training/conditioning drills (1–2 h/day, 4 days/week), and spring football practice/resistance training (1–3 h/day, 4–5 day/week). Athletic coaches, athletic trainers, and/or research assistants supervised all training sessions. Training averaged of 121 ± 67 min per session with an average intensity of 3.3 ± 1 on a 1–5 scale where 1 was equivalent to a walk-through practice prior to games and 5 was equivalent to game competition. Environmental conditions during training and competition ranged from 8–40°C (mean 24.7 ± 9°C) and 19–98% relative humidity (49.3 ± 17%).

*Statistical analysis*

The numbers of injuries observed in creatine users and non-creatine users was determined for each injury category monitored during the study. The percentage of injuries observed for creatine users vs. non-users was calculated by dividing the number injuries observed among creatine users by the total number of injuries observed in a given category. The total number of injuries for all categories was determined by adding number of injuries observed for all categories for the creatine users and all athletes. The overall incidence of injuries was then calculated by dividing the total number of injuries observed for creatine users by the total number of injuries observed for all athletes. The percentage of injuries for creatine users was descriptively compared to the creatine use rate among the athletes (45–56%) in order to determine whether athletes who took creatine during training experienced a disproportionately higher incidence of injuries in comparison to athletes who did not take creatine. Data are presented as percentages of injuries for creatine users compared to non-users. It should be noted that the data in this investigation are presented in descriptive form because some subject's reported multiple injuries in the same injury category thus violating the statistical assumption of mutual exclusiveness regarding all other parametric and non-parametric statistical procedures.

## Results

Table 1 presents the injuries treated by the athletic training staff during the three-year safety study for the creatine and non-creatine groups. Results revealed that the incidence of cramping, heat/dehydration, muscle tightness, non-contact joint injuries, contact injuries, illness, number of missed practices due to injury, players lost for the season, and total injuries/missed practices were generally proportional or lower than the creatine use rate among players during the course of the three year safety study (39%).

*Table 1.* Observed injury rates during three years of NCAA Division IA college football training/games in which creatine use rate was 45–56%

| Treated Injury | Number of injuries (Creatine users/ total injuries) | Percentage of injuries for creatine users |
|---|---|---|
| Cramping | 37/96 | 39% |
| Heat/dDehydration | 8/28 | 36% |
| Muscle tightness | 18/42 | 43% |
| Muscle pulls/strains | 25/51 | 49% |
| Non-contact joint injuries | 44/132 | 33% |
| Contact injuries | 39/104 | 44% |
| Illness | 12/27 | 44% |
| Missed practices | 19/41 | 46% |
| Players lost for season | 3/8 | 38% |
| Total injuries/missed practices | 205/529 | 39% |

## Discussion

Anecdotal reports have suggested that creatine supplementation may promote dehydration, cramping, and musculoskeletal injury [2, 4, 38, 40]. Since many of these reports have emanated from athletic trainers and coaches, they are commonly reported as side effects from creatine supplementation [2, 4, 40]. As a result, some trainers and coaches have restricted availability of creatine to their athletes (particularly during intense training periods performed in the heat) and some have warned against the use of creatine until more long-term data demonstrated safety in athletes. In addition, some athletic organizations (e.g. NCAA) have banned allowing teams to 'provide' creatine to their athletes citing safety and fairness issues although athletes are still allowed to take creatine. Results of this long-term safety study indicate that creatine supplementation (~ 5 g/day) does not increase the incidence of dehydration, cramping, and/or muscle injury among Division IA college football players during intense training in very hot and humid environments. Moreover, that creatine supplementation did not cause a consistent pattern of 'unknown' side effects. These findings add to the growing body of evidence indicating that creatine supplementation does not increase the incidence of anecdotally reported side effects and/or cause unknown health problems [10, 13, 14, 18, 33–35].

One of the most commonly reported anecdotal side effects associated with creatine supplementation has been that creatine may increase the incidence of dehydration, muscle cramping, and/or heat tolerance. In this regard, some have suggested that since creatine supplementation may increase work capacity, athletes who take creatine during training in hot and humid environments may experience a greater rate of dehydration, muscle cramping, and/or heat illness [40]. Another theory suggests that since creatine has been suggested to promote fluid retention, it may alter electrolyte status and thereby promote muscle cramping by interfering with the muscle's contraction/relaxation mechanisms [2, 4, 40]. Over the last few years, a number of studies have examined the effects of creatine supplementation on hydration status, electrolyte levels, and dehydration during exercise performed in the heat [20–23, 41–43]. Results of these studies have indicated that creatine does not promote dehydration, alter electrolyte levels, or increase thermal stress. In addition, there has been no evidence that creatine supplementation promotes muscle cramping among athletes [21, 23, 37, 44–50]. In fact, recent studies have indicated that creatine supplementation may actually promote hydration [21, 23, 41], reduce thermal stress during exercise in the heat [23, 42], and/or reduce incidence of injury [11, 17, 31, 33, 37, 44–47, 49–52]. The results of the present study support these findings in that the incidence of cramping (39%) and dehydration (36%) observed among creatine users was lower than the creatine use rate among these athletes (i.e. 45–56%).

Anecdotal reports have also suggested that creatine may promote a higher incidence of muscle injuries such as muscle strains and pulls [2, 4, 40]. Proponents of this theory have postulated that since creatine supplementation may promote rapid increases in strength and body mass, the athlete may be more predisposed to additional stress placed on muscles, bones, joints, ligaments, and connective tissues. Results of the present study indicate that creatine supplementation does not increase the incidence of muscle injuries in athletes. Specifically, the incidence of muscle tightness (43%), muscle pulls/strains (49%), non-contact (33%) and contact injuries (44%) among creatine users were similar or lower than the creatine use rate among these athletes (i.e. 45–56%). There was also no evidence of a greater proportion of individual injuries (e.g. hamstring pulls, groin pulls, etc.). Moreover, the incidence of illness (44%), missed practices (46%), players lost with season ending injuries (38%), and total injuries observed (39%) were similar or lower than the creatine use rate among these athletes. If creatine supplementation increased the incidence of these problems, the incidence of injury in creatine users should have been markedly higher than the creatine use rate among these players. Yet, the incidence of injury among athletes who took creatine during training was similar or lower than the observed use rate. Based on these results, one could argue that creatine supplementation may have allowed the athletes to tolerate training to a greater degree and thereby lessened the incidence of injury. These findings support recent reports that creatine supplementation during training may lessen injury rates among athletes [11, 17, 31, 33, 37, 44–47, 49–52] and/or hasten recovery following immobilization injury [32].

In summary, results of this study indicate that creatine supplementation during training does not increase the incidence of dehydration, cramping, and/or injury in college football players. Although athletes who take creatine during intense training may experience some of these problems, it appears that the incidence of these problems is similar or lower than the creatine use rate among athletes. Hopefully, these findings will help dispel anecdotal myths suggesting that creatine supplementation may increase the prevalence of dehydration, cramping and/or injury among athletes. Further, we hope that these findings may help professionals involved in the training and/or medical supervision of athletes (i.e. athletic coaches, athletic trainers, researchers, certified strength and conditioning coaches, nutritional consultants, administrators, athletic governing bodies) better examine the methods employed to train and/or manage athletes (i.e. 3-a-day training in extreme climates, exhaustive conditioning drills, hydration practices, etc.). In this regard, it appears that the type and conditions that athletes are asked to train and/or compete may place them at a greater risk of dehydration, cramping and/or injury than anecdotally associating these problems with creatine supplementation.

## Acknowledgements

We would like to thank the athletes who participated in this study, the laboratory assistants in Exercise and Sport Nutrition Laboratory (ESNL) who assisted in data collection and analysis, and the coaches who allowed us to conduct this study. This study was funded in part by the ESNL at The University of Memphis, MetaResponse Sciences/ImagiNutrition (Laguna Niguel, CA, USA), and Experimental and Applied Sciences (Golden, CO, USA). Supplements used in this study were provided by Experimental and Applied Sciences, Degussa Bioactives/Traco Labs (Champaign, IL, USA), and MetaResponse Modifiers (San Clemente, CA, USA). Investigators independently collected, analyzed and interpreted data from this study and have no financial interest in the outcome of results reported. Presentation of results in this study does not constitute endorsement by authors or their institutions of the supplements investigated.

## References

1. Williams MH, Kreider R, Branch JD: Creatine: In: The Power Supplement. Human Kinetics Publishers, Champaign, IL, 1999
2. Kreider R: Creatine supplementation: Analysis of ergogenic value, medical safety, and concerns. J Exerc Physiol Online 1: 7–18, 1998. Available: http://www.css.edu/users/tboone2/asep/jan3.htm
3. Kraemer WJ, Volek JS: Creatine supplementation. Its role in human performance. Clin Sports Med 18: 651–666, 1999
4. Terjung RL, Clarkson P, Eichner ER, Greenhaff PL, Hespel PJ, Israel RG, Kraemer WJ, Meyer RA, Spriet LL, Tarnopolsky MA, Wagenmakers AJ, Williams MH: American College of Sports Medicine roundtable. The physiological and health effects of oral creatine supplementation. Med Sci Sports Exerc 32: 706–717, 2000

5. Vandenberghe K, Goris M, Van Hecke P, Van Leemputte M, Vangerven L, Hespel P: Long-term creatine intake is beneficial to muscle performance during resistance training. J Appl Physiol 83:2055–2063, 1997
6. Guerrero-Ontiveros ML, Wallimann T: Creatine supplementation in health and disease. Effects of chronic creatine ingestion *in vivo*: Down-regulation of the expression of creatine transporter isoforms in skeletal muscle. Mol Cell Biochem 184: 427–437, 1998
7. Tarnopolsky MA, Parise G, Fu MH, Brose A, Mahoney D, Wallimann T: Creatine monohydrate supplementation at the recommended dosage does not affect creatine transporter content during resistance exercise training in either the young or the elderly. Paper presented at the 6th International Conference on Guanidino Compounds in Biology and Medicine, Cincinatti, OH, September 3, 2001
8. Kuehl K, Soehler S, Kulacki K, Goldberg L, Elliot D, Bennett W, Haddock B: Effects of oral creatine monohydrate supplementation on renal function in adults. Med Sci Sports Exerc 32: S168, 2000
9. Kreider R, Ransom J, Rasmussen C, Hunt J, Melton C, Stroud T, Cantler E, Milnor P: Creatine supplementation during pre-season football training does not affect markers of renal function. FASEB J 13: A543, 1999
10. Kreider R, Melton C, Rasmussen C, Greenwood M, Lancaster S, Cantler E, Milnor P, Almada AL: Effects of long-term creatine supplementation on renal function and muscle and liver enzyme efflux. Med Sci Sports Exerc 33: S207, 2001
11. Rasmussen C, Kreider R, Melton C, Ransom J, Stroud T, Cantler E, Greenwood M, Milnor P: Long-term creatine supplementation during football training does not affect markers of renal stress. J Strength Cond Res 13: 431, 1999
12. Almada AL, Kreider R, Melton C, Rasmussen C, Lundberg J, Ransom J, Greenwood M, Stroud T, Cantler E, Milnor P, Fox J: Long-term creatine supplementation does not affect markers of renal stress in athletes. J Strength Cond Res 14: 359, 2000
13. Robinson TM, Sewell DA, Casey A, Steenge G, Greenhaff PL: Dietary creatine supplementation does not affect some haematological indices, or indices of muscle damage and hepatic and renal function. Br J Sports Med 34: 284–288, 2000
14. Poortmans JR, Francaux M: Long-term oral creatine supplementation does not impair renal function in healthy athletes. Med Sci Sports Exerc 31: 1108–1110, 1999
15. Poortmans JR, Auquier H, Renaut V, Durussel A, Saugy M, Brisson GR: Effect of short-term creatine supplementation on renal responses in men. Eur J Appl Physiol Occup Physiol 76: 566–567, 1997
16. Ransom J, Kreider R, Hunt J, Melton C, Rasmussen C, Stroud T, Cantler E, Milnor P: Effects of creatine supplementation during training on markers of catabolism and muscle and liver enzymes. Med Sci Sports Exerc 31: S265, 1999
17. Ransom J, Kreider R, Rasmussen C, Melton C, Stroud T, Cantler E, Greenwood M, Milnor P: Effects of long-term creatine supplementation during training on markers of catabolism and enzyme efflux. J Str Cond Res 13: 431, 1999
18. Schilling BK, Stone MH, Utter A, Kearney JT, Johnson M, Coglianese R, Smith L, O'Bryant HS, Fry AC, Starks M, Keith R, Stone ME: Creatine supplementation and health variables: A retrospective study. Med Sci Sports Exerc 33: 183–188, 2001
19. Kreider RB, Ferreira M, Wilson M, Grindstaff P, Plisk S, Reinardy J, Cantler E, Almada AL: Effects of creatine supplementation on body composition, strength, and sprint performance. Med Sci Sports Exerc 30: 73–82, 1998
20. Rasmussen C, Kreider R, Ransom J, Hunt J, Melton C, Stroud T, Cantler E, Milnor P: Creatine supplementation during pre-season football training does not affect fluid or electrolyte status. Med Sci Sports Exerc 31: S299, 1999
21. Hulver MW, Campbell A, Haff G, Schroeder C, Comeau M, Potteiger JA: The effects of creatine supplementation on total body fluids, performance, and muscle cramping during exercise. Med Sci Sports Exerc 32: S133, 2000
22. Oopik V, Paasuke M, Timpmann S, Medijainen L, Ereline J, Smirnova T: Effect of creatine supplementation during rapid body mass reduction on metabolism and isokinetic muscle performance capacity. Eur J Appl Physiol Occup Physiol 78: 83–92, 1998
23. Volek JS, Mazzetti SA, Farquhar WB, Barnes BR, Gomez, Al, Kraemer WJ: Physiological responses to short-term exercise in the heat after creatine loading. Med Sci Sports Exerc 33: 1101–1108, 2001
24. Mihic S, MacDonald JR, McKenzie S, Tarnopolsky MA: Acute creatine loading increases fat-free mass, but does not affect blood pressure, plasma creatinine, or CK activity in men and women. Med Sci Sports Exerc 32: 291–296, 2000
25. Peeters BM, Lantz CD, Mayhew JL: Effect of oral creatine monohydrate and creatine phosphate supplementation on maximal strength indices, body composition, and blood pressure. J Strength Cond Res 13: 3–9, 1999
26. Earnest CP, Snell P, Rodriguez R, Almada A, Mitchell TL: The effect of creatine monohyerate ingestion on anaerobic power indices, muscular strength and body composition. Acta Physiol Scand 153: 207–209, 1995
27. Earnest CP, Almada A, Mitchell TL: High-performance capillary electrophoresis-pure creatine monohydrate reduced blood lipids in men and women. Clinical Science 91: 113–118, 1996
28. Bemben MG, Bemben DA, Loftiss DD, Knehans AW: Creatine supplementation during resistance training in college football athletes. Med Sci Sports Exerc 33: 1667–1673, 2001
29. Bermon S, Venembre P, Sachet C, Valour S, Dolisi C: Effects of creatine monohydrate ingestion in sedentary and weight-trained older adults. Acta Physiol Scand 164: 147–155, 1998
30. Eijnde BO, Hespel P: Short-term creatine supplementation does not alter the hormonal response to resistance training. Med Sci Sports Exerc 33: 449–453, 2001
31. Greenwood M, Kreider R, Melton C, Rasmussen C, Lundberg J, Stroud T, Cantler E, Milnor P, Almada AL: Short- and long-term creatine supplementation does not affect hematological markers of health. J Strength Cond Res 14: 362–363, 2000
32. Hespel P, Eijnde BO, Van Leemputte M, Urso B, Greenhaff PL, Labarque V, Dymarkowski S, Van Hecke P, Richter EA: Oral creatine supplementation facilitates the rehabilitation of disuse atrophy and alters the expression of muscle myogenic factors in humans. J Physiol 536: 625–633, 2001
33. Kreider R, Rasmussen C, Melton C, Greenwood M, Stroud T, Ransom J, Cantler E, Milnor P, Almada AL: Long-term creatine supplementation does not adversely affect clinical markers of health. Med Sci Sports Exerc 32: S134, 2000
34. Stone MH, Schilling BK, Fry AC, Johnson M, Keith RE, Kearney JT, Coglianese RH, Stone ME, Utter A, Smith L, O'Bryant HS: A retrospective study of long-term creatine supplementation on blood markers of health. J Strength Cond Res 13: 433, 1999
35. Sipila I, Rapola J, Simell O, Vannas A: Supplementary creatine as a treatment for gyrate atrophy of the choroid and retina. New Engl J Med 304: 867–870, 1981
36. Vannas-Sulonen K, Sipila I, Vannas A, Simell O, Rapola J: Gyrate atrophy of the choroid and retina. A five-year follow-up of creatine supplementation. Ophthalmology 92: 1719–1727, 1985
37. Greenwood M, Farris J, Kreider R, Greenwood L, Byars A: Creatine supplementation patterns and perceived effects in select division I collegiate athletes. Clin J Sport Med 10: 191–194, 2000
38. Juhn MS, O'Kane JW, Vinci DM: Oral creatine supplementation in male collegiate athletes: A survey of dosing habits and side effects. J Am Diet Assoc 99: 593–595, 1999

39. LaBotz M, Smith BW: Creatine supplement use in an NCAA Division I athletic program. Clin J Sport Med 9: 167–169, 1999
40. Juhn MS, Tarnopolsky M: Potential side effects of oral creatine supplementation: A critical review. Clin J Sport Med 8: 298–304, 1998
41. Ziegenfuss TN, Lowery LM, Lemon PWR: Acute fluid volume changes in men during three days of creatine supplementation. J Exerc Physiol Online 1: 1–9, 1998. Available: http://www.css.edu/users/tboone2/asep/jan3.htm
42. Kern M, Podewils LJ, Vukovich M, Buono MJ: Physiological response to exercise in the heat following creatine supplementation. J Exerc Physiol Online 4: 18–27, 2001. Available: http://www.css.edu/users/tboone2/asep/Kern.pdf
43. Papadopoulos C, Imamura R, Bandon LJ: The effect of creatine supplementation on repeated bouts of high-intensity exercise in the heat. Med Sci Sports Exerc 33: S203, 2001
44. Greenwood M, Kreider R, Rasmussen C, Ransom J, Melton C, Stroud T, Cantler E, Milnor P: Creatine supplementation does not increase incidence of cramping or injury during college football training II. J Strength Cond Res 13: 425–426, 1999
45. Greenwood M, Greenwood LD, Kreider R, Byars A: Effects of creatine supplementation on the incidence of cramping/injury during college football three a day training. Med Sci Sports Exerc 32: S136, 2000
46. Greenwood LD, Greenwood M, Kreider R, Byars A: Effects of creatine supplementation on the incidence of cramping/injury during collegiate fall baseball. Med Sci Sports Exerc 32: S136, 2000
47. Hunt J, Kreider R, Melton C, Ransom J, Rasmussen C, Stroud T, Cantler E, Milnor P: Creatine does not increase incidence of cramping or injury during pre-season college football training II. Med Sci Sports Exerc 31: S355, 1999
48. Kreider R, Rasmussen C, Ransom J, Almada AL: Effects of creatine supplementation during training on the incidence of muscle cramping, injuries and GI distress. J Strength Cond Res 12: 275, 1998
49. Kreider R, Melton C, Ransom J, Rasmussen C, Stroud T, Cantler E, Greenwood M, Milnor P: Creatine supplementation does not increase incidence of cramping or injury during college football training I. J Strength Cond Res 13: 428, 1999
50. Kreider R, Melton C, Hunt J, Rasmussen C, Ransom J, Stroud T, Cantler E, Milnor P: Creatine does not increase incidence of cramping or injury during pre-season college football training I. Med Sci Sports Exerc 31: S355, 1999
51. Ortega Gallo PA, Dimeo F, Batista J, Bazan F, Betchakian L, Garcia Cambon C, Griffa J: Creatine supplementation in soccer players, effects in body composition and incidence of sport-related injuries. Med Sci Sports Exerc 32: S134, 2000
52. Greenwood M, Kreider R, Greenwood LD, Comeau M, Brown LE, Stahura K, Byars A: Perceived health status and side-effects associated with creatine supplementation during the collegiate baseball season. Med Sci Sports Exerc 33: S205, 2001

*Molecular and Cellular Biochemistry* **244:** 89–94, 2003.


# Effects of creatine supplementation on performance and training adaptations

Richard B. Kreider

*Exercise and Sport Nutrition Laboratory, Center for Exercise, Nutrition and Preventive Health Research, Department of Health, Human Performance and Recreation, Baylor University, Waco, TX, USA*

## Abstract

Creatine has become a popular nutritional supplement among athletes. Recent research has also suggested that there may be a number of potential therapeutic uses of creatine. This paper reviews the available research that has examined the potential ergogenic value of creatine supplementation on exercise performance and training adaptations. Review of the literature indicates that over 500 research studies have evaluated the effects of creatine supplementation on muscle physiology and/or exercise capacity in healthy, trained, and various diseased populations. Short-term creatine supplementation (e.g. 20 g/day for 5–7 days) has typically been reported to increase total creatine content by 10–30% and phosphocreatine stores by 10–40%. Of the approximately 300 studies that have evaluated the potential ergogenic value of creatine supplementation, about 70% of these studies report statistically significant results while remaining studies generally report non-significant gains in performance. No study reports a statistically significant ergolytic effect. For example, short-term creatine supplementation has been reported to improve maximal power/strength (5–15%), work performed during sets of maximal effort muscle contractions (5–15%), single-effort sprint performance (1–5%), and work performed during repetitive sprint performance (5–15%). Moreover, creatine supplementation during training has been reported to promote significantly greater gains in strength, fat free mass, and performance primarily of high intensity exercise tasks. Although not all studies report significant results, the preponderance of scientific evidence indicates that creatine supplementation appears to be a generally effective nutritional ergogenic aid for a variety of exercise tasks in a number of athletic and clinical populations. (Mol Cell Biochem **244**: 89–94, 2003)

*Key words*: sport nutrition, ergogenic aids, exercise, training, phosphocreatine

## Introduction

An ergogenic aid is a technique or practice that serves to increase performance capacity, the efficiency to perform work, the ability to recover from exercise, and/or the quality of training thereby promoting greater training adaptations. [1] When evaluating the potential ergogenic value of a proposed aid, it is important to evaluate the theoretical rationale, the scientific evidence that the proposed aid affects exercise metabolism and/or performance, whether studies have incorporated an appropriate research design (e.g. double blind, placebo controlled, randomized clinical trial), and the reliability of the experimental methods employed. It is also important to examine whether a proposed ergogenic aid is safe for a given population. Based on a thorough analysis of the literature, it is then possible to make conclusions regarding the ergogenic value and safety of a proposed aid [1].

In the case of creatine, it has been well established that increasing dietary availability of creatine serves to increase total creatine (TC) and phosphocreatine (PC) concentrations in the muscle [2–9]. Moreover, that availability of creatine and PC play a significant role in contributing to energy metabolism particularly during intense exercise. For example, creatine supplementation (e.g. 20 g/day × 5 days) has been reported to increase muscle TC and PC typically by 15–40% [10–12]. Theoretically, increasing the availability of PC would enhance cellular bioenergetics of the phosphagen system that is involved in high-intensity exercise performance [7, 11, 13] as well as the shuttling of high-energy phosphates between the mitochondria and cytosol via the creatine phosphate shut-

*Address for offprints*: R.B. Kreider, Exercise and Sport Nutrition Laboratory, Center for Exercise, Nutrition and Preventive Health Research, Department of Health, Human Performance and Recreation, Baylor University, P.O. Box 97313, Waco, TX 76798-7313, USA (E-mail: Richard_Kreider@baylor.edu)

tle that may enhance both anaerobic and aerobic capacity [14, 15].

Over the last several years, a number of reviews were published examining the potential ergogenic value of creatine supplementation [6, 7, 10–13, 16–19]. These reviews generally concluded that creatine supplementation serves to increase muscle TC and PC content. In addition, that creatine may improve performance primarily during short-duration, high intensity exercise. However, there was less evidence that creatine supplementation enhanced exercise performance during moderate to high-intensity prolonged exercise. In addition, there were some questions whether results observed in laboratory settings would transfer to performance on the field, whether performance changes observed would enhance training adaptations, and whether long-term creatine supplementation was safe. Since these reviews, a number of research studies have been published evaluating the effects of creatine supplementation on performance and training adaptations in a variety of populations. The purpose of this paper is to examine the most recent research that has examined the effects of short-term creatine supplementation on exercise performance and whether creatine supplementation during training can serve as a safe and effective ergogenic aid for athletes.

## Effects of short-term creatine supplementation on performance

Numerous studies have examined the effects of short-term creatine supplementation (5–7 days) on exercise performance. As described in a number of reviews, the majority of initial studies suggested that creatine supplementation can significantly increases strength, power, sprint performance, and/or work performed during multiple sets of maximal effort muscle contractions [6, 7, 10–13, 16, 20]. More recent studies have supported these initial observations. For example, Volek *et al.* [21] reported that creatine supplementation (25 g/day for 7 days) resulted in a significant increases in the amount of work performed during five sets of bench press and jump squats in comparison to a placebo group. Urbanski *et al.* [22] reported that creatine supplementation (20 g/day × 5 days) increased maximal isometric knee extension strength and time to fatigue. Tarnopolsky *et al.* [23] reported creatine supplementation (20 g/day × 4 days) increased peak cycling power, dorsi-flexion maximal voluntary contractions (MVC) torque, and lactate in men and women with no apparent gender effects. Moreover, Wiroth *et al.* [24] reported that creatine supplementation (15 g/day × 5 days) significantly improved maximal power and work performed during 5 × 10-sec cycling sprints with 60-sec rest recovery in younger and older subjects. These findings and many others support prior reports indicating that creatine supplementation can improve performance when evaluated in controlled laboratory and testing settings.

Some have criticized this type of early creatine research suggesting that although performance gains have been observed in controlled laboratory settings, it was less clear whether these changes would improve athletic performance on the field [17, 19]. Since then, a number of studies have attempted to evaluate the effects of creatine supplementation on field performance. These studies have generally indicated that short-term creatine supplementation may improve high intensity, short-duration performance in various athletic tasks. For example, Skare *et al.* [25] reported that creatine supplementation (20 g/day) decreased 100-m sprint times and reduced the total time of 6 × 60-m sprints in a group of well-trained adolescent competitive runners. Mujika *et al.* [26] reported that creatine supplementation (20 g/day × 6 days) improved repeated sprint performance (6 × 15 m sprints with 30-sec recovery) and limited the decay in jumping ability in 17 highly trained soccer players. Similarly, Theodorou *et al.* [27] reported that creatine supplementation (25 g/day × 4 days) significantly improved mean interval performance times in 22 elite swimmers. These recent preliminary findings and many others suggest that creatine supplementation can improve performance of athletes in a variety of sport-related field activities [28–41].

Since creatine supplementation may affect shuttling of high-energy phosphates between the cytosol and mitochondria, some have suggested that creatine supplementation may affect performance during more prolonged exercise bouts. Recent studies also provide some support this contention. For example, Earnest *et al.* [42] reported that creatine supplementation (20 g/day × 4 days and 10 g/day × 6 days) improved cumulative run time to exhaustion in two runs lasting approximately 90-sec each. Smith *et al.* [43] reported that creatine supplementation (20 g/day × 5 days) increased work rate during exercise bouts lasting between 90–600 sec primarily at the shorter, more intense exercise bouts. Nelson *et al.* [44] found that creatine supplementation (20 g/day × 7 days) decreased submaximal heart rate and oxygen uptake ($VO_2$), while increasing ventilatory anaerobic threshold (VANT) and total time to exhaustion during a maximal exercise test in 36 trained adults. Rico-Sanz *et al.* [45] reported that creatine supplementation (20 g/day × 5 days) increased time to exhaustion (29.9 ± 3.8 to 36.5 ± 5.7 min) while reducing ammonia levels (a marker of adenine nucleotide degradation) when cycling at 30 and 90% of maximum until exhaustion. Finally, Preen *et al.* [46] evaluated the effects of ingesting creatine (20 g/day × 5 days) on resting and post-exercise TC and PC as well as performance of an 80-min intermittent sprint test (10 sets of 5–6 × 6-sec sprints with varying recovery intervals). The authors reported that creatine increased resting and post-exercise TC and PC, mean work performed, and total work performed during 6 × 6-sec sets with 54- and 84-sec recovery. In addition, work performed during 5 × 6-sec sprints with 24-sec recovery tended to be greater (p = 0.056). Collectively, these findings support contentions that creatine supplementation may provide ergogenic benefit for more prolonged exercise bouts involving both anaerobic and aerobic energy systems.

However, as with previous creatine research, not all of the recent studies have found that creatine supplementation enhances exercise performance. For example, McKenna *et al.* [47] reported that creatine supplementation (30 g/day × 5 days) did not affect 5 × 10-sec sprints with rest intervals of 180, 50, and 20-sec in 14 untrained subjects. Gilliam *et al.* [48] found that creatine supplementation (20 g/day × 5 days) did not affect isokinetic knee extension performance during 5 × 30 MVC in 23 untrained subjects. Deutekom *et al.* [49] reported that creatine (20 g/day × 6 days) increased body mass but did not affect muscle activation, fatigue, and/or recovery from electrical stimulation of the quadriceps or maximal exercise performance during sprint cycling in 23 well-trained rowers. Similarly, Edwards *et al.* [50] reported that creatine (20 g/day × 6 days) did not affect running fatigue to exhaustion following performing 4 × 15-sec sprints in 21 moderately active subjects. However, ammonia levels were lower following creatine supplementation suggesting that may have lessened the degree of adenine nucleotide degradation and improved metabolic efficiency. In another study, Op't Eijnde *et al.* [51] reported that creatine (20 g/day × 5 days) did not enhance stroke performance or 70-m agility sprint performance in well-trained tennis players. Finally, Finn *et al.* [2] reported that although creatine supplementation (20 g/day × 5 days) increased TC content and 1-sec relative peak power in 16 triathletes, no significant effects were observed in repetitive cycling sprint performance (4 × 20-sec sprints with 20-sec rest recovery).

In my view, when one examines all of the available literature on creatine supplementation, the following conclusions can be drawn. First, although some intra-subject variability has been reported, the vast majority of studies available to date (> 90%) indicate that short-term creatine supplementation significantly increases TC and PC content as determined by assessing muscle biopsies, urinary whole body creatine retention, and/or magnetic resonance spectroscopy (MRS) [4, 6, 9, 10, 12, 52, 53]. Consequently, it is clear that creatine supplementation enhances the potential to perform high intensity exercise much like carbohydrate loading enhances the potential to perform endurance exercise to exhaustion. Overall, approximately 70% of short-term studies on creatine supplementation report some ergogenic benefit particularly during high- intensity, repetitive exercise [10, 12]. These benefits have been primarily found when performing laboratory tests that have good test-to-test reliability [23]. However, as described above, a number of recent studies have indicated that creatine supplementation can also improve performance in field type events like running, soccer, and swimming. It is also interesting to note that over the last few years, the percentage of studies reporting ergogenic benefit from creatine supplementation has risen to 80–85% presumably due to a greater understanding of how to properly design studies to assess the ergogenic value of creatine supplementation. Benefits have been reported in untrained, trained, and diseased children, adolescents, adults, and elderly populations [10, 12, 54]. Studies reporting no significant effects of creatine supplementation have generally observed small but non-significant improvements in performance (i.e. 1–7%). It should be noted that no study has reported a statistically significant ergolytic (negative) effect from creatine supplementation. Studies that have reported no significant benefit of creatine supplementation often have low statistical power, have evaluated performance tests with large test-to-test reliability, and/or have not incorporated appropriate experimental controls. Consequently, it is my view that the preponderance of evidence indicates that short-term creatine supplementation enhances performance in a variety of laboratory and on-field exercise tasks.

## Effects of creatine supplementation on training adaptations

Theoretically, increasing the ability to perform high-intensity exercise may lead to greater training adaptations over time. Consequently, a number of studies have evaluated the effects of creatine supplementation on training adaptations. For example, Vandenberghe *et al.* [55] reported that in comparison to a placebo group, creatine supplementation (20 g/day × 4 days; 5 g/day × 65 days) during 10-weeks of training in women increased TC and PC, maximal strength (20–25%), maximal intermittent exercise capacity of the arm flexors (10–25%), and fat free mass (FFM) by 60%. In addition, the researchers reported that creatine supplementation during 10-weeks of detraining helped maintain training adaptations to a greater degree. Kelly *et al.* [56] reported that 26-days of creatine supplementation (20 g/day × 4 days; 5 g/day × 22 days) significantly increased body mass, FFM, three repetition maximum (RM) on the bench press, and the number of repetitions performed in the bench press over a series of sets in 18 power lifters. Noonan *et al.* [57] reported that creatine supplementation (20 g/day × 5 days; 100 or 300 mg/kg/day of FFM × 51 days) in conjunction with resistance and speed/agility training significantly improved 40-yard dash time and bench press strength in 39 college athletes. Kreider *et al.* [58] reported that creatine supplementation (15.75 g/day × 28 days) during off-season college football training promoted greater gains in FFM and repetitive sprint performance in comparison to subjects ingesting a placebo. Likewise, Stone *et al.* [38] reported that 5-weeks of creatine ingestion (~ 10 or 20 g/day with and without pyruvate) promoted significantly greater increases in body mass, FFM, 1 RM bench press, combined 1 RM squat and bench press, vertical jump power output, and peak rate of force development during in-season training in 42 Division IAA college football players.

Volek *et al.* [8] reported that 12-weeks of creatine supplementation (25 g/day × 7 days; 5 g/day × 77 days) during periodized resistance training increased muscle TC and PC, FFM, type I, IIa, and IIb muscle fiber diameter, bench press and squat 1 RM, and lifting volume (weeks 5–8) in 19 resistance trained athletes. Peters *et al.* [59] reported that creatine monohydrate and creatine phosphate supplementation (20 g/day × 3 days; 10 g/day × 39 days) during training significantly increased body mass, FFM, and 1-RM bench press in 35 resistance-trained males. Kirksey *et al.* [60] found that creatine supplementation (0.3 g/kg/day × 42 days) during off-season training promoted greater gains in vertical jump height and power, sprint cycling performance, and FFM in 36 Division IAA male and female track and field athletes. Pearson *et al.* [61] reported that creatine supplementation (5 g/day × 10 weeks) during resistance training promoted greater gains in strength, power, and body mass with no change in percent body fat in 16 Division IA college football players during summer conditioning. Moreover, Jones *et al.* [32] reported that creatine (20 g/day × 5 days; 5 g/day × 10 weeks) promoted greater gains in sprint performance (5 × 15-sec with 15-sec recovery) and average on-ice sprint performance (6 × 80-m sprints) in 16 elite ice-hockey players. Becque *et al.* [62] found that creatine supplementation (20 g/day × 5 days; 2 g/day × 37 days) during strength training led to greater gains in arm flexor muscular strength, upper arm muscle area, and FFM than strength training alone in 23 resistance trained athletes.

Additionally, Burke *et al.* [63] reported that low dose creatine supplementation (7.7 g/day × 21 days) during training promoted greater gains in total work until fatigue, peak force, peak power, and fatigue resistance in 41 college athletes. Brenner *et al.* [64] reported that creatine supplementation (20 g/day × 7 days; 2 g/day × 28 days) significantly improved upper-body strength gain and decreased percent body fat in 16 female college lacrosse players during pre-season training. Larson-Meyer *et al.* [34] reported that creatine supplementation (15 g/day × 7 days; 5 g/day × 84 days) promoted greater gains in bench press and squat maximal strength with no differences in FFM during off-season training in 14 female college soccer players. Interestingly, Jowko *et al.* [65] recently reported that creatine supplementation (20 g/day × 7 days; 10 g/day × 14 days) significantly increased FFM and cumulative strength gains during training in 40 subjects initiating training. Additional gains were observed when 3 g/day of calcium beta-hydroxy-beta-methylbutyrate (HMB) was co-ingested with creatine.

In a very interesting experimental design, Stevenson *et al.* [66] evaluated the effects of creatine supplementation (20 g/day × 7 days; 5 g/day × 49 days) on volitional and electrical stimulated training in 18 resistance trained subjects. Subjects participated in a traditional resistance training program as well as an electromyostimulation (EMS) training program (i.e. 3–5 sets × 10 eccentric and concentric contractions performed twice per week on one leg). The researchers found that creatine supplementation did not affect mechanical or hypertrophic responses to the EMS training. However, magnetic resonance imaging (MRI) determined cross-sectional area of the traditionally trained but non-electrically stimulated leg was significantly greater in the creatine group. Finally, Willougby *et al.* [9] recently reported that in comparison to controls, creatine supplementation (6 g/day × 12 weeks) during resistance training (6–8 repetitions at 85–90%; 3 × weeks) significantly increased total body mass, FFM, and thigh volume, 1 RM strength, myofibrillar protein content, Type I, IIa, and IIx myosin heavy chain (MHC) mRNA expression, and MHC protein expression. This study provides strong evidence that creatine supplementation during intense resistance training leads to greater gains in strength and muscle mass.

In my view, after evaluating the available data on the effects of creatine supplementation on training adaptations, the following conclusions can be drawn. Studies that evaluated the effects of creatine supplementation on muscle TC and PC stores described in the present review as well as the majority of previous studies reviewed elsewhere indicate that creatine loading increases TC and PC. Creatine supplementation during training is typically associated with a 0.5–2 kg greater increase in body mass and/or FFM. Although it has been hypothesized that the initial weight gain associated with creatine supplementation may be due to fluid retention, a number of studies indicate that long-term creatine supplementation increases FFM and/or muscle fiber diameter with no disproportional increase in total body water. These findings suggest that the weight gain observed during training appears to be muscle mass. About 90% of long-term training studies report some ergogenic benefit with gains typically 10–100% greater than controls. Improvements have been reported in untrained and trained adolescents, adults, and elderly populations. No clinically significant side effects have been reported in these studies even though many of them involved intense training in a variety of exercise conditions. These findings suggest that creatine supplementation during training serves to enhance training adaptations. Moreover, these beneficial changes may offer some therapeutic benefit for a variety of pathologies involving muscle weakness and/or muscle wasting.

## Conclusions

Creatine appears to be an effective and safe nutritional ergogenic aid to improve high intensity exercise performance and/or training adaptations in a variety of sports. Although more research on the potential ergogenic value of creatine for specific athletic populations may be useful, it is my view that the most promising area of future research will be to examine potential therapeutic benefit for various clinical populations.

## References


1. Leutholtz B, Kreider RB: Exercise and sport nutrition. In: T. Wilson, N. Temple (eds). Nutritional Health. Humana Press, Inc., Totowa, NJ, 2001, pp 207–239
2. Finn JP, Ebert TR, Withers RT, Carey MF, Mackay M, Phillips JW, Febbraio MA: Effect of creatine supplementation on metabolism and performance in humans during intermittent sprint cycling. Eur J Appl Physiol 84: 238–243, 2001
3. Harris RC, Soderlund K, Hultman E: Elevation of creatine in resting and exercised muscle of normal subjects by creatine supplementation. Clin Sci (Colch) 83: 367–374., 1992
4. Greenhaff PL, Bodin K, Soderlund K, Hultman E: Effect of oral creatine supplementation on skeletal muscle phosphocreatine resynthesis. Am J Physiol 266: E725–E730, 1994
5. Greenhaff P, Constantin-Teodosiu D, Casey A, Hultman E: The effect of oral creatine supplementation on skeletal muscle ATP degradation during repeated bouts of maximal voluntary exercise in man. J Physiol 476: 84P, 1994
6. Hultman E, Soderlund K, Timmons JA, Cederblad G, Greenhaff PL: Muscle creatine loading in men. J Appl Physiol 81: 232–237, 1996
7. Williams MH, Branch JD: Creatine supplementation and exercise performance: An update. J Am Coll Nutr 17: 216–234, 1998
8. Volek JS, Duncan ND, Mazzetti SA, Staron RS, Putukian M, Gomez AL, Pearson DR, Fink WJ, Kraemer WJ: Performance and muscle fiber adaptations to creatine supplementation and heavy resistance training. Med Sci Sports Exerc 31: 1147–1156, 1999
9. Willoughby DS, Rosene J: Effects of oral creatine and resistance training on myosin heavy chain expression. Med Sci Sports Exerc 33: 1674–1681, 2001
10. Kreider RB: Creatine supplementation in exercise and sport. In: J. Driskell, I. Wolinsky (eds). Energy-Yielding Macronutrients and Energy Metabolism in Sports Nutrition. CRC Press LLC, Boca Raton, FL, 1999, pp 213–242
11. Kreider R: Creatine supplementation: Analysis of ergogenic value, medical safety, and concerns. J Exerc Physiol Online 1: 7–18, 1998. Available: http://www.css.edu/users/tboone2/asep/jan3.htm
12. Williams MH, Kreider R, Branch JD: In: Creatine: The Power Supplement. Human Kinetics Publishers, Champaign, IL, 1999
13. Kraemer WJ, Volek JS: Creatine supplementation. Its role in human performance. Clin Sports Med 18: 651–666, 1999
14. Bessman S, Savabi F: The role of phosphocreatine energy shuttle in exercise and muscle hypertrophy. In: M.A. Conway, J.F. Clark (eds). Creatine and Creatine Phosphate: Scientific and Clinical Perspectives. Academic Press, San Diego, CA, 1988, pp 185–198
15. Wallimann T, Dolder M, Schlattner U, Eder M, Hornemann T, O'Gorman E, Ruck A, Brdiczka D: Some new aspects of creatine kinase (CK): Compartmentation, structure, function and regulation for cellular and mitochondrial bioenergetics and physiology. Biofactors 8: 229–234, 1998
16. Greenhaff P: Creatine supplementation and implications for exercise performance. In: A. Jeudendrup, M. Brouns, F. Brouns (eds). Advances in Training and Nutrition for Endurance Sports. Novartis Nutrition Research Unit, Maastricht, 1997
17. Juhn MS, Tarnopolsky M: Oral creatine supplementation and athletic performance: A critical review. Clin J Sport Med 8: 286–297., 1998
18. Graham AS, Hatton RC: Creatine: A review of efficacy and safety. J Am Pharm Assoc (Wash) 39: 803–810, 1999
19. Mujika I, Padilla S: Creatine supplementation as an ergogenic acid for sports performance in highly trained athletes: A critical review. Int J Sports Med 18: 491–496, 1997
20. Balsom PD, Soderlund K, Ekblom B: Creatine in humans with special reference to creatine supplementation. Sports Med 18: 268–280, 1994
21. Volek JS, Kraemer WJ, Bush JA, Boetes M, Incledon T, Clark KL, Lynch JM: Creatine supplementation enhances muscular performance during high-intensity resistance exercise. J Am Diet Assoc 97: 765–770, 1997
22. Urbanski RL, Vincent WJ, Yaspelkis BB III: Creatine supplementation differentially affects maximal isometric strength and time to fatigue in large and small muscle groups. Int J Sport Nutr 9: 136–145, 1999
23. Tarnopolsky MA, MacLennan DP: Creatine monohydrate supplementation enhances high-intensity exercise performance in males and females. Int J Sport Nutr Exerc Metab 10: 452–463, 2000
24. Wiroth JB, Bermon S, Andrei S, Dalloz E, Hebuterne X, Dolisi C: Effects of oral creatine supplementation on maximal pedalling performance in older adults. Eur J Appl Physiol 84: 533–539, 2001
25. Skare OC, Skadberg, Wisnes AR: Creatine supplementation improves sprint performance in male sprinters. Scand J Med Sci Sports 11: 96–102, 2001
26. Mujika I, Padilla S, Ibanez J, Izquierdo M, Gorostiaga E: Creatine supplementation and sprint performance in soccer players. Med Sci Sports Exerc 32: 518–525, 2000
27. Theodorou AS, Cooke CB, King RF, Hood C, Denison T, Wainwright BG, Havenetidis K: The effect of longer-term creatine supplementation on elite swimming performance after an acute creatine loading. J Sports Sci 17: 853–859, 1999
28. Aaserud R, Gramvik P, Olsen SR, Jensen J: Creatine supplementation delays onset of fatigue during repeated bouts of sprint running. Scand J Med Sci Sports 8: 247–251, 1998
29. Cornish SM, Chilibeck PD, Burke DG, Whelan HK: The effect of creatine monohydrate supplementation on sprint skating in hockey players. Med Sci Sports Exerc 32: S135, 2000
30. Cox GR, Burke LM, Mujika I, Tumilty D: Acute creatine supplementation and performance during a field test simulating match play in elite female soccer players. Med Sci Sports Exerc 33: S204, 2001
31. Crowder T, Jensen N, Richmond S, Viogts J, Sweeney B, McIntyre G, Thompson B: Influence of creatine type and diet on strength and body composition of collegiate lightweight football players. Med Science Sports Exerc 30: S264, 1998
32. Jones AM, Atter T, Georg KP: Oral creatine supplementation improves multiple sprint performance in elite ice-hockey players. J Sports Med Phys Fitness 39: 189–196, 1999
33. Larson DE, Hunter GR, Trowbridge CA, Turk JC, Harbin PA, Torman SL: Creatine supplementation and performance during off-season training in female soccer players. Med Sci Sports Exerc 30: S264, 1998
34. Larson-Meyer DE, Hunter GR, Trowbridge CA, Turk JC, Ernest JM, Torman SL, Harbin PA: The effect of creatine supplementation on muscle strength and body composition during off-season training in female soccer players. J Strength Cond Res 14: 434–442, 2000
35. Lefavi RG, Mcmillan JL, Kahn PJ, Crosby JF, DiGioacchino RF, Streater JA: Effects of creatine monohydrate on performance of college baseball and basketball players. J Strength Cond Res 12: 275, 1998

36. Noonan B, French J, Street G: Creatine supplementation and multiple skating task performance in Division I hockey players. Med Sci Sports Exerc 30: S310, 1998
37. Sasaki H, Hiruma E, Aoyama R: Effects of creatine loading on muscular strength and endurance in female softball players. Med Sci Sports Exerc 33: S207, 2001
38. Stone MH, Sanborn K, Smith LL, O'Bryant HS, Hoke T, Utter AC, Johnson RL, Boros R, Hruby J, Pierce KC, Stone ME, Garner B: Effects of in-season (5 weeks) creatine and pyruvate supplementation on anaerobic performance and body composition in American football players. Int J Sport Nutr 9: 146–165, 1999
39. Stout J, Echerson J, Noonan D, Moore G, Cullen D: The effects of a supplement designed to augment creatine uptake on exercise performance and fat free mass in football players. Med Sci Sports Exerc 29: S251, 1997
40. Selsby JT, Beckett KD, Devor ST, Dern M: Swim performance following creatine supplementation in division III athletes. Med Sci Sports Exerc 33: S206, 2001
41. Romer LM, Barrington JP, Jeukendrup AE: Effects of oral creatine supplementation on high intensity, intermittent exercise performance in competitive squash players. Int J Sports Med 22: 546–552., 2001
42. Earnest CP, Almada A, Mitchell TL: Effects of creatine monohydrate ingestion on intermediate duration anaerobic treadmill running to exhaustion. J Strength Cond Res 11: 234–238, 1997
43. Smith JC, Stephens DP, Hall EL, Jackson AW, Earnest CP: Effect of oral creatine ingestion on parameters of the work rate-time relationship and time to exhaustion in high-intensity cycling. Eur J Appl Physiol Occup Physiol 77: 360–365, 1998
44. Nelson AG, Day R, Glickman-Weiss EL, Hegsted M, Kokkonen J, Sampson B: Creatine supplementation alters the response to a graded cycle ergometer test. Eur J Appl Physiol 83: 89–94, 2000
45. Rico-Sanz J, Mendez Marco MT: Creatine enhances oxygen uptake and performance during alternating intensity exercise. Med Sci Sports Exerc 32: 379–385, 2000
46. Preen D, Dawson B, Goodman C, Lawrence S, Beilby J, Ching S: Effect of creatine loading on long-term sprint exercise performance and metabolism. Med Sci Sports Exerc 33: 814–821, 2001
47. McKenna MJ, Morton J, Selig SE, Snow RJ: Creatine supplementation increases muscle total creatine but not maximal intermittent exercise performance. J Appl Physiol 87: 2244–2252, 1999
48. Gilliam JD, Hohzorn C, Martin D, Trimble MH: Effect of oral creatine supplementation on isokinetic torque production. Med Sci Sports Exerc 32: 993–996, 2000
49. Deutekom M, Beltman JG, de Ruiter CJ, de Koning JJ, de Haan A: No acute effects of short-term creatine supplementation on muscle properties and sprint performance. Eur J Appl Physiol 82: 223–229, 2000
50. Edwards MR, Rhodes EC, McKenzie DC, Belcastro AN: The effect of creatine supplementation on anaerobic performance in moderately active men. J Strength Cond Res 15: 357–361, 2000
51. Eijnde BO, Vergauwen L, Hespel P: Creatine loading does not impact on stroke performance in tennis. Int J Sports Med 22: 76–80, 2001
52. Greenhaff PL, Casey A, Short AH, Harris R, Soderlund K, Hultman E: Influence of oral creatine supplementation of muscle torque during repeated bouts of maximal voluntary exercise in man. Clin Sci (Colch) 84: 565–571., 1993
53. Volek JS, Duncan ND, Mazzetti SA, Putukian M, Gomez AL, Staron RS, Kraemer WJ: Performance and muscle fiber adaptations to 12 weeks of creatine supplementation and heavy resistance training. Med Sci Sports Exerc 31: 1999
54. Tarnopolsky MA: Potential benefits of creatine monohydrate supplementation in the elderly. Curr Opin Clin Nutr Metab Care 3: 497–502, 2000
55. Vandenberghe K, Goris M, Van Hecke P, Van Leemputte M, Vangerven L, Hespel P: Long-term creatine intake is beneficial to muscle performance during resistance training. J Appl Physiol 83: 2055–2063, 1997
56. Kelly VG, Jenkins DG: Effect of oral creatine supplementation on near-maximal strength and repeated sets of high-intensity bench press exercise. J Strength Cond Res 12: 109–115, 1998
57. Noonan D, Berg K, Latin RW, Wagner JC, Reimers K: Effects of varying dosages of oral creatine relative to fat free body mass on strength and body composition. J Strength Cond Res 12: 104–108, 1998
58. Kreider RB, Ferreira M, Wilson M, Grindstaff P, Plisk S, Reinardy J, Cantler E, Almada AL: Effects of creatine supplementation on body composition, strength, and sprint performance. Med Sci Sports Exerc 30: 73–82, 1998
59. Peeters BM, Lantz CD, Mayhew JL: Effect of oral creatine monohydrate and creatine phosphate supplementation on maximal strength indices, body composition, and blood pressure. J Strength Cond Res 13: 3–9, 1999
60. Kirksey KB, Stone MH, Warren BJ, Johnson RL, Stone M, Haff G, Williams FE, Proulx C: The effects of 6 weeks of creatine monohydrate supplementation on performance measures and body composition in collegiate track and field athletes. J Strength Cond Res 13: 148–156, 1999
61. Pearson DR, Hamby DG, Russel W, Harris T: Long-term effects of creatine monohydrate on strength and power. J Strength Cond Res 13: 187–192, 1999
62. Becque MD, Lochmann JD, Melrose DR: Effects of oral creatine supplementation on muscular strength and body composition. Med Sci Sports Exerc 32: 654–658, 2000
63. Burke DG, Silver S, Holt LE, Smith Palmer T, Culligan CJ, Chilibeck PD: The effect of continuous low dose creatine supplementation on force, power, and total work. Int J Sport Nutr Exerc Metab 10: 235–244, 2000
64. Brenner M, Walberg Rankin J, Sebolt D: The effect of creatine supplementation during resistance training in women. J Strength Cond Res 14: 207–213, 2000
65. Jowko E, Ostaszewski P, Jank M, Sacharuk J, Zieniewicz A, Wilczak J, Nissen S: Creatine and beta-hydroxy-beta-methylbutyrate (HMB) additively increase lean body mass and muscle strength during a weight-training program. Nutrition 17: 558–566, 2001
66. Stevenson SW, Dudley GA: Dietary creatine supplementation and muscular adaptation to resistive overload. Med Sci Sports Exerc 33: 1304–1310, 2001

*Molecular and Cellular Biochemistry* **244:** 95–104, 2003.


# Long-term creatine supplementation does not significantly affect clinical markers of health in athletes

Richard B. Kreider,[1,5] Charles Melton,[1] Christopher J. Rasmussen,[5] Michael Greenwood,[2] Stacy Lancaster,[1] Edward C. Cantler,[1] Pervis Milnor[3] and Anthony L. Almada[4]

[1]*Exercise and Sport Nutrition Laboratory, Department of Human Movement Sciences and Education, The University of Memphis, Memphis, TN;* [2]*Department of HPESS, Arkansas State University, AR;* [3]*The Milnor Clinic, Memphis, TN;* [4]*MetaResponse Sciences Inc., Laguna Niguel, CA;* [5]*Exercise and Sport Nutrition Laboratory, Center for Exercise, Nutrition and Preventative Health Research, Department of Health, Human Performance and Recreation, Baylor University, Waco, TX, USA*

## Abstract

Creatine has been reported to be an effective ergogenic aid for athletes. However, concerns have been raised regarding the long-term safety of creatine supplementation. This study examined the effects of long-term creatine supplementation on a 69-item panel of serum, whole blood, and urinary markers of clinical health status in athletes. Over a 21-month period, 98 Division IA college football players were administered in an open label manner creatine or non-creatine containing supplements following training sessions. Subjects who ingested creatine were administered 15.75 g/day of creatine monohydrate for 5 days and an average of 5 g/day thereafter in 5–10 g/day doses. Fasting blood and 24-h urine samples were collected at 0, 1, 1.5, 4, 6, 10, 12, 17, and 21 months of training. A comprehensive quantitative clinical chemistry panel was determined on serum and whole blood samples (metabolic markers, muscle and liver enzymes, electrolytes, lipid profiles, hematological markers, and lymphocytes). In addition, urine samples were quantitatively and qualitative analyzed to assess clinical status and renal function. At the end of the study, subjects were categorized into groups that did not take creatine ($n = 44$) and subjects who took creatine for 0–6 months (mean $4.4 \pm 1.8$ months, $n = 12$), 7–12 months (mean $9.3 \pm 2.0$ months, $n = 25$), and 12–21 months (mean $19.3 \pm 2.4$ months, $n = 17$). Baseline and the subjects' final blood and urine samples were analyzed by MANOVA and $2 \times 2$ repeated measures ANOVA univariate tests. MANOVA revealed no significant differences ($p = 0.51$) among groups in the 54-item panel of quantitative blood and urine markers assessed. Univariate analysis revealed no clinically significant interactions among groups in markers of clinical status. In addition, no apparent differences were observed among groups in the 15-item panel of qualitative urine markers. Results indicate that long-term creatine supplementation (up to 21-months) does not appear to adversely effect markers of health status in athletes undergoing intense training in comparison to athletes who do not take creatine. (Mol Cell Biochem **244**: 95–104, 2003)

*Key words*: ergogenic aids, nutrition, safety, exercise, renal function, muscle, metabolism

## Introduction

Creatine is a naturally occurring amino acid that is obtained from the diet and/or synthesized endogenously from the amino acids glycine, arginine, and methionine [1, 2]. Approximately 95% of creatine is stored in skeletal muscle while the remaining 5% is stored the heart, brain, and testes [3]. Of this, approximately two thirds of creatine is stored as phos-

*Address for offprints*: R.B. Kreider, Exercise and Sport Nutrition Laboratory, Center for Exercise, Nutrition and Preventive Health Research, Department of Health, Human Performance and Recreation, Baylor University, P.O. Box 97313, Waco, TX 76798-7313, USA (E-mail: Richard_Kreider@baylor.edu)

phocreatine with the remaining creatine comprising the free creatine pool [4]. Creatine supplementation (e.g. 20 g/day × 5 days) has been consistently reported to increase muscle creatine and phosphocreatine typically by 15–40% [4–6]. Theoretically, increasing the availability of phosphocreatine would enhance cellular bioenergetics of the phosphagen system [1, 2, 5–7] as well as the shuttling of high-energy phosphates between the mitochondria and cytosol via the creatine phosphate shuttle [8–10].

In support of this contention, approximately 70% of studies that have evaluated the potential ergogenic value of creatine supplementation have reported significant improvements in performance particularly those involving high-intensity exercise and/or training [2, 3, 7]. There is also recent evidence that creatine supplementation may provide therapeutic benefit for patients with a variety of metabolic disorders [11–15], neuromuscular diseases [16–21], as well hasten recovery following immobilization [22]. These findings have indicated that creatine may serve as a promising ergogenic aid for athletes as well as may offer some clinical benefit for certain populations.

The only side effect that has been consistently reported has been weight gain which may be a desired effect for many athletes and patient populations [1–3, 7, 23]. Despite this apparent safety record, concerns have been raised in the popular media and scientific community regarding the safety of creatine supplementation [23–28]. In this regard, concerns have been raised that creatine supplementation may promote long-term suppression of creatine synthesis, increase renal stress, promote muscle and liver damage, alter fluid and electrolyte status, and/or cause unknown long-term side effects [26, 27]. In addition, creatine has been anecdotally suggested to increase gastrointestinal upset, cause diarrhea, promote cramping and dehydration, and increase the incidence of musculoskeletal injury [2, 23, 27].

Recent studies have attempted to determine the validity of these anecdotally reported concerns. For example, a number of studies have evaluated the effects of short-term creatine supplementation (e.g. 5 days–12 weeks) on thermal stress/dehydration [29–33], cramping [31, 34–43], electrolyte status [31, 40, 44–47], renal stress [48–55], muscle trauma [34–37, 39, 41, 43, 56, 57], and/or general markers of clinical health/safety [55, 58–65]. Several researchers have also attempted to retrospectively compare the medical status of self-reported creatine users (up to 5 years) to non-users [66, 67]. Other studies have attempted to use questionnaires to assess the prevalence of these potential side effects [38, 68–73]. However, most of these questionnaire based studies did not compare side effects of creatine users to non-creatine using controls and/or attempt to determine whether the side effects commonly reported in the media influenced responses to these questionnaires. Although results of these studies have consistently indicated that creatine supplementation does not appear to cause any of the anecdotally reported side effects, additional long-term research is warranted [1, 2, 26, 27]. The purpose of this study was to examine the short and long-term medical safety of creatine supplementation among athletes in comparison to athletes who did not take creatine during training and competition.

## Materials and methods

### *Subjects*

One hundred and sixteen National Collegiate Athletic Association (NCAA) Division IA college football players volunteered to participate in this study over a 2 year period. Subjects were informed as to the experimental procedures and signed informed consent statements in adherence with the Internal Review Board for use of human subjects in research at The University of Memphis and the American College of Sports Medicine. Of these, 98 subjects donated pre- and at least one subsequent blood and urine sample during the course of the study. Descriptively (means ± S.D.), subjects were 19.2 ± 2 years (range 18–23 years), 185 ± 8 cm (range 173–191 cm), 101 ± 18 kg (range 70–148 kg) and included a balanced representation of athletes from each position on the team (i.e. backs, receivers, tight ends, linebackers, lineman, and kickers).

### *Methods and procedures*

Subjects were recruited to participate in this study during pre-season training prior to the 1998 and 1999 seasons. Approximately 65 subjects volunteered to participate during the first year and about 40 subjects volunteered to participate in the second year of the study. All subjects underwent pre-season medical examinations and were cleared to participate in football according to NCAA criteria. Subjects who volunteered to participate in the study chose whether they wanted to take creatine or non-creatine containing supplements during training. Subjects who chose to take creatine were administered in an open label manner 15.75 g/day of creatine monohydrate for 5 days and an average of 5 g/day thereafter in 5–10 g doses following supervised training sessions. Creatine was added to sports drinks or carbohydrate/protein drinks that were offered to players following training sessions, practices, and games by research assistants working with the strength and conditioning staff. Supplement intake was monitored and recorded in order to document creatine intake. When subjects were not on campus, they were provided creatine to take on their own and self-reported compliance. If for some reason a subject fell behind in taking creatine, subjects were administered up to 10 g/day in order to catch them up to an average of 5 g/day.

Fasting blood and 24-h urine samples were collected on as many athletes willing to provide samples at 0, 1, 1.5, 4, 6, 10, 12, 17, and 21 months of training (typically 30–55 per testing session). Collection of blood and urine samples coincided with the athletes reporting for summer school/preseason training (0 months), pre-fall football camp (1 month), post-fall football camp (1.5 months), post-season (4 months), start of spring semester/winter conditioning (6 months), and following spring football practice at the end of the spring semester (10 months) in the first year of supplementation. In the second year, blood and urine samples were obtained prior to fall football camp (12 months), at the end of the football season (17 months), and at of the end of the spring semester (21 months). Subjects who began the study in the second year donated blood and urine samples prior to the fall football season (0 months), after the football season (4 months), and/or at the end of the second semester (9 months).

Training consisted of summer resistance training/conditioning drills (1–2 h/day, 4 days/week), fall football camp (3–6 h/day, 6 days/week), practicing/competing during football season (2–4 h/day, 6 days/week) off-season resistance training/conditioning drills (1–2 h/day, 4 days/week), and spring football practice/resistance training (1–3 h/day, 4–5 days/week). Coaches, athletic trainers, and/or research assistants supervised all training sessions and games. Training duration, type, and general intensity as well as environmental conditions were recorded. Training averaged of 118 ± 68 min per session with an average intensity of 3.3 ± 1 on a 1–5 scale where 1 was equivalent to a walk-through practice prior to games and 5 was equivalent to game competition. Environmental conditions during training and competition ranged from 8–37°C (mean 24.2 ± 8°C) and 20–98% relative humidity (52.2 ± 16%). Injuries and medical conditions treated by athletic training/medical staff were recorded to assess medical status throughout the study. These data were reported in a companion paper presented at the 6th International Meeting on Guanidino Compounds in Biology and Medicine and published in the Journal of Molecular and Cellular Biochemistry [39].

Subjects observed an overnight 8-h fast prior to donating blood samples. Blood samples were obtained via venipuncture from an antecubital vein in the forearm using standard phlebotomy procedures between 6:00–9:00 am during each assessment period. Blood samples were collected into three 10 mL serum separation tubes (SST) and one 5 mL anticoagulant tube containing K3 (EDTA). The SST's were centrifuged at 5,000 rev/min for 10-min using a Biofuge 17R centrifuge (Heraeus Inc., Germany). Serum from two SST was transferred into microcentrifuge tubes and frozen at –80°C for subsequent analysis. Serum from the remaining SST was transferred into a 10 ml plain sterile tube. The plain and EDTA tubes were refrigerated and shipped overnight in cold containers to SmithKline Beecham Clinical Laboratories (Ann Arbor, MI, USA) for standard clinical analysis. A complete metabolic clinical chemistry panel was run on serum samples using the Olympus AU5200 automated chemistry analyzer (Melville, NY, USA) following standard clinical procedures. Cell blood counts with percent differentials and platelet determination were run on whole blood samples using a Coulter STKS automated analyzer using standard procedures (Coulter Inc., Hialeah, FL, USA). These analyzers were calibrated daily to controls according to manufacturers recommendations and federal guidelines for clinical diagnostic laboratories. Test to test reliability of performing these assays ranged from 2–8% for individual assays with an average variation of ±3%. Samples were run in duplicate to verify results if the observed values were outside control values and/or clinical norms according to standard clinical procedures.

Urine samples were collected in 24-h collection containers according to standard procedures. Urine volume was recorded and approximately 10 ml of urine was placed into a urine preservative tube, refrigerated, and shipped overnight in a cold container to SmithKline Beecham Clinical Laboratories to have a 15-item urinalysis performed using the Clinitek Atlas® automated urine chemistry analyzer (Bayer Diagnostics, Tarrytown, NY, USA). This analyzer was calibrated to controls according to manufacturers recommendations and federal guidelines for clinical diagnostic laboratories. In addition, approximately 10 ml of urine was pipetted into a plain sterile transfer tube and stored at –80°C for subsequent analysis. A microcentrifuge of frozen serum and a frozen urine transfer tube was shipped on dry ice to Nichols Institute (San Juan Capistrano, CA, USA) or to the Department of Biomedical Sciences at Queen's Medical Centre (Nottingham, UK) for determination of plasma creatinine, urine creatinine, and creatinine clearance using high performance liquid chromatography (HPLC) according to previously published procedures [6, 52, 53, 55].

### *Statistical analysis*

At the end of the study, subjects were categorized as non-creatine users (n = 44); subjects who took creatine for 0–6 months (mean 4.4 ± 1.8 months, n = 12); subjects who took creatine for 7–12 months (mean 9.3 ± 2.0 months, n = 25); and subjects who took creatine for 12–21 months (mean 19.3 ± 2.4 months, n = 17). The subjects' baseline and final blood and urine samples were analyzed by MANOVA using SPSS for Windows Version 10.05 software (SPSS, Inc., Chicago, IL, USA). In addition, 2 × 2 repeated measures ANOVA univariate tests were performed on all dependent variables. Data were considered statistically significant when the probability of Type I error was 0.05 or less. Data are presented as means ± S.D.

## Results

MANOVA revealed no significant differences (p = 0.51) between creatine users and non-users in the quantitative panel of blood and urine markers assessed. Since the clinical safety of creatine supplementation is of interest to the scientific and medical community, we have presented the means and standard deviations from the univariate repeated measures ANOVA tests performed on serum metabolic markers, muscle and liver enzymes, electrolytes, blood lipids, hematological markers, and quantitative urinary markers in Tables 1–6, respectively. No significant differences (p > 0.05) were observed among groups in any of the quantitative markers analyzed with the exception that significant interactions were observed in sodium, chloride, and hematocrit. However, as Tables 3 and 5 indicate, the differences observed among groups for sodium (141–142 meq/L), chloride (104–105 meq/L), and hematocrit (43–455%) were small, within normal ranges, and therefore do not appear to be of any physiological or clinical significance. No apparent differences were observed among groups in the qualitative urinary assessment of color, appearance, glucose, bilirubin, ketones, hemoglobin, total protein, nitrates, leukocyte esterase, white blood cells, red blood cells, epithelial cells, bacteria, amorphic cyrstals, or calcium oxalate.

*Table 1.* Metabolic markers for the creatine and non-creatine supplemented subjects. The pre-value is listed at the top and the post value is listed at the bottom of each cell

| Variable | Non-creatine control (n = 44) | Creatine 0–6 M (n = 12) | Creatine 7–12 M (n = 25) | Creatine 12–21 M (n = 17) | Univariate interaction |
|---|---|---|---|---|---|
| Glucose | 87 ± 11 | 90 ± 14 | 88 ± 15 | 84 ± 15 | I = 0.16 |
| (mg/dl) | 85 ± 6 | 84 ± 12 | 84 ± 10 | 89 ± 9 | |
| Total protein | 7.3 ± 0.4 | 7.2 ± 0.7 | 7.3 ± 0.03 | 7.3 ± 0.4 | I = 0.15 |
| (g/dl) | 7.3 ± 0.3 | 7.4 ± 0.3 | 7.2 ± 0.3 | 7.4 ± 0.4 | |
| Albumin | 4.4 ± 0.2 | 4.4 ± 0.3 | 4.4 ± 0.2 | 4.4 ± 0.2 | I = 0.37 |
| (g/dl) | 4.4 ± 0.2 | 4.5 ± 0.1 | 4.4 ± 0.2 | 4.4 ± 0.2 | |
| Globulin | 2.9 ± 0.4 | 2.9 ± 0.3 | 2.9 ± 0.3 | 2.9 ± 0.2 | I = 0.55 |
| (g/dl) | 2.9 ± 0.4 | 2.9 ± 0.3 | 2.8 ± 0.3 | 3.0 ± 0.3 | |
| Albumin/globulin | 1.55 ± 0.3 | 1.55 ± 0.2 | 1.56 ± 0.2 | 1.52 ± 0.2 | I = 0.75 |
| ratio | 1.53 ± 0.2 | 1.56 ± 0.1 | 1.59 ± 0.2 | 1.48 ± 0.1 | |
| Creatinine | 1.23 ± 0.1 | 1.29 ± 0.2 | 1.26 ± 0.1 | 1.16 ± 0.2 | I = 0.56 |
| (mg/dl) | 1.35 ± 0.1 | 1.41 ± 0.2 | 1.42 ± 0.2 | 1.35 ± 0.2 | |
| Blood urea nitrogen | 15.2 ± 3.8 | 15.2 ± 3.0 | 15.5 ± 3.8 | 15.6 ± 3.7 | I = 0.85 |
| (BUN) (mg/dl) | 15.0 ± 2.9 | 15.9 ± 3.4 | 15.2 ± 2.6 | 15.6 ± 3.5 | |
| BUN/creatinine | 12.5 ± 3.3 | 12.0 ± 2.7 | 12.5 ± 3.5 | 13.8 ± 4.4 | I = 0.52 |
| ratio | 11.2 ± 2.2 | 11.5 ± 3.1 | 10.7 ± 1.8 | 11.7 ± 3.2 | |
| Uric acid | 5.5 ± 1.1 | 5.5 ± 1.0 | 5.7 ± 1.5 | 5.3 ± 0.9 | I = 0.78 |
| (mg/dl) | 5.4 ± 1.1 | 5.6 ± 1.5 | 5.5 ± 0.9 | 4.9 ± 1.0 | |

Data are means ± S.D.

*Table 2.* Muscle and liver enzymes for the creatine and non-creatine supplemented subjects. The pre-value is listed at the top and the post value is listed at the bottom of each cell

| Variable | Non-creatine control (n = 44) | Creatine 0–6 M (n = 12) | Creatine 7–12 M (n = 25) | Creatine 12–21 M (n = 17) | Univariate interaction |
|---|---|---|---|---|---|
| Creatine kinase | 796 ± 1124 | 427 ± 235 | 579 ± 373 | 862 ± 1821 | I = 0.19 |
| (U/l) | 433 ± 146 | 469 ± 250 | 659 ± 545 | 355 ± 248 | |
| Lactate dehydrogenase | 180 ± 56 | 159 ± 91 | 159 ± 43 | 177 ± 63 | I = 0.33 |
| (U/l) | 152 ± 21 | 151 ± 40 | 149 ± 38 | 143 ± 25 | |
| Aspartate aminotransferase | 34 ± 24 | 26 ± 11 | 28 ± 8 | 32 ± 22 | I = 0.08 |
| (U/l) | 27 ± 4 | 31 ± 19 | 30 ± 8 | 25 ± 6 | |
| Alanine aminotransferase | 27 ± 11 | 22 ± 10 | 28 ± 11 | 27 ± 14 | I = 0.45 |
| (U/l) | 25 ± 8 | 26 ± 13 | 26 ± 12 | | |
| Alkaline phosphatase | 91 ± 29 | 91 ± 23 | 93 ± 25 | 101 ± 27 | I = 0.09 |
| (U/l) | 93 ± 20 | 94 ± 31 | 100 ± 19 | 93 ± 17 | |

Data are means ± S.D.

*Table 3.* Serum electrolyte levels for the creatine and non-creatine supplemented subjects. The pre-value is listed at the top and the post value is listed at the bottom of each cell

| Variable | Non-creatine control (n = 44) | Creatine 0–6 M (n = 12) | Creatine 7–12 M (n = 25) | Creatine 12–21 M (n = 17) | Univariate interaction |
|---|---|---|---|---|---|
| Sodium | 141 ± 1 | 142 ± 3 | 142 ± 2 | 141 ± 2 | I = 0.01 |
| (meq/L) | 141 ± 1 | 141 ± 1 | 141 ± 1 | 140 ± 1 | |
| Chloride | 104 ± 2 | 105 ± 3 | 105 ± 2 | 105 ± 2 | I = 0.01 |
| (meq/L) | 103 ± 2 | 104 ± 2 | 103 ± 2 | 102 ± 2 | |
| Potassium | 4.5 ± 0.3 | 4.6 ± 0.5 | 4.7 ± 0.6 | 4.6 ± 0.3 | I = 0.33 |
| (meq/L) | 4.4 ± 0.3 | 4.7 ± 0.7 | 4.5 ± 0.4 | 4.4 ± 0.2 | |
| Phosphorus | 4.3 ± 0.7 | 4.3 ± 0.6 | 3.9 ± 0.7 | 4.1 ± 0.5 | I = 0.17 |
| (meq/L) | 4.5 ± 0.5 | 4.1 ± 0.4 | 4.3 ± 0.5 | 4.5 ± 0.8 | |

Data are means ± S.D.

*Table 4.* Blood lipid profiles for the creatine and non-creatine supplemented subjects. The pre-value is listed at the top and the post value is listed at the bottom of each cell

| Variable | Non-creatine control (n = 44) | Creatine 0–6 M ( n = 12) | Creatine 7–12 M ( n = 25) | Creatine 12–21 M (n = 17) | Univariate interaction |
|---|---|---|---|---|---|
| Cholesterol | 163 ± 29 | 174 ± 28 | 176 ± 27 | 164 ± 22 | I= 0.29 |
| (mg/dl) | 172 ± 19 | 175 ± 22 | 179 ± 27 | 180 ± 26 | |
| HDL | 47 ± 10 | 46 ±7 | 45 ± 10 | 50 ± 13 | I = 0.95 |
| (mg/dl) | 48 ± 13 | 47 ± 6 | 46 ±7 | 49 ± 11 | |
| LDL | 98 ± 24 | 111 ± 26 | 111 ± 22 | 98 ± 18 | I = 0.35 |
| (mg/dl) | 106 ± 17 | 111 ± 22 | 111 ± 21 | 108 ± 25 | |
| CHL/HDL | 3.6 ± 1.1 | 3.8 ± 0.7 | 4.0 ± 0.9 | 3.5 ± 0.9 | I= 0.46 |
| ratio | 3.8 ± 1.0 | 3.8 ± 0.6 | 4.0 ± 0.9 | 3.8 ± 0.8 | |
| Triglycerides | 92 ± 55 | 91 ± 40 | 97 ± 37 | 83 ± 34 | I = 0.19 |
| (mg/dl) | 97 ± 44 | 86 ± 31 | 108 ± 56 | 110 ± 48 | |

Data are means ± S.D.

# Discussion

Results of the present study indicate that short and long-term creatine supplementation (up to 21 months) does not appear to adversely effect clinical markers of health status in a large number of athletes undergoing intense training in comparison to athletes who do not take creatine. These findings provide the strongest evidence to date that long-term creatine supplementation does not appear to pose a health risk for athletes. In addition, these findings support previous reports from short-term studies (5 days–12 weeks) and long-term retrospective studies (up to 5 years) that creatine supplementation does not pose a health risk in apparently healthy individuals, athletes, or patient populations [38, 52, 53, 55, 58, 59, 64–66, 74]. The following discusses the results of the present study in consideration of concerns raised regarding the safety of creatine supplementation.

Concerns have been raised that creatine supplementation may increase renal stress and/or impair renal function. These concerns have been primarily fueled by reports of four case studies of possible renal dysfunction in individuals believed to have been taking creatine [75–78]. In each instance, elevations in serum creatinine (e.g. 1.5–1.7 mg/dl) were initially used to diagnose renal stress. Although the conclusions drawn about these case studies have been criticized [63, 79, 80] because these individuals had pre-existing kidney disease [77], may have been misdiagnosed [75], and/or apparently one subject ingested liquid creatine with only 25 mg of creatine per serving indicating that creatine could not have been related to the renal dysfunction observed [78], they have nevertheless raised concerns regarding the long-term safety of creatine supplementation.

Several studies have reported that creatine supplementation during training may increase serum creatinine levels (e.g. from 1.1 to 1.3–1.5 mg/dl). However, since creatine is naturally degraded to creatinine, the increased serum creatinine has been suggested to be due to a greater turnover of creatine following creatine loading and/or due to an ability to maintain a greater training volume/intensity following creatine supplementation [2, 44, 63]. Several recent studies have evaluated the effects of creatine supplementation on renal function by assessing urinary creatinine clearance [31, 48, 51–54] and/or using iohexol infusion techniques to assess glomular filtration [51]. These studies found that creatine

*Table 5.* Hematological markers for the creatine and non-creatine supplemented subjects. The pre-value is listed at the top and the post value is listed at the bottom of each cell

| Variable | Non-creatine control (n = 44) | Creatine 0–6 M (n = 12) | Creatine 7–12 M (n = 25) | Creatine 12–21 M (n = 17) | Univariate interaction |
|---|---|---|---|---|---|
| White blood cells | 5.5 ± 1.4 | 5.2 ± 1.0 | 6.0 ± 1.3 | 5.4 ± 1.3 | I = 0.38 |
| (thous/cum) | 5.7 ± 1.0 | 5.0 ± 0.8 | 6.0 ± 1.2 | 5.9 ± 1 1 | |
| Red blood cells | 5.0 ± 0.3 | 4.9 ± 0.2 | 5.1 ± 0.4 | 5.0 ± 0.3 | I = 0.28 |
| (mil/cum) | 5.1 ± 0.3 | 5.1 ± 0.3 | 5.1 ± 0.3 | 5.1 ± 0.3 | |
| Hemoglobin | 15.0 ± 1.1 | 15.0 ± 0.7 | 15.1 ± 1.0 | 14.6 ± 0.5 | I = 0.22 |
| (g/dl) | 15.1 ± 0.9 | 15.3 ± 0.9 | 15.3 ± 0.7 | 15.2 ± 0.5 | |
| Hematocrit | 45 ± 3 | 43 ± 3 | 45 ± 3 | 44 ± 2 | I = 0.03 |
| (%) | 45 ± 2 | 45 ± 2 | 46 ± 2 | 45 ± 1 | |
| Total bilirubin | 0.21 ± 0.16 | 0.22 ± 0.12 | 0.19 ± 0.01 | 0.22 ± 0.13 | I = 0.22 |
| (mg/dl) | 0.16 ± 0.01 | 0.22 ± 0.01 | 0.13 ± 0.004 | 0.14 ± 0.007 | |
| Mean corpuscle | 90 ± 5 | 89 ± 5 | 88 ± 5 | 88 ± 4 | I = 0.34 |
| volume (fl) | 89 ± 4 | 90 ± 3 | 89 ± 4 | 87 ± 4 | |
| Mean corpuscle | 30.0 ± 2 | 30.3 ± 1 | 29.7 ± 2 | 29.4 ± 1 | I = 0.87 |
| hemoglobin (pg) | 30.2 ± 1 | 30.7 ± 1 | 29.9 ± 2 | 29.5 ± 1 | |
| Mean corpuscle | 30.6 ± 0.6 | 34.0 ± 0.6 | 33.5 ± 0.6 | 33.4 ± 0.6 | I = 0.55 |
| hemoglobin content (%) | 33.7 ± 0.6 | 34.0 ± 0.4 | 33.5 ± 0.7 | 33.5 ± 0.7 | |
| Red cell dimension | 12.8 ± 0.7 | 12.3 ± 0.8 | 12.8 ± 0.8 | 12.8 ± 0.8 | I = 0.86 |
| width (%) | 12.6 ± 0.6 | 12.2 ± 0.7 | 12.6 ± 0.5 | 12.6 ± 0.7 | |
| Platelets | 223 ± 38 | 205 ± 30 | 212 ± 48 | 226 ± 55 | I = 0.89 |
| (thous/cum) | 227 ± 34 | 209 ± 33 | 219 ± 44 | 237 ± 51 | |
| Mean platelet volume | 9.8 ± 1.0 | 9.3 ± 1.1 | 9.9 ± 0.9 | 10.3 ± 2.9 | I = 0.27 |
| (fl) | 9.5 ± 0.7 | 9.1 ± 1.1 | 10.0 ± 0.9 | 9.8 ± 1.6 | |
| Neutrophils | 47.7 ± 12 | 48.2 ± 10 | 51.7 ± 12 | 51.0 ± 11 | I = 0.72 |
| (%) | 47.6 ± 10 | 49.6 ± 5 | 54.3 ± 9 | 49.4 ± 7 | |
| Lymphocyges | 38.7 ± 11 | 39.4 ± 11 | 35.5 ± 11 | 35.6 ± 9 | I=0.36 |
| (%) | 39.6 ± 10 | 37.5 ± 6 | 33.2 ± 7 | 38.1 ± 6 | |
| Monocytes | 0.48 ± 0.2 | 0.40 ± 0.2 | 0.46 ± 0.2 | 0.44 ± 0.1 | I = 0.91 |
| (thous/cum) | 0.47 ± 0.2 | 0.41 ± 0.2 | 0.44 ± 0.1 | 0.46 ± 0.1 | |
| Eosonophils | 0.24 ± 0.2 | 0.20 ± 0.1 | 0.26 ± 0.1 | 0.22 ± 0.2 | I= 0.98 |
| (thous/cum) | 0.24 ± 0.1 | 0.20 ± 0.08 | 0.27 ± 0.2 | 0.22 ± 0.1 | |
| Basophils | 0.037 ± 0.04 | 0.040 ± 0.02 | 0.036 ± 0.03 | 0.036 ± 0.04 | I = 0.59 |
| (thous/cum) | 0.034 ± 0.02 | 0.025 ± 0.02 | 0.032 ± 0.02 | 0.043 ± 0.04 | |
| Neutrophil/ | 1.46 ± 0.9 | 1.39 ± 0.8 | 1.71 ± 1.0 | 1.57 ± 0.6 | I = 0.69 |
| lymphocyte ratio | 1.32 ± 0.5 | 1.38 ± 0.4 | 1.80 ± 0.8 | 1.36 ± 0.6 | |

Data are means ± S.D.

*Table 6.* Quantitative urine markers for the creatine and non-creatine supplemented subjects. The pre-value is listed at the top and the post value is listed at the bottom of each cell

| Variable | Non-creatine control (n = 44) | Creatine 0–6 M ( n = 12) | Creatine 7–12 M ( n = 25) | Creatine 12–21 M (n = 17) | Univariate interaction |
|---|---|---|---|---|---|
| Volume | 1.50 ± 0.6 | 1.38 ± 0.8 | 1.67 ± 0.7 | 1.38 ± 0.5 | I = 0.38 |
| (l) | 1.22 ± 0.3 | 1.05 ± 0.4 | 1.27 ± 0.6 | 1.37 ± 0.7 | |
| Specific gravity | 1.023 ± 0.006 | 1.024 ± 0.009 | 1.020 ± 0.006 | 1.019 ± 0.005 | I = 0.24 |
| | 1.026 ± 0.004 | 1.024 ± 0.007 | 1.025 ± 0.005 | 1.025 ± 0.007 | |
| pH | 6.3 ± 0.6 | 6.1 ± 0.5 | 6.3 ± 0.6 | 6.1 ± 0.4 | I = 0.36 |
| | 6.1 ± 0.3 | 6.2 ± 0.4 | 6.2 ± 0.4 | 6.1 ± 0.6 | |
| Creatinine | 2 .82 ± 1.6 | 2.45 ± 1.36 | 2.55 ± 1.2 | 2.67 ± 1.1 | I = 0.76 |
| (g/24 h) | 2.30 ± 0.9 | 2.13 ± 0.8 | 2.47 ± 1.2 | 2.27 ± 1.8 | |
| Creatinine clearance | 269 ± 241 | 171 ± 117 | 234 ± 165 | 213 ± 150 | I = 0.69 |
| (ml/min) | 162 ± 100 | 120 ± 63 | 168 ± 165 | 177 ± 185 | |

Data are means ± S.D.

supplementation has no apparent impact on renal function. Results of the present study support these findings in that no significant differences were observed among creatine and non-creatine users in serum creatinine, urinary creatinine excretion, or creatinine clearance.

Interestingly, baseline creatinine levels in the present study were at the upper end of normal for untrained individuals in all groups (i.e. 0.5–1.2 mg/dl) and the post-values were above normal values (1.35–1.42 mg/dl) in all groups. Although no significant differences were observed among creatine users and controls, most athletes had serum creatinine levels between 1.2–1.7 mg/dl particularly during more intense training periods. If one only used serum creatinine to diagnose renal stress, one could infer that many of these athletes were experiencing renal stress (regardless of whether they were taking creatine or not). Yet, no significant differences were observed among groups in creatinine clearance and values were within or exceeded norms (i.e. > 75–150 ml/min). It is also interesting to note that baseline creatinine clearance values (typically obtained prior to fall football season) were above norms (171–269 ml/min) and that post creatinine clearance values analysis (typically obtained at the end of the football season or at the end of the second academic semester) decreased in all groups from pre- to post analysis (suggesting a decreased renal function) but remained within or slightly above clinical norms (120–177 ml/min). These findings indicate that renal function may vary among college football players possibly due to their large body mass and/or the type and volume of training they are engaged. Consequently, it is our view that care must be taken when interpreting individual changes in serum creatinine and/or creatinine clearance in these types of athletes and inferring that creatine supplementation may have been related to changes in these renal markers when large variations in these variables are apparently normal for this population.

Concerns have also been raised that creatine supplementation may increase muscle and/or liver damage [2, 27, 55]. This concern has been based on an initial report suggested that athletes taking creatine during training may experience slightly elevated muscle and/or liver enzymes [44]. Although the levels reported were within normal values for athletes, some have suggested that creatine may increase muscle and/ or liver damage. Results of the present study indicated that athletes engaged in intense training have creatine kinase (CK) levels above clinical norms for untrained individuals (i.e. > 225 IU/L). However, the mean values observed were within normal ranges for athletes engaged in intense training (i.e. typically 250–1,000 IU/L) and no significant differences were observed among creatine and non-creatine users in CK values. In addition, no significant differences were observed among creatine and non-creatine supplemented groups in lactate dehydrogenase (LDH), aspartate aminotransferase (AST), or alanine aminotransferase (ALT) values and all of these values were within normal ranges (i.e. LDH 100–250; AST and ALT < 55 IU/L) for non-athletes. These findings indicate that although these athletes may have had elevated CK levels, there does not appear to be any difference in muscle and liver enzyme efflux among athletes who do and do not take creatine during intense training and competition.

Creatine supplementation has also been suggested to alter fluid balance and/or electrolyte status. The basis of this concern was from initial reports suggesting that urine output may decrease slightly during creatine loading thereby suggesting that short-term creatine supplementation may increase fluid retention [6]. Although subsequent studies have been unable to demonstrate a disproportionate increase in total body water following creatine supplementation [31, 40, 44, 46], the potential increase in fluid retention has been theorized to dilute electrolytes and predispose athletes to cramping. Results of the present study do not support this hypothesis. Although significant interactions were observed among groups in sodium and chloride levels, differences among groups were negligible (i.e. < 1 meq/L) and of no physiological or clinical significance. In addition, no significant differences were observed among groups in potassium, calcium, phosphorus, urine output, or urine specific gravity. Moreover, as reported in our companion paper to this study, creatine supplementation did not increase the incidence of dehydration or muscle cramping in these athletes monitored over a 3-year period [39]. These findings support results of previous studies indicating that creatine supplementation does not increase thermal stress or promote dehydration [29–33, 46, 47, 81], cramping [31, 34–43], or alter electrolyte status [31, 40, 44, 45].

Earnest and colleagues [82] reported that creatine supplementation lowered blood lipids in a group of subjects with high triglycerides. Since then, several researchers have examined whether creatine supplementation affects blood lipid profiles [44, 83–85]. One theory for this phenomenon is that creatine supplementation may enhance the quality of training thereby accentuating the positive effects of exercise on blood lipid profiles. Potentially, a creatine induced reduction in cholesterol and/or triglyceride levels may have significant health benefits. Although we previously reported that blood lipid profiles were improved during the initial 42-days of this study as subjects went through pre-season training and football camp [86], no significant differences were observed among creatine and non-creatine users in our one-year analysis [62] or in the present overall analysis. These findings suggest that the possible influence of creatine on lipid profiles in athletes with normal blood lipids is either transient or non-existent. Present findings support other reports indicating creatine supplementation does not appear to affect blood lipid levels in athletes [55, 67, 84, 85]. However, additional research should examine the potential influence of creatine supplementation on lipid profiles particularly in individuals with elevated cholesterol and/or triglycerides.

The last major concern about creatine supplementation that we would like to address is that creatine supplementation may cause unknown long-term side effects. Results of the present study do not support these contentions. In this regard, we saw no evidence that short- or long-term creatine supplementation caused any clinically significant change in serum metabolic markers, muscle and liver enzyme efflux, serum electrolytes, blood lipid profiles, red and white whole blood cell hematology, or quantitative and qualitative urinary markers. These findings support other reports indicating that long-term creatine supplementation (up to 5 years) did not appear to cause clinically significant side effects in various patient populations and/or retrospective analysis of athletes [52, 55, 64–66]. Although research should continue to evaluate the health consequences of creatine supplementation (particularly in adolescents and various patient populations), results of this study indicate that creatine supplementation (~ 5 g/day for up to 21 months) appears to be a safe nutritional supplement for athletes engaged in intense training and competition.

## Acknowledgements

We would like to thank the athletes who participated in this study, the laboratory assistants in Exercise and Sport Nutrition Laboratory (ESNL), and the coaches who allowed us to conduct this study. In addition, we would like to thank Dr. Paul Greenhaff and his associates from University of Nottingham for analyzing serum and urine creatinine for this study.

This study was funded in part by the ESNL, MetaResponse Sciences (Laguna Niguel, CA, USA), and Experimental and Applied Sciences (Golden, CO, USA). Supplements used in this study were provided by Experimental and Applied Sciences, degussa Bioactives/Traco Labs (Champaign, IL, USA), and MetaResponse Modifiers (San Clemente, CA, USA). Investigators from The University of Memphis independently collected, analyzed and interpreted data from this study and have no financial interest in the outcome of results reported. Presentation of results in this study does not constitute endorsement by The University of Memphis of the supplements investigated.

## References

1. Williams MH, Branch JD: Creatine supplementation and exercise performance: An update. J Am Coll Nutr 17: 216–234, 1998
2. Kreider R: Creatine supplementation: Analysis of ergogenic value, medical safety, and concerns. J Exerc Physiol Online 1: 7–18, 1998. Available: http://www.css.edu/users/tboone2/asep/jan3.htm
3. Williams MH, Kreider R, Branch JD: In: Creatine: The Power Supplement. Human Kinetics Publishers, Champaign, IL, 1999
4. Balsom PD, Soderlund K, Ekblom B: Creatine in humans with special reference to creatine supplementation. Sports Med 18: 268–280, 1994
5. Greenhaff P: The nutritional biochemistry of creatine. J Nutr Biochem 11: 610–618, 1997
6. Hultman E, Soderlund K, Timmons JA, Cederblad G, Greenhaff PL: Muscle creatine loading in men. J Appl Physiol 81: 232–237, 1996
7. Kraemer WJ, Volek JS: Creatine supplementation. Its role in human performance. Clin Sports Med 18: 651–666, 1999
8. Bessman S, Savabi F: The role of phosphocreatine energy shuttle in exercise and muscle hypertrophy. In: M.A. Conway, J. F. Clark (eds). Creatine and Creatine Phosphate: Scientific and Clinical Perspectives. Academic Press, San Diego, CA, 1988, pp 185–198
9. Ma TM, Friedman DL, Roberts R: Creatine phosphate shuttle pathway in tissues with dynamic energy demand. In: M.A. Conway, J.F. Clark (eds). Creatine and Creatine Phosphate: Scientific and Clinical Perspectives. Academic Press, San Diego, CA, 1996, pp 17–32.
10. Wallimann T, Dolder M, Schlattner U, Eder M, Hornemann T, O'Gorman E, Ruck A, Brdiczka D: Some new aspects of creatine kinase (CK): Compartmentation, structure, function and regulation for cellular and mitochondrial bioenergetics and physiology. Biofactors 8: 229–234, 1998
11. Vorgerd M, Grehl T, Jager M, Muller K, Freitag G, Patzold T, Bruns N, Fabian K, Tegenthoff M, Mortier W, Luttmann A, Zange J, Malin JP: Creatine therapy in myophosphorylase deficiency (McArdle disease): A placebo-controlled crossover trial. Arch Neurol 57: 956–963, 2000
12. Leuzzi V, Bianchi MC, Tosetti M, Carducci C, Cerquiglini CA, Cioni G, Antonozzi I: Brain creatine depletion: Guanidinoacetate methyltransferase deficiency (improving with creatine supplementation). Neurology 55: 1407–1409, 2000
13. Stockler S, Hanefeld F: Guanidinoacetate methyltransferase deficiency: A newly recognized inborn error of creatine biosynthesis. Wiener Klinische Wochenschrift 109: 86–88, 1997
14. Stockler S, Hanefeld F, Frahm J: Creatine replacement therapy in guanidinoacetate methyltransferase deficiency, a novel inborn error of metabolism. Lancet 21: 789–790, 1996
15. Ganesan V, Johnson A, Connelly A, Eckhardt S, Surtees RA: Guanidinoacetate methyltransferase deficiency: New clinical features. Pediatr Neurol 17: 155–157, 1997
16. Willer B, Stucki G, Hoppeler H, Bruhlmann P, Krahenbuhl S: Effects of creatine supplementation on muscle weakness in patients with rheumatoid arthritis. Rheumatology (Oxford) 39: 293–298, 2000
17. Felber S, Skladal D, Wyss M, Kremser C, Koller A, Sperl W: Oral creatine supplementation in Duchenne muscular dystrophy: A clinical and 31P magnetic resonance spectroscopy study. Neurol Res 22: 145–150, 2000
18. Valtonen M, Nanto-Salonen K, Jaaskelainen S, Heinanen K, Alanen A, Heinonen OJ, Lundbom N, Erkintalo M, Simell O: Central nervous system involvement in gyrate atrophy of the choroid and retina with hyperornithinaemia. J Inherit Metab Dis 22: 855–866, 1999
19. Tarnopolsky MA, Parise G: Direct measurement of high-energy phosphate compounds in patients with neuromuscular disease. Muscle Nerve 22: 1228–1233, 1999
20. Heinanen K, Nanto-Salonen K, Komu M, Erkintalo M, Alanen A, Heinonen OJ, Pulkki K, Nikoskelainen E, Sipila I, Simell O: Creatine corrects muscle 31P spectrum in gyrate atrophy with hyperornithinaemia. Eur J Clin Invest 29: 1060–1065, 1999
21. Tarnopolsky MA, Roy BD, MacDonald JR: A randomized, controlled trial of creatine monohydrate in patients with mitochondrial cytopathies. Muscle Nerve 20: 1502–1509, 1997
22. Hespel P, Eijnde BO, Van Leemputte M, Urso B, Greenhaff PL, Labarque V, Dymarkowski S, Van Hecke P, Richter EA: Oral creatine

supplementation facilitates the rehabilitation of disuse atrophy and alters the expression of muscle myogenic factors in humans. J Physiol 536: 625–633, 2001
23. Terjung RL, Clarkson P, Eichner ER, Greenhaff PL, Hespel PJ, Israel RG, Kraemer WJ, Meyer RA, Spriet LL, Tarnopolsky MA, Wagenmakers AJ, Williams MH: American College of Sports Medicine roundtable. The physiological and health effects of oral creatine supplementation. Med Sci Sports Exerc 32: 706–717, 2000
24. Benzi G: Is there a rationale for the use of creatine either as nutritional supplementation or drug administration in humans participating in a sport? Pharmacol Res 41: 255–264, 2000
25. Benzi G, Ceci A: Creatine as nutritional supplementation and medicinal product. J Sports Med Phys Fitness 41: 1–10, 2001
26. Graham AS, Hatton RC: Creatine: A review of efficacy and safety. J Am Pharm Assoc (Wash) 39: 803–810; quiz 875–807, 1999
27. Juhn MS, Tarnopolsky M: Potential side effects of oral creatine supplementation: A critical review. Clin J Sport Med 8: 298–304, 1998
28. Yu PH, Deng Y: Potential cytotoxic effect of chronic administration of creatine, a nutrition supplement to augment athletic performance. Med Hypotheses 54: 726–728, 2000
29. Papadopoulos C, Imamura R, Bandon LJ: The effect of creatine supplementation on repeated bouts of high-intensity exercise in the heat. Med Sci Sports Exerc 33: S203, 2001
30. Kern M, Podewils LJ, Vukovich M, Buono MJ: Physiological response to exercise in the heat following creatine supplementation. J Exerc Physiol Online 4: 18–27, 2001. Available:http://www.css.edu/users/tboone12/asep/Kern.pdf
31. Volek JS, Mazzetti SA, Farquhar WB, Barnes BR, Gomez AL, Kraemer WJ: Physiological responses to short-term exercise in the heat after creatine loading. Med Sci Sports Exerc 33: 1101–1108, 2001
32. McArthur PD, Webster MJ, Body JC, May RA, Eschbach LD, Eimer AF, Angelosoulos TJ, Zoeller RF, Krebs GV: Creatine supplementation and acute dehydration. Med Sci Sports Exerc 31: S263, 1999
33. Webster MJ, Vogel RA, Erdmann LD, Clark RD: Creatine supplementation: Effect on exercise performance at two levels of acute dehydration. Med Sci Sports Exerc 31: S263, 1999
34. Greenwood M, Kreider R, Rasmussen C, Ransom J, Melton C, Stroud T, Cantler E, Milnor P: Creatine supplementation does not increase incidence of cramping or injury during college football training II. J Strength Cond Res 13: 425–426, 1999
35. Kreider R, Melton C, Hunt J, Rasmussen C, Ransom J, Stroud T, Cantler E, Milnor P: Creatine does not increase incidence of cramping or injury during pre-season college football training I. Med Sci Sports Exerc 31: S355, 1999
36. Greenwood M, Greenwood LD, Kreider R, Byars A: Effects of creatine supplementation on the incidence of cramping/injury during college football three a day training. Med Sci Sports Exerc 32: S136, 2000
37. Greenwood LD, Greenwood M, Kreider R, Byars A: Effects of creatine supplementation on the incidence of cramping/injury during collegiate fall baseball. Med Sci Sports Exerc 32: S136, 2000
38. Greenwood M, Farris J, Kreider R, Greenwood L, Byars A: Creatine supplementation patterns and perceived effects in select division I collegiate athletes. Clin J Sport Med 10: 191–194, 2000
39. Greenwood M, Kreider R, Melton C, Rasmussen C, Lancaster S, Cantler E, Milnor P, Almada AL: Creatine supplementation during college football training does not increase the incidence of cramping or injury. Abstracts of 6th International Conference on Guanidino Compounds in Biology and Medicine, Cincinnati, OH, September 1, 2001
40. Hulver MW, Campbell A, Haff G, Schroeder C, Comeau M, Potteiger JA: The effects of creatine supplementation on total body fluids, performance, and muscle cramping during exercise. Med Sci Sports Exerc 32: S133, 2000
41. Hunt J, Kreider R, Melton C, Ransom J, Rasmussen C, Stroud T, Cantler E, Milnor P: Creatine does not increase incidence of cramping or injury during pre-season college football training II. Med Sci Sports Exerc 31: S355, 1999
42. Kreider R, Rasmussen C, Ransom J, Almada AL: Effects of creatine supplementation during training on the incidence of muscle cramping, injuries and GI distress. J Strength Cond Res 12: 275, 1998
43. Kreider R, Melton C, Ransom J, Rasmussen C, Stroud T, Cantler E, Greenwood M, Milnor P: Creatine supplementation does not increase incidence of cramping or injury during college football training I. J Strength Cond Res 13: 428, 1999
44. Almada AL, Kreider R, Melton C, Rasmussen C, Lundberg J, Ransom J, Greenwood M, Stroud T, Cantler E, Milnor P, Fox J: Long-term creatine supplementation does not affect markers of renal stress in athletes. J Strength Cond Res 14: 359, 2000
45. Earnest CP, Almada A, Mitchell TL: Influence of chronic creatine supplementation on hepatorenal function. FASEB J 10: A790, 1996
46. Kreider R, Ransom J, Rasmussen C, Hunt J, Melton C, Stroud T, Cantler E, Milnor P: Creatine supplementation during pre-season football training does not affect markers of renal function. FASEB J 13: A543, 1999
47. Kuehl K, Soehler S, Kulacki K, Goldberg L, Elliot D, Bennett W, Haddock B: Effects of oral creatine monohydrate supplementation on renal function in adults. Med Sci Sports Exerc 32: S168, 2000
48. Poortmans JR, Francaux M: Long-term oral creatine supplementation does not impair renal function in healthy athletes. Med Sci Sports Exerc 31: 1108–1110, 1999
49. Poortmans JR, Auquier H, Renaut V, Durussel A, Saugy M, Brisson GR: Effect of short-term creatine supplementation on renal responses in men. Eur J Appl Physiol Occup Physiol 76: 566–567, 1997
50. Rasmussen C, Kreider R, Melton C, Ransom J, Stroud T, Cantler E, Greenwood M, Milnor P: Long-term creatine supplementation during football training does not affect markers of renal stress. J Strength Cond Res 13: 431, 1999
51. Robinson TM, Sewell DA, Casey A, Steenge G, Greenhaff PL: Dietary creatine supplementation does not affect some haematological indices, or indices of muscle damage and hepatic and renal function. Br J Sports Med 34: 284–288, 2000
52. Bemben MG, Bemben DA, Loftiss DD, Knehans AW: Creatine supplementation during resistance training in college football athletes. Med Sci Sports Exerc 33: 1667–1673, 2001
53. Bermon S, Venembre P, Sachet C, Valour S, Dolisi C: Effects of creatine monohydrate ingestion in sedentary and weight-trained older adults. Acta Physiol Scand 164: 147–155, 1998
54. Eijnde BO, Hespel P: Short-term creatine supplementation does not alter the hormonal response to resistance training. Med Sci Sports Exerc 33: 449–453, 2001
55. Greenwood M, Kreider R, Melton C, Rasmussen C, Lundberg J, Stroud T, Cantler E, Milnor P, Almada AL: Short- and long-term creatine supplementation does not affect hematological markers of health. J Strength Cond Res 14: 362–363, 2000
56. Kreider R, Rasmussen C, Melton C, Greenwood M, Stroud T, Ransom J, Cantler E, Milnor P, Almada AL: Long-term creatine supplementation does not adversely affect clinical markers of health. Med Sci Sports Exerc 32: S134, 2000
57. Poortmans JR, Francaux M: Adverse effects of creatine supplementation: fact or fiction? Sports Med 30: 155–170, 2000
58. Sipila I, Rapola J, Simell O, Vannas A: Supplementary creatine as a treatment for gyrate atrophy of the choroid and retina. New Engl J Med 304: 867–870, 1981
59. Vannas-Sulonen K, Sipila I, Vannas A, Simell O, Rapola J: Gyrate atrophy of the choroid and retina. A five-year follow-up of creatine supplementation. Ophthalmology 92: 1719–1727, 1985

60. Schilling BK, Stone MH, Utter A, Kearney JT, Johnson M, Coglianese R, Smith L, O'Bryant HS, Fry AC, Starks M, Keith R, Stone ME: Creatine supplementation and health variables: a retrospective study. Med Sci Sports Exerc 33: 183–188, 2001
61. Stone MH, Schilling BK, Fry AC, Johnson M, Keith RE, Kearney JT, Coglianese RH, Stone ME, Utter A, Smith L, O'Bryant HS: A retrospective study of long-term creatine supplementation on blood markers of health. J Strength Cond Res 13: 433, 1999
62. Juhn MS, O'Kane JW, Vinci DM: Oral creatine supplementation in male collegiate athletes: a survey of dosing habits and side effects. J Am Diet Assoc 99: 593–595, 1999
63. Ray TR, Eck JC, Covington LA, Murphy RB, Williams R, Knudtson J: Use of oral creatine as an ergogenic aid for increased sports performance: Perceptions of adolescent athletes. South Med J 94: 608–612, 2001
64. Sheppard HL, Raichada SM, Kouri KM, Stenson-Bar-Maor L, Branch JD: Use of creatine and other supplements by members of civilian and military health clubs: Across-sectional survey. Int J Sport Nutr Exerc Metab 10: 245–259, 2000
65. Smith J, Dahm DL: Creatine use among a select population of high school athletes. Mayo Clin Proc 75: 1257–1263, 2000
66. Stephens MB, Olsen C: Ergogenic supplements and health risk behaviors. J Fam Pract 50: 696–699, 2001
67. Metzl JD, Small E, Levine SR, Gershel JC: Creatine use among young athletes. Pediatrics 108: 421–425, 2001
68. Greenwood M, Kreider R, Greenwood LD, Comeau M, Brown LE, Stahura K, Byars A: Perceived health status and side-effects associated with creatine supplementation during the collegiate baseball season. Med Sci Sports Exerc 33: S205, 2001
69. Kuehl K, Goldber L, Elliot D: Renal insufficiency after creatine supplementation in a college football athlete. Med Sci Sports Exerc 30: S235, 1998
70. Pritchard NR, Kalra PA: Renal dysfunction accompanying oral creatine supplements. Lancet 351: 1252–1253, 1998
71. Koshy KM, Griswold E, Schneeberger EE: Interstitial nephritis in a patient taking creatine. N Engl J Med 340: 814–815, 1999
72. Loud K, Rozycki A, Chobanian M: Creatine nephropathy – Lacrosse. Med Sci Sports Exerc 33: S10, 2001
73. Poortmans JR, Francaux M: Renal dysfunction accompanying oral creatine supplements. Lancet 352: 234, 1998
74. Greenhaff P: Renal dysfunction accompanying oral creatine supplements. Lancet 352: 233–234, 1998
75. Kreider RB, Ferreira M, Wilson M, Grindstaff P, Plisk S, Reinardy J, Cantler E, Almada AL: Effects of creatine supplementation on body composition, strength, and sprint performance. Med Sci Sports Exerc 30: 73–82, 1998
76. Ziegenfuss TN, Lowery LM, Lemon PWR: Acute fluid volume changes in men during three days of creatine supplementation. J Exerc Physiol Online 1: 1–9, 1998. Available: http://www.css.edu/users/tboone2/asep/jan3.htm
77. Rasmussen C, Kreider R, Ransom J, Hunt J, Melton C, Stroud T, Cantler E, Milnor P: Creatine supplementation during pre-season football training does not affect fluid or electrolyte status. Med Sci Sports Exerc 31: S299, 1999
78. Earnest CP, Almada A, Mitchell TL: High-performance capillary electrophoresis-pure creatine monohydrate reduced blood lipids in men and women. Clin Sci 91: 113–118, 1996
79. Arciero PJ, Hannibal III N, Hamed J, Gentile C: Effect of creatine on resting metabolic rate, body composition, strength, and blood cholesterol. Med Sci Sports Exerc 31: S264, 1999
80. Dulacki K, Kuehl K, Koehler S, Elliot D, Goldberg L: Effects of creatine on body composition, strength, maximal aerobic capacity, cholesterol and gender. Med Sci Sports Exerc 32: S135, 2000
81. Volek JS, Duncan ND, Mazzetti SA, Putukian M, Gomez AL, Kraemer WJ: No effect of heavy resistance training and creatine supplementation on blood lipids. Int J Sport Nutr Exerc Metab 10: 144–156, 2000
82. Melton C, Kreider R, Rasmussen C, Ransom J, Hunt J, Stroud T, Cantler E, Milnor P: Effects of ingesting creatine containing supplements during training on blood lipid profiles. FASEB J 13: A559, 1999

*Molecular and Cellular Biochemistry* **244:** 105–112, 2003.


# Creatine supplementation enhances anaerobic ATP synthesis during a single 10 sec maximal handgrip exercise

Yuko Kurosawa,[1] Takafumi Hamaoka,[1] Toshihito Katsumura,[1] Masasuke Kuwamori,[2] Naoto Kimura,[3] Takayuki Sako[4] and Britton Chance[5]

[1]*Department of Preventive Medicine and Public Health, Tokyo Medical University, Shinjuku-ku, Tokyo;* [2]*Department of Health and Physical Education, School of Commerce, Meiji University, Tokyo;* [3]*Department of Preventive Medicine and Public Health, Nippon Sports Science University, Tokyo;* [4]*Exercise Physiology, Department of Food and Nutrition, Japan Women's University, Tokyo, Japan;* [5]*Department of Biochemistry and Biophysics, University of Pennsylvania, Philadelphia, PA, USA*

## Abstract

Forearm muscles of twelve healthy male subjects (age = 22.3 ± 1.1 years (mean ± S.E.)) were examined during a 10 sec maximal dynamic handgrip exercise (Ex10) using 31-phosphorus magnetic resonance spectroscopy before and after ingestion with 30 g creatine (Cr) monohydrate or placebo per day for 14 days. Cr supplementation produced a 11.5 ± 4.6% increase in the resting muscle phosphocreatine (PCr) concentration and a 65.0 ± 4.2% increase in the PCr degradation during Ex10. ATP synthesis rate through PCr hydrolysis and total anaerobic ATP synthesis rate during Ex10 increased from 0.64 ± 0.08 (pre-value) to 0.86 ± 0.14 mmol/kg ww/sec (post-value, $p < 0.05$) and from 0.97 ± 0.16 (pre-value) to 1.33 ± 0.27 mmol/kg ww/sec (post-value, $p < 0.05$), respectively. An increase in total anaerobic ATP synthesis during Ex10 after Cr supplementation positively correlated with the increase in ATP synthesis through PCr hydrolysis. Cr supplementation produced a 15.1 ± 3.8% increase in the mean power output during Ex10. There was no significant difference in the mean power output per unit of total anaerobic ATP synthesis during Ex10 between before and after Cr supplementation. ATP synthesis rate through PCr hydrolysis positively correlated with mean power output during Ex10 in all twelve subjects after treatment ($r = 0.58$, $p < 0.05$). The results suggest that Cr supplementation enhanced PCr degradation during Ex10. It is strongly indicated that an improvement in performance during Ex10 was associated with the increased PCr availability for the synthesis of ATP. (Mol Cell Biochem **244**: 105–112, 2003)

*Key words*: creatine supplementation, phosphocreatine, anaerobic ATP synthesis, muscle power, magnetic resonance spectroscopy, dynamic handgrip exercise

## Introduction

Hultman *et al.* [1] reported in their study that, the force declined at approximately the same rate as the ATP turnover during high-intensity muscle contractions. In the early phase of maximal contractions, the rapid rephosphorylation of ADP from phosphocreatine (PCr) via the creatine kinase reaction contributes a substantial fraction of ATP synthesis [2]. Because PCr is the fastest substrate capable of supplying ATP to the contracting muscle, and the maximal ATP synthesis rate through PCr hydrolysis is greater than that from glycolysis or aerobic pathways [3, 4]. In addition, net PCr hydrolysis

*Address for offprints*: Y. Kurosawa, Department of Preventive Medicine and Public Health, Tokyo Medical University, 6-1-1 Shinjuku, Shinjuku-ku, Tokyo 160-8402, Japan (E-mail: kuro@tokyo-med.ac.jp)

consumes hydrogen ions and therefore may also act as a buffer in intracellular acidosis [5]. Therefore, muscle PCr availability during maximal exercise is important for the continuation of muscle force production [6, 7].

Because muscle PCr availability during maximal intensity exercise is affected by muscle PCr concentration before exercise [8], increasing muscle PCr via creatine (Cr) supplementation might benefit the continuation of muscle force production [9]. Several previous studies, using a variety of experimental models have examined the effects of Cr supplementation on exercise performance [9–12]. Some of these studies have demonstrated that Cr supplementation improves exercise performance [9–11]. However, only a few studies have simultaneously monitored exercise performance and muscle energy metabolism during exercise before and after Cr supplementation [10, 11]. The effects of Cr supplementation on PCr breakdown rate and exercise performance during a ultrashort maximal exercise has yet to be investigated. Thus, we chose a single 10 sec maximal exercise since PCr plays an important role in synthesizing ATP in this type of exercise [13]. It is also possible that, Cr supplementation affects glycolysis during exercise, because the products of PCr hydrolysis (i.e. Cr and Pi) activate this metabolic pathway [14]. The effects of Cr supplementation on the rate of ATP synthesis through PCr hydrolysis and glycolysis during maximal exercise has not been sufficiently examined.

In this study, we evaluated the rate of ATP synthesis through PCr hydrolysis and glycolysis and mean power output during a 10 sec maximal dynamic handgrip exercise using 31-phosphorus magnetic resonance spectroscopy ($^{31}$P-MRS) before and after Cr supplementation. We also evaluated the relationship between the total anaerobic ATP synthesis rate and mean power output during a 10 sec maximal exercise before and after Cr supplementation.

## Materials and methods

### *Subjects*

Twelve healthy male subjects (age: 22.3 ± 1.1 years; height: 172.5 ± 1.5 cm; weight: 66.1 ± 2.0 kg (mean ± S.E.)) participated in this study. Prior to the experiment, all subjects were fully informed of the nature of the protocol and signed a written consent form. They completed brief medical histories and to ensure their medical safety we sampled venous blood and urine both before and after Cr and placebo treatments. We confirmed that the subjects would not experience any side effects on their liver function, renal function or electrolyte balance following Cr supplementation. The subjects were categorized as physically active, meaning they engaged in some type of recreational exercise, but were not highly trained. During the course of the study the subjects continued with their normal daily activities and no restraints were placed on their diet or general activities.

### *Creatine supplementation*

Subjects were randomly assigned to one of two experimental groups: (A) Cr (n = 6), and (B) placebo (n = 6). Subjects in Group A consumed 30 g (5 g × 6 times) of Cr monohydrate (Phosphagen™, Experimental and Applied Sciences, Golden, CO, USA) and subjects in Group B consumed 30 g of placebo (glucose) on a daily basis for 14 consecutive days. The subjects received 5 g Cr monohydrate or placebo in 300 ml warm water 6 times with a 2 h interval. The subjects were blinded as to whether they were receiving the placebo or Cr.

### *Exercise testing*

One day before and 1 day after 14 days of either treatment (Cr or placebo intake), exercise testing was carried out to evaluate mean power output and muscle bioenergetics. While in an upright position, the subjects in both treatment groups performed dynamic handgrip contractions with a handgrip ergometer in a $^{31}$P-MRS magnet bore. This ergometer was equipped with a weight-loaded system made of nonmagnetic materials inside a 2.0-T $^{31}$P-MRS apparatus (BEM250/80, Otsuka electronics, Japan). The details of these systems were reported in a previous paper [15].

In our pilot study, we examined maximal power output during a 10 sec dynamic handgrip exercise with maximal effort using a variety of weight-load intensities. The maximal power output was recorded at 40% of MVC. After the pilot study, we chose a weight that exhibited 40% of MVC during a 10 sec dynamic handgrip exercise. Before we began the experimental trials the subjects participated in 2–3 exercise practice sessions to familiarize themselves with the experimental procedures. They were instructed to give their maximal effort for 10 sec by gripping the apparatus as hard as possible. In principal experiment, the subjects lifted the mass a distance of 2 cm, adjusted to a 40% of maximal voluntary contraction (MVC) during 10 sec (Ex10). The mean power output was determined from the measured average power output. The power output was measured from the tension and displacement applied to a pulley. Before treatment, the mean power output for all of the subjects was 9.2 ± 0.9 watts. After the treatments, the MVC, irrespective of treatment, did not change significantly. Therefore, the difference in the absolute weight utilized for exercise testing between before and after treatment in both group, was also not significant.

*$^{31}P$ magnetic resonance spectroscopy ($^{31}P$-MRS)*

$^{31}P$ spectra were obtained in the finger flexor muscles using a 2.0-T superconducting magnet with a 26-cm diameter horizontal bore. A 30-mm diameter surface coil was tuned to either the proton (85.44 MHz) or the phosphorus (34.58 MHz) frequency. The field was shimmed using the proton signal from the intramuscular $H_2O$ (proton half width was < 0.3 ppm). $^{31}P$ spectra were acquired using a 2 sec interpulse delay (pulse width 60 μsec, sweep width 5,000 Hz, 1,024 data points, no zero filling). The data for 5 scans were averaged to produce a single spectrum. The deconvoluted time-domain signal was filtered exponentially (5 Hz line broadening) to further improve the signal-to-noise ratio and was Fourier transformed for spectral analysis. The peak areas of PCr, Pi, and β-ATP were calculated by the Vivospec spectrometer curve-fitting program (Otsuka Electronics, Japan). Baseline correction was also performed. Due to incomplete relaxation during the 2 sec repetition period, these values were corrected for partial saturation (correction factor of 1.330 for PCr, 1.081 for Pi, and 1.184 for β-ATP). Correction factors were determined by comparing the partially relaxed spectrum (2 sec repetition period) with the fully relaxed spectrum (20 sec repetition period) of a single subject who was separate from this study. It has been demonstrated that the MRS and biochemical measurements of skeletal muscle ATP concentration are similar [16]. Individual differences of muscle ATP concentration measured biochemically were relatively small [17]. Before and after Cr supplementation there were no significant differences in muscle ATP levels [12, 18]. Thus, PCr and Pi concentrations were calculated by multiplying the ratio of the saturation corrected peak areas of PCr and Pi to β-ATP by the ATP concentration. Metabolite levels are represented as millimoles per kg wet weight. Because there was no data available for the direct chemical analysis of forearm muscles, the ATP concentration reported from muscle biopsies of the human vastus lateralis (ATP = 5.5 mmol/kg wet wt) was used [17]. As a result, in this study the resting forearm muscle PCr concentration before treatment was slightly higher compared with the previously published data that used leg muscles measured by $^{31}P$-MRS [11]. The chemical shift of Pi, with PCr set at zero, was used to calculate muscle pH [19]. The change in muscle PCr and Pi concentrations and pH at rest and during Ex10 were measured. $^{31}P$ spectra were averaged for 15 min at rest and for 10 sec during Ex10.

*ATP turnover rate*

Kemp *et al.* [20] quantitatively investigated human muscle bioenergetics during dynamic exercise at three intensities. They conducted their study under both ischemic and non-ischemic conditions using $^{31}P$-MRS. In their study, they confirmed that the anaerobic ATP synthesis in the initial phase of non-ischemic exercise, as calculated from changes in pH and PCr concentration, was almost identical to the ATP synthesis rate in the initial phase of ischemic exercise. In the initial phase of non-ischemic exercise, proton efflux was considered to be negligible and oxidative ATP synthesis was small. The equations they used in the initial phase of non-ischemic exercise to evaluate anaerobic ATP synthesis are found here [20]:

Total anaerobic ATP synthesis rate (F) = PCr depletion rate (D) + glycolysis rate (L)

$$F = D + L = (-\Delta PCr/\Delta t) + (3/2)\,(\phi D - \beta \Delta pH/\Delta t)$$

ΔPCr: Difference in PCr from steady state (mmol/liter);
Δt: Difference in time from steady state (min);
$\phi$: $[1/1 + 10^{(pH - 6.75)}]$;
β: Buffer capacity (mmol/liter per pH unit);
ΔpH: Difference in pH from steady state

*Statistics*

Data are expressed as means ± S.E. All analyses were conducted using the StatView J4.02 statistical software package (Abacus Concepts, Berkeley, CA, USA). Data were analyzed in four stages. First, the relationship between resting muscle PCr concentration and mean power output during Ex10 was evaluated using Pearson's correlation analysis. This analysis was conducted for all twelve subjects before they received treatment. Second, baseline values for muscle PCr concentration, Pi concentration, muscle pH, grip strength and mean power output were analyzed with Student's unpaired *t*-tests between Cr and placebo groups. Changes in muscle PCr concentration, Pi concentration, muscle pH, grip strength and mean power output before and after treatments were analyzed with Student's paired *t*-tests by treatment group. Third, to examine the effect of Cr supplementation on the total and fractional anaerobic ATP synthesis rate through PCr hydrolysis and glycolysis during Ex10 repeated measures ANOVAs with treatment (Cr vs. placebo) and time (before vs. after treatment) as factors were conducted. After these analyses, Student's paired *t*-tests were performed by treatment group. Furthermore, to clarify the relationship between individual changes in total anaerobic ATP synthesis rate and individual changes in ATP synthesis rate through PCr hydrolysis after Cr supplementation, Pearson's correlation analysis was conducted. Finally, mean power output per unit of total anaerobic ATP synthesis before and after treatments was analyzed with Student's paired *t*-tests by treatment group. The relationship between ATP synthesis rate through PCr hydrolysis and mean power output of all twelve subjects after treatments was evaluated using Pearson's correlation analysis. All statistical significance was set at a type 1 error level of 0.05 (two-tailed).

## Results

*Muscle energy metabolism*

*Muscle PCr concentration*

Figure 1 illustrates the resting PCr concentration of the forearm muscles, as measured by $^{31}$P-MRS, before and after 14 days of Cr and placebo treatment. There was no significant difference in the baseline values for the Cr and placebo subjects (22.1 ± 0.7 and 23.1 ± 1.4 mmol/kg ww, respectively; Fig. 1). After Cr supplementation, resting muscle PCr concentration increased by 11.5 ± 4.6% ($p < 0.01$).

Figure 2 illustrates the concentration of forearm muscle PCr and the degradation of muscle PCr during Ex10 measured by $^{31}$P-MRS before and after Cr and placebo treatments. PCr breakdown rate during Ex10 increased significantly after Cr supplementation (from 0.63 ± 0.08 to 1.09 ± 0.27 mmol/kg ww/sec, $p < 0.05$).

*Muscle Pi concentration*

Cr supplementation did not affect resting muscle Pi concentration. The muscle Pi increment during Ex10 did not change significantly following Cr supplementation (from 0.77 ± 0.13 to 0.94 ± 0.17 mmol/kg ww/sec, $p = 0.40$).

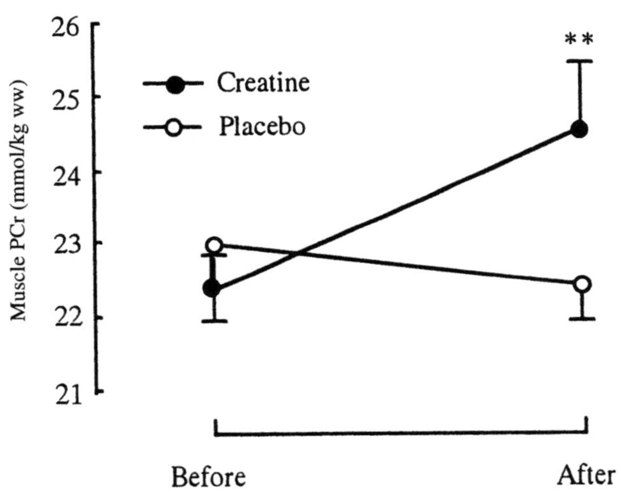


*Fig. 1.* Resting phosphocreatine (PCr) concentration of the forearm muscles measured by 31-phosphorus magnetic resonance spectroscopy ($^{31}$P-MRS) before and after creatine (Cr) supplementation (30 g/day, filled symbols, n = 6) and placebo (open symbols, n = 6) for 14 days. Values are means ± S.E. **p < 0.01, significant difference before and after Cr supplementation.

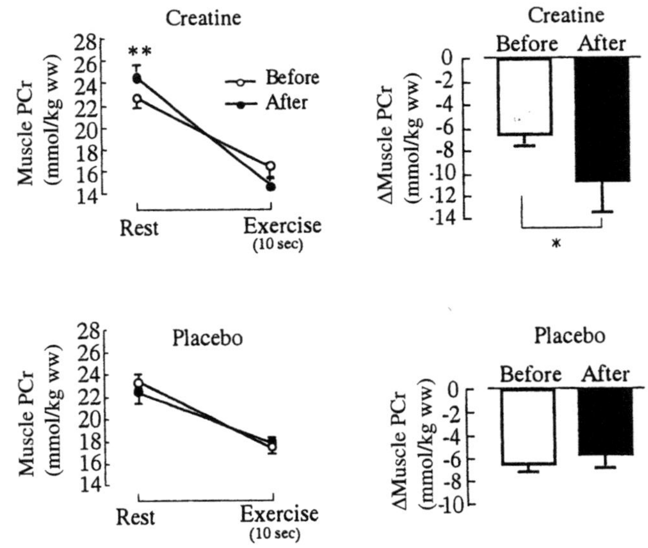


*Fig. 2.* Forearm muscle PCr concentration and change in forearm muscle PCr concentration during dynamic handgrip exercise with 10 sec maximal effort measured by $^{31}$P-MRS before (open symbols and bars) and after (filled symbols and bars) Cr supplementation (30 g/day, n = 6) and placebo (n = 6) for 14 days. Values are means ± S.E. **p < 0.01, significant difference before and after Cr supplementation; *p < 0.05, significant difference before and after Cr supplementation.

*Muscle pH*

Before and after Cr and placebo treatments, the resting muscle pH did not change significantly (Cr: 7.01 ± 0.01 and 7.01 ± 0.01, placebo: 7.03 ± 0.02 and 7.05 ± 0.01, before and after treatments, respectively). The muscle pH during Ex10, before and after both treatments, was also not significantly altered (Cr: 7.02 ± 0.03 and 7.00 ± 0.05, placebo: 7.02 ± 0.02 and 7.07 ± 0.05, before and after treatments, respectively).

*ATP synthesis rate*

Figure 3 illustrates the rate of ATP synthesis through PCr hydrolysis and glycolysis during Ex10 before and after Cr and placebo treatments. Irrespective of treatment, the ATP synthesis rate through PCr hydrolysis was always greater than that generated from glycolysis. The ATP synthesis rate through PCr hydrolysis was significantly increased following Cr supplementation (from 0.64 ± 0.08 to 0.86 ± 0.14 mmol/kg ww/sec, $p < 0.05$). The ATP synthesis rate through glycolysis increased in five subjects and slightly decreased in one subject after Cr supplementation, but this was not statistically significant (n = 6, $p = 0.13$). The analytical results for the 'ATP-synthesis-increased' five subjects, however, was significant (n = 5, $p = 0.02$). The total anaerobic ATP synthesis rate increased significantly after Cr supplementation (from 0.97 ± 0.16 to 1.33 ± 0.27 mmol/kg ww/sec, $p < 0.05$).

Figure 4 illustrates an increase in the total anaerobic ATP synthesis rate during Ex10, following Cr supplementation, which correlated with an increase in the ATP synthesis rate

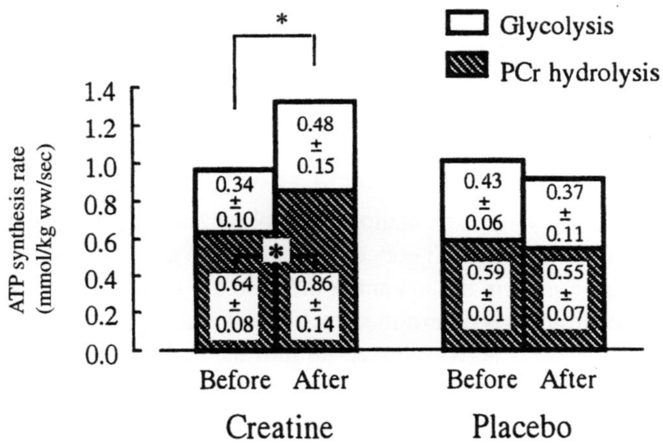


*Fig. 3.* ATP synthesis rate through PCr hydrolysis (hatched bars) and glycolysis (open bars) during dynamic handgrip exercise with 10 sec maximal effort before and after Cr supplementation (30 g/day, n = 6) and placebo (n = 6) for 14 days. Values are means ± S.E. *$p < 0.05$, significant differences before and after Cr supplementation on ATP synthesis rate through PCr hydrolysis and total anaerobic ATP synthesis rate.

through PCr hydrolysis during Ex10 ($r = 0.87$, $p < 0.05$). A linear correlation was found between the increasing total anaerobic ATP synthesis rate and the increasing ATP synthesis rate through glycolysis following Cr supplementation. This correlation was statistically significant among the 'ATP-synthesis-increased' five subjects ($n = 5$, $r = 0.86$, $p < 0.05$), but not among all six subjects ($n = 6$, $r = 0.72$, $p = 0.09$).

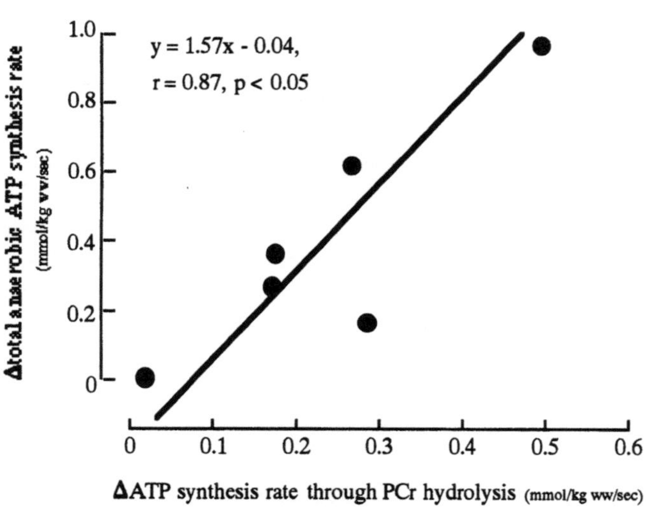


*Fig. 4.* Relationship between individual changes in ATP synthesis rate through PCr hydrolysis and individual changes in total anaerobic ATP synthesis rate during dynamic handgrip exercise with 10 sec maximal effort in the Cr supplementation group (n = 6), before and after Cr supplementation for 14 days (6 × 5 g/day).

*Mean power output*

The mean power output during Ex10 is illustrated in Fig. 5. There were no significant differences in the baseline values of mean power output during Ex10 between the Cr and placebo subjects (8.8 ± 1.2 and 9.4 ± 0.9 watts, respectively). Although there is a large interindividual variation in mean power output increase, all subjects showed an improvement in mean power output after Cr supplementation (+ 15.1 ± 3.8%, $p < 0.05$).

*Mean power output and ATP synthesis rate*

*Mean power output per unit of total anaerobic ATP synthesis*

Irrespective of treatment, the mean power output per unit of total anaerobic ATP synthesis during Ex10 did not change significantly after treatments (Cr: 9.9 ± 1.6 and 9.3 ± 1.9 J/mmol/kg ww, placebo: 9.1 ± 1.3 and 9.9 ± 0.9 J/mmol/kg ww, before and after treatments, respectively). Mean power output was enhanced significantly after Cr supplementation, but mean power output per unit of total anaerobic ATP synthesis during Ex10, was not affected by Cr supplementation.

*ATP synthesis rate through PCr hydrolysis – mean power output relationships*

Figure 6 illustrates the ATP synthesis rate through PCr hydrolysis during Ex10 refers to mean power output in all twelve subjects after treatment. It is evident that the ATP synthesis rate through PCr hydrolysis positively correlated with mean power output during Ex10 ($r = 0.58$, $p < 0.05$).

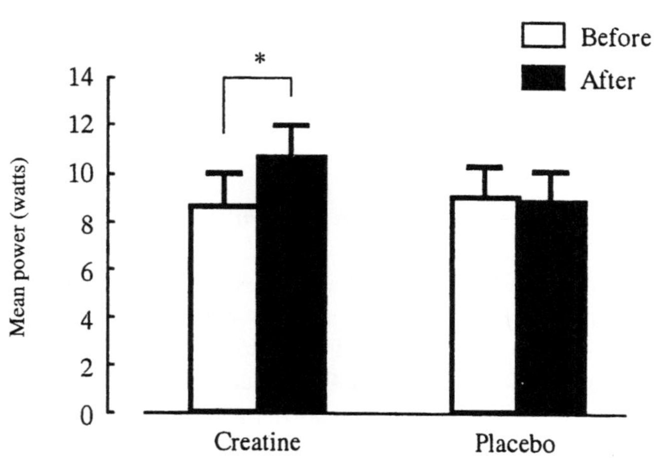


*Fig. 5.* Mean power output during dynamic handgrip exercise with 10 sec maximal effort before (open bars) and after (filled bars) Cr supplementation (30 g/day, n = 6) and placebo (n = 6) for 14 days. Values are means ± S.E. *$p < 0.05$, significant difference before and after Cr supplementation.

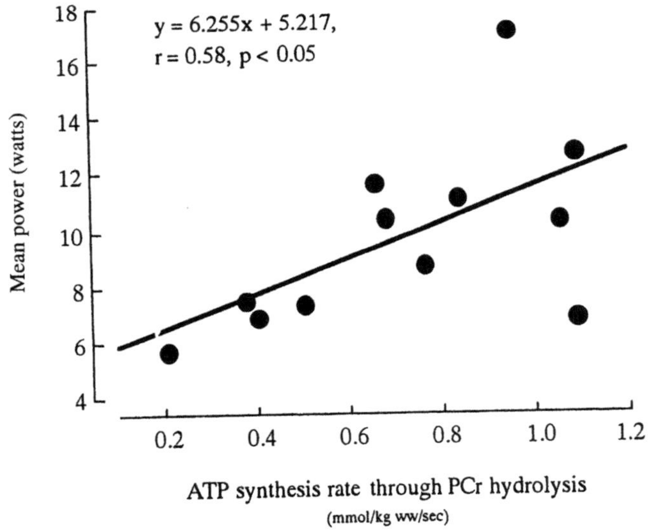


*Fig. 6.* Relationship between ATP synthesis rate through PCr hydrolysis and mean power output during dynamic handgrip exercise with 10 sec maximal effort of all 12 subjects after treatments.

## Discussion

This is the first study to investigate, using $^{31}$P-MRS, the effects of Cr supplementation on ATP synthesis rate through PCr hydrolysis and glycolysis and exercise performance during Ex10. The main finding is that Cr supplementation enhances the ATP synthesis rate through PCr hydrolysis and the total anaerobic ATP synthesis rate during Ex10. In this study, the increase in total anaerobic ATP synthesis during Ex10 positively correlated with an increase in ATP synthesis through PCr hydrolysis. Cr supplementation also produced an increase in mean power output during Ex10. The ATP synthesis rate through PCr hydrolysis positively correlated with mean power output during Ex10. There was no significant difference in the mean power output per unit of total anaerobic ATP synthesis during Ex10 between before and after Cr supplementation.

*Muscle PCr – mean power relationship*

During short-term maximal intensity exercise, the inability of human skeletal muscle to maintain a high rate of anaerobic ATP synthesis through PCr hydrolysis affects muscle power output [6, 7]. Muscle PCr concentration before exercise affects muscle PCr breakdown rate during maximal intensity exercise [8]. In agreement with these findings, the result of the present experiment demonstrates that resting muscle PCr before treatment of all twelve subjects relates to muscle power output during Ex10 ($r = 0.64$, $p < 0.05$, not shown in result). This data indicates the importance of increasing muscle PCr concentration with regard to enhancing muscle power.

*Resting muscle PCr*

This study's supplementation protocol successfully increased the intramuscular PCr concentration by 11.5 ± 4.6% (Fig. 1). Because muscle tissue cannot produce Cr, total cellular Cr concentration is determined by the cell's ability to take up Cr [21]. Cr is actively taken up into the cell via a specific transport process [21], and the activity of the transporter is affected by blood Cr concentration and other factors as well [21]. Unlike our study, which examined forearm muscle most previous studies evaluated the effects of Cr supplementation on leg muscle PCr concentration with supplementation lasting 5–6 days. We conducted the pilot study to examine the amplitude and time course of the resting forearm muscle PCr concentration in our protocol. In our pilot study, which utilized the same supplementation protocol, the PCr concentration of the resting forearm muscle in one subject gradually increased from the start of Cr supplementation until the 9th day (data not shown). Therefore, for more effective results we chose a supplementation period of 14 days and confirmed that Cr supplementation produced a 11.5 ± 4.6% increase in the resting forearm muscle PCr concentration. These results indicate that Cr supplementation increases PCr concentration in forearm muscle by the same level as in leg muscles that were measured by biopsy technique (10% [9] and 9% [22] increase) and $^{31}$P-MRS (16% [11] and 15% [10] increase).

*Muscle energy metabolism and mean power output during exercise*

After Cr supplementation, PCr degradation during Ex10 increased by 65.0 ± 4.2% (Fig. 2), and the ATP synthesis rate through PCr hydrolysis, as evaluated from changes in muscle PCr concentration and muscle pH (20), increased significantly (Fig. 3). The total anaerobic ATP synthesis rate also increased significantly (Fig. 3), and this positively correlated with an enhanced ATP synthesis rate through PCr hydrolysis (Fig. 4). In previous studies, it was found that muscle PCr breakdown rate during exercise is affected by exercise intensity [23] and duration [13], PCr concentration [8, 9] and muscle fiber types [24]. Casey *et al.* [8] performed two bouts of 30 sec maximal isokinetic cycling with a 4 min interval to evaluate thigh muscle PCr utilization in Type I and II fibers using the biopsy method. They demonstrated that muscle PCr utilization during the first exercise bout was greater in Type II fibers, when compared with Type I fibers. In the second bout, however, muscle PCr utilization during exercise was not significantly different between Type I and II fibers. They reported that this was likely to have been a reduction in substrate availability, resulted from the relatively low restoration of ATP and PCr in Type II fiber during recovery between bout 1 and 2. In another study, Casey *et al.* [9] reported that an increase in muscle PCr concentration after Cr supplementation in Type II fibers positively correlates with an increase in muscle PCr degradation during exercise. Therefore, according to this study the pre-exercise level of PCr, especially in Type II fibers is an important determining factor of PCr utilization during short-term, high intensity exercise. This is because recruitment of Type II fibers has more advantage than Type I fibers for ATP utilization and ATP synthesis during this type of exercise. In this study we tested forearm muscles which is expected to have a larger proportion of Type II fibers (average is 60%, ranged from 35–85%) than leg muscles [25–27]. This is one of the reasons why we clearly detected a significant increase in PCr degradation during Ex10 after Cr supplementation.

It is speculated that PCr plays an important role during short-term, maximal exercise. In this study, we used Ex10 to clearly demonstrate the effect of an increased PCr level on ATP synthesis through PCr degradation. We detected a significant increase in muscle PCr degradation during Ex10 after Cr supplementation. Bogdanis *et al.* [13] examined PCr breakdown during the former and latter half of a 20 sec maximal cycle ergometer sprint with the biopsy method. This group reported that PCr breakdown during the latter half (10 sec) of the sprint was one third of the breakdown during the former 10 sec of the sprint (15 vs. 45 mmol/kg dry muscle). This indicates a large difference in the PCr breakdown rate between the former and latter 10 sec maximal sprint exercise. In contrast, Casey *et al.* [9] did not detect a significant increase in muscle PCr degradation during a 30 sec exercise in both Type I and II fibers after Cr supplementation. This might be due to the longer duration of exercise a 30 sec maximal cycling exercise.

In our study, the percentage of ATP synthesis through PCr hydrolysis during Ex10 was 61% before treatment. The percentage of ATP synthesis through PCr hydrolysis evaluated in this study is in agreement with the Hultman *et al.* data [1] that was obtained by biopsy methods. Their data showed that ATP synthesis through PCr hydrolysis was 60% of the total anaerobic amount at the 10 sec after the start of intense electrical stimulation to the leg muscle. In this study, we used $^{31}$P-MRS for evaluating the ATP synthesis rate from changes in muscle PCr and pH. In a previous study [28], it is reported that post-exercise muscle pH was overestimated with the MRS measurement. Muscle pH was 6.82 with chemical analysis of the biopsy samples, and 6.66 with the MRS measurement at the end of exercise [28]. In our study, however, even if the rate of glycolysis was overestimated, the ATP synthesis rate through glycolysis was always smaller than that produced from PCr hydrolysis, confirming that PCr is a major provider of anaerobic ATP during Ex10.

After Cr supplementation, all subjects showed an improvement in mean power output (+15.1 ± 3.8%). However, calculated mean power output per unit of total anaerobic ATP synthesis during Ex10, an indicator of ATP cost of contraction, did not change significantly. This finding agrees with a previous study in which, following Cr supplementation, no change in ATP cost was observed in muscle contractions [29]. This mean power positively correlated with the ATP synthesis rate through PCr hydrolysis ($r = 0.58$, $p < 0.05$) in this study. The marked enhancement in muscle power output during Ex10, which has been associated with Cr supplementation, is likely a result of enhanced PCr availability due to an increase in resting muscle PCr. These data suggest that the ability to synthesize ATP, especially through PCr hydrolysis, plays a key role in regulating mean power output during short-term, maximal intensity exercise. This conclusion is in agreement with previous studies [6, 7].

We examined the contribution of glycolysis to the total anaerobic ATP synthesis before and after Cr supplementation. In five subjects we found a significant increase in the ATP synthesis rate through glycolysis and a significant correlation between the increasing total anaerobic ATP synthesis rate and the increasing ATP synthesis rate through glycolysis after Cr supplementation. This inconsistency in statistical significance might be due to a type II error resulting from a small sample size. We believe that glycolysis plays an important role in increasing total anaerobic ATP synthesis after Cr supplementation. In a previous study, muscle lactate accumulation immediately after intense, electrically stimulated contraction was greater after Cr supplementation in responders when compared with the pre-value using the biopsy method [30]. This research group speculated that glycolysis made a greater contribution to ATP synthesis during exercise after Cr supplementation [30], although the type and length of exercise differed from ours. The increase in Pi, the products of PCr hydrolysis, activates glycogenolysis through stimulation of phosphorylase, and activates glycolysis through stimulation of phosphofructokinase [14]. It can also be deduced that an increase in muscle PCr after Cr supplementation could raise the metabolic buffering capacity of muscle [5]. In another study, however, muscle lactate accumulation immediately after 30 sec maximal exercise did not change significantly after Cr supplementation [9]. ATP synthesis through glycolysis during exercise is affected by exercise intensity and duration, substrate availability, muscle fiber type and other factors. While it has been shown that an increase in Pi levels may play a role in regulating glycolytic activity in humans [14], further studies are needed to evaluate whether Cr supplementation activates glycolytic pathways using a variety of exercise models.

In conclusion, a daily dose of 30 g Cr for 14 days improved ATP synthesis through PCr hydrolysis and mean power output during short-term, maximal exercise. An increase in total anaerobic ATP synthesis during short-term, maximal exercise positively correlated with an increase in ATP synthesis through PCr hydrolysis. Cr supplementation did not affect the ATP cost of contraction during short-term, maximal exercise. These data strongly indicate that an improvement in performance during short-term, maximal exercise after Cr supplementation was associated with the increased PCr availability for the synthesis of ATP.

## Acknowledgements

We thank the entire staff of the Department of Preventive Medicine and Public Health, Tokyo Medical University, especially Toshio Kimura, Gregg Goldstein and Sawako Wakui, for excellent collaboration in this investigation, and the Experimental and Applied Sciences, Incorporated for their donation of Cr supplements.

## References


1. Hultman E, Sjoholm H: Energy metabolism and contraction force of human skeletal muscle *in situ* during electrical stimulation. J Physiol 345: 525–532, 1983
2. Hultman E, Greenhaff PL, Ren JM, Soderlund K: Energy metabolism and fatigue during intense muscle contraction. Biochem Soc Trans 19: 347–353, 1991
3. Hultman E, Sjoholm H: Substrate availability. In: H.G. Knuttgen, J.A. Vogel, J. Poortmans (eds). Biochemistry of Exercise. Human Kinetics, Champaign, 1983, pp 63–75
4. Hultman E, Bergstrom M, Spriet LL, Soderlund K: Energy metabolism and fatigue. In: A.W. Taylor, P.D. Gollnick, H.J. Green *et al.* (eds). Biochemistry of Exercise VII. Human Kinetics, Champaign, 1990, pp 73–92
5. Hultman E, Sahlin K: Acid-base balance during exercise. In: R.S. Hutton, D.I. Miller (eds). Exercise and Sports Science Review. Franklin Institute Press, Philadelphia, 1980, pp 41–128
6. Katz A, Sahlin K, Henriksson J: Muscle ATP turnover rate during isometric contraction in humans. J Appl Physiol 60: 1839–1842, 1986
7. Hultman E, Bergstrom J, Anderson NM: Breakdown and resynthesis of phosphorylcreatine and adenosine triphosphate in connection with muscular work in man. Scand J Clin Lab Invest 19: 56–66, 1967
8. Casey A, Constantin-Teodosiu D, Howell S, Hultman E, Greenhaff PL: Metabolic response of type I and II muscle fibers during repeated bouts of maximal exercise in humans. Am J Physiol 271: E38–E43, 1996
9. Casey A, Constantin-Teodosiu D, Howell S, Hultman E, Greenhaff PL: Creatine ingestion favorably affects performance and muscle metabolism during maximal exercise in humans. Am J Physiol 271: E31–E37, 1996
10. Smith SA, Montain SJ, Matott RP, Zientara GP, Jolesz FA, Fielding RA: Creatine supplementation and age influence muscle metabolism during exercise. J Appl Physiol 85: 1349–1356, 1998
11. Vandenberghe K, Van Hecke P, Van Leemputte M, Vanstapel F, Hespel P: Phosphocreatine resynthesis is not affected by creatine loading. Med Sci Sports Exerc 31: 236–242, 1999
12. Snow RJ, McKenna MJ, Selig SE, Kemp J, Stathis CG, Zhao S: Effect of creatine supplementation on sprint exercise performance and muscle metabolism. J Appl Physiol 84: 1667–1673, 1998
13. Bogdanis GC, Nevill ME, Lakomy HK, Boobis LH: Power output and muscle metabolism during and following recovery from 10 and 20 sec of maximal sprint exercise in humans. Acta Physiol Scand 163: 261–272, 1998
14. Meyer RA, Sweeney HL, Kushmerick MJ: A simple analysis of the 'phosphocreatine shuttle'. Am J Physiol 246: C365–C377, 1984
15. Sako T, Hamaoka T, Higuchi H, Kurosawa Y, Katsumura T: Validity of NIR spectroscopy for quantitatively measuring muscle oxidative metabolic rate in exercise. J Appl Physiol 90: 338–344, 2001
16. Bangsbo J, Johansen L, Quistorff B, Saltin B: NMR and analytic biochemical evaluation of CrP and nucleotides in the human calf during muscle contraction. J Appl Physiol 74: 2034–2039, 1993
17. Harris RC, Hultman E, Nordesjo LO: Glycogen, glycolytic intermediates and high-energy phosphates determined in biopsy samples of musculus quadriceps femoris of man at rest: Methods and variance of values. Scand J Clin Lab Invest 33: 109–120, 1974
18. Harris RC, Soderlund K, Hultman E: Elevation of creatine in resting and exercised muscle of normal subjects by creatine supplementation. Clin Sci (Lond) 83: 367–374, 1992
19. Kushmerick MJ, Meyer RA: Chemical changes in rat leg muscle by phosphorus nuclear magnetic resonance. Am J Physiol 248: C542–C549, 1985
20. Kemp GJ, Thompson CH, Barnes PR, Radda GK: Comparisons of ATP turnover in human muscle during ischemic and aerobic exercise using $^{31}$P magnetic resonance spectroscopy. Magn Reson Med 31: 248–258, 1994
21. Clark JF, Odoom J, Tracey I, Dunn J, Boehm EA, Paternostro G, Radda GK: Creatine transport. In: M.A. Conway, J.F. Clark (eds). Creatine and Creatine Phosphate. Academic Press, San Diego, 1996, pp 41–46
22. Hultman E, Soderlund K, Timmons JA, Cederblad G, Greenhaff PL: Muscle creatine loading in men. J Appl Physiol 81: 232–237, 1996
23. Hamaoka T, Iwane H, Shimomitsu T, Katsumura T, Murase N, Nishio S, Osada T, Kurosawa Y, Chance B: Noninvasive measures of oxidative metabolism on working human muscles by near-infrared spectroscopy. J Appl Physiol 81: 1410–1417, 1996
24. Soderlund K, Greenhaff PL, Hultman E: Energy metabolism in type I and type II human muscle fibres during short term electrical stimulation at different frequencies. Acta Physiol Scand 144: 15–22, 1992
25. Mizuno M, Secher NH, Quistorff B: $^{31}$P-NMR spectroscopy, rsEMG, and histochemical fiber types of human wrist flexor muscles. J Appl Physiol 76: 531–538, 1994
26. Barstow TJ, Buchthal SD, Zanconato S, Cooper DM: Changes in potential controllers of human skeletal muscle respiration during incremental calf exercise. J Appl Physiol 77: 2169–2176, 1994
27. Nishida M, Nishijima H, Yonezawa K, Sato I, Anzai T, Okita K, Yasuda H: Phosphorus-31 magnetic resonance spectroscopy of forearm flexor muscles in student rowers using an exercise protocol adjusted for differences in cross-sectional muscle area. Eur J Appl Physiol Occup Physiol 64: 528–533, 1992
28. Constantin-Teodosiu D, Greenhaff PL, McIntyre DB, Round JM, Jones DA: Anaerobic energy production in human skeletal muscle in intense contraction: A comparison of $^{31}$P magnetic resonance spectroscopy and biochemical techniques. Exp Physiol 82: 593–601, 1997
29. Smith SA, Montain SJ, Matott RP, Zientara GP, Jolesz FA, Fielding RA: Effects of creatine supplementation on the energy cost of muscle contraction: A $^{31}$P-MRS study. J Appl Physiol 87: 116–123, 1999
30. Greenhaff PL, Bodin K, Soderlund K, Hultman E: Effect of oral creatine supplementation on skeletal muscle phosphocreatine resynthesis. Am J Physiol 266: E725–E730, 1994

*Molecular and Cellular Biochemistry* **244:** 113–118, 2003.


# Association of ecNOS gene polymorphisms with end stage renal diseases

Sohji Nagase,[1] Hiromichi Suzuki,[1] Ying Wang,[1] Shuichi Kikuchi,[1] Aki Hirayama,[1] Atsushi Ueda,[1] Kenji Takada,[1] Takaaki Oteki,[1] Mami Obara,[1] Kazumasa Aoyagi[2] and Akio Koyama[1]
[1]*Institute of Clinical Medicine, University of Tsukuba, Tsukuba;* [2]*Tsukuba College of Technology, Tsukuba, Japan*

## Abstract

Nitric oxide (NO) is a very potent regulator of intrarenal hemodynamics and is thought to be an important factor in the deterioration of renal function. Several polymorphisms of the endothelial NO synthase (eNOS) gene have been reported. For instance, tandem 27-bp repeats in intron 4 of the eNOS gene are polymorphic, i.e. eNOS4*a* allele has 4 and eNOS4*b* has 5 tandem repeats, and the association between eNOS4*a* and myocardial infarction has been reported. In addition, a missense Glu298Asp mutation in exon 7 of the eNOS gene is reported to be a risk factor for hypertension or myocardial infarction. In this study, we investigated the frequencies of these 2 polymorphisms of eNOS gene in patients with end-stage renal diseases (ESRD), and compared them with those of healthy subjects.

Genomic DNA was obtained from regularly hemodialyzed patients and healthy volunteers. The allele frequencies of eNOS4*a* and eNOS4*b* in intron 4 were analyzed by PCR and the missense Glu298Asp mutation in exon 7 were determined by PCR FMLP analysis.

The allele frequency of eNOS4*a* (eNOS4*a*/*b* and eNOS4*a*/*a*) in non-diabetic group is significantly higher than that in healthy controls (27.3% vs. 19.0%, $p = 0.01$) though there is no significant difference between diabetic group and healthy controls. On the other hand, the frequencies of missense Glu298Asp mutation in both non-diabetic and diabetic groups are significantly higher than that in healthy controls (22.5% in non-diabetic, 20.8% in diabetic and 7.4% in control group, $p = 0.002$: non-diabetic vs. control, $p = 0.01$: diabetic vs. control).

This study clarified that the polymorphisms in intron 4 and exon 7 of eNOS gene are the genetic risk factors for ESRD. The polymorphisms in intron may change the transcriptional activity and those in exon may alter the 3 dimensional structure of the enzyme, and may affect the progression of renal diseases via decreased NO synthesis. Further study is required to clarify the detailed mechanisms. (Mol Cell Biochem **244**: 113–118, 2003)

*Key words*: endothelial nitric oxide synthase, end stage renal disease, gene polymorphism, 27-bp repeats in intron 4, Glu298Asp mutation, polymerase chain reaction

## Introduction

Among the various factors involved in the deterioration of renal function, changes in hemodynamics are thought to be important. Nitric oxide (NO), an extensively studied endothelium-derived relaxing factor, is reported to be a very potent regulator of intrarenal hemodynamics [1, 2]. NO is produced from L-arginine by nitric-oxide synthase (NOS). The endothelial NOS (eNOS) is encoded by a gene located on chromosome 7q35-36 comprising 26 exons that span 21 kilobases (kb) [3, 4] and expressed in endothelium. Recently, several studies have shown that the polymorphisms of eNOS are related to the hemodynamics. Though the highly polymorphic(CA)n repeats in intron 13 and 2 biallelic markers in intron 18 have been shown to be not associated with essential hypertension [5], Wang *et al.* [6] showed that polymor-

*Address for offprints*: S. Nagase, Division of Nephrology, Institute of Clinical Medicine, University of Tsukuba, 1-1-1 Ten-nodai, Tsukuba, Ibaraki 305-8575, Japan (E-mail: sohji-n@md.tsukuba.ac.jp)

phism of the variable number of tandem repeats (VNTR) in intron 4 of eNOS gene is associated with coronary artery disease among smokers. They identified 2 alleles in eNOS intron 4, the larger of which has 5 tandem 27-bp repeats (GAAGTCTAGACCTGCTGC(A/G)GGGGTGAG). The first 3 repeats had A and the last 2 had G at the 19th base of the 27-bp repeats, respectively. However, the smaller one has only 4 repeats, in which the first 2 repeats had A and the last 2 had G at the 19th base of the repeat, respectively. They denoted these 2 alleles as eNOS4*a* for 4 repeats and eNOS4*b* for 5 repeats. Ichihara *et al.* observed an association between the eNOS4*a* allele and myocardial infarction in both a smoking and non-smoking Japanese population [7]. Adding to this, a missense polymorphism of Glu298Asp within exon 7 of the eNOS gene was reported to be associated with essential hypertension, vasospastic angina and myocardial infarction in a Japanese population [8–10]. As both coronary lesions and chronic renal failure are basically vascular disorders, we speculate that the polymorphism in eNOS gene might have some relevance to progression in chronic renal failure. In this view point, we studied the frequencies of VNTR in intron 4 and Glu298Asp mutation in exon 7 of eNOS gene in patients with end-stage renal diseases (ESRD), and compared them with those of healthy subjects.

## I. Variable number of tandem repeats (VNTR) in intron 4

### *Subjects and methods*

A total of 302 hemodialysis (HD) patients (183 men and 119 women, mean age 57.1 years) and 248 genetically unrelated healthy Japanese subjects (179 men and 69 women, mean age 49.3 years) were entered in this study. All of the patients and normal controls were Japanese. Informed consent was obtained from each person enrolled in this study. Underlying diseases and other clinical features are shown in Table 1. We compared the frequencies of eNOS gene polymorphism among all the patients, the non-diabetic group (all cases except diabetic nephropathy, n = 231), the diabetic group (n = 71), the chronic glomerulonephritis (CGN) group (n = 133) and the healthy controls.

DNA was extracted from peripheral blood using a DNA extraction kit (Wako Pure Chemical, Osaka, Japan) and was stored at 4°C until analysis. eNOS genotypes were determined by the polymerase chain reaction (PCR) as described by Wang *et al.* [6]. Briefly, the oligonucleotide primers (the forward primer was 5′-AGGCCCTATGGTAGTGCCTTT-3′, and the reverse primer was 5′-TCTCTTAGTGCTGTGGTC-AC-3′) that flank the region of the 27-bp direct repeat in eNOS intron 4 were used for DNA amplification. Each reaction mixture was heated to 94°C for 4 min for denaturation and underwent 35 cycles at 94°C for 1 min, finally annealing at 56°C for 1 min with an extension at 72°C for 2 min, and a final extension at 74°C for 7 min. The PCR products were analyzed by 3% NuSieve 3:1 agarose gel electrophoresis (FMC BioProducts, Rockland, ME, USA), and fragments were visualized by ethidium bromide staining and ultraviolet transillumination.

### *Statistical analyses*

A chi-square test was used to test observed vs. expected type frequencies, assuming the Hardy-Weinberg equilibrium. To correct for the contribution of age and sex, we performed multiple logistic regression analysis to explore the relationship between the eNOS gene polymorphism and ESRD, and also for the effect of eNOS intron 4 variant as a risk factor for ESRD. The analysis was performed by means of SPSS Advanced Statistics 6.1 for Macintosh (SPSS Japan Inc. Japan). The intron 4 genotypes were coded as dummy variables in which 0 stood for *bb*, 1 for *ba* or *aa*. The odds ratio and 95% confidence interval (CI) were also calculated. A p value

*Table 1.* Underlying diseases of haemodialysis patients and age of study groups

| | Numbers (males/females) | Age, years (mean ± S.D.) | |
|---|---|---|---|
| | | At start of dialysis | At present |
| All patients | 302 (183/119) | 48.7 ± 16.0 | 57.1 ± 14.0 |
| Non-DM total | 231 (135/96) | 45.5 ± 16.4 | 55.6 ± 14.8 |
| Chronic glomerulonephritis (CGN) | 133 (80/53) | 42.3 ± 15.3 | 53.4 ± 13.5 |
| Polycystic kidney | 18 (10/8) | 52.8 ± 10.2 | 58.8 ± 10.9 |
| Nephrosclerosis | 13 (7/6) | 66.4 ± 14.4 | 73.8 ± 12.2 |
| Systemic lupus erythematosus | 7 (1/6) | 35.3 ± 11.2 | 40.7 ± 11.8 |
| Others | 18 (10/8) | 41.4 ± 19.4 | 51.4 ± 16.9 |
| Unclear | 42 (27/15) | 51.8 ± 15.8 | 60.0 ± 15.7 |
| DM | 71 (48/23) | 57.0 ± 9.9 | 62.0 ± 9.2 |
| Control | 248 (179/69) | | 49.3 ± 8.8 |

DM – diabetic nephropathy; S.D. – standard deviation.

of < 0.05 was considered statistically significant. These analyses were basically the same in the study of part II.

*Results*

PCR analysis of genomic DNA generated fragments of 393 or 420 bp corresponding to the eNOS4*a* and eNOS4*b* alleles, respectively [6]. As there were few subjects who were homozygous for eNOS*a*, we considered subjects who were both homo-and heterozygous for eNOS*a* for comparison with subjects having only the eNOS*b* genotype. The frequencies of eNOS4*b/b*, eNOS4*b/a*, eNOS4*a/a* were 81.0% (201/248), 19.0% (47/248), 0.0% (0/248) in the control group, 74.8% (226/302), 23.5% (71/302), 1.7% (5/302) in all the patients, 72.7% (168/231), 25.1% (58/231), 2.2% (5/231) in the non-diabetic group, 70.7% (94/133), 26.3% (35/133), 3.0% (4/133) in the CGN group and 81.7% (58/71), 18.3% (13/71), 0.0% (0/71) in the diabetic group, respectively (Table 2). The genotype frequencies in controls were in agreement with those predicted by the Hardy-Weinberg equilibrium ($x^2 = 2.71$; df = 1; $p > 0.05$).

The multiple logistic regression analysis for correcting the contribution of age and sex revealed that the frequency of the eNOS4*a* (eNOS4*b/a* and eNOS4*a/a*) in all the patients, in the non-diabetic group and in the CGN group were significantly higher than that in the control group ($p = 0.021$; $p = 0.0096$; $p = 0.010$, respectively). In contrast, there was no significant difference in the frequencies of eNOS genotypes between the diabetic group and the control group ($p = 0.81$).

The patients of the non-diabetic group and the diabetic group had been under the treatment with HD on an average of 10 and 5 years respectively when this study was performed. In order to examine the survival bias in analyzing the relationship between eNOS intron 4 gene polymorphism and the progression of chronic renal failure, we divided the non-diabetic patients into 2 groups one of which had been receiving dialysis for under 10 years and the other for over 10 years, since the mean duration under treatment was 10 years. A similar division into 2 groups was made regarding diabetic nephropathy where one group was on treatment under 5 years and the other was over 5 years since the mean duration of HD was 5 years. We compared the 2 groups with regard to the frequencies of eNOS intron 4 gene polymorphism (shown in Table 3) by means of the chi-square test. Statistical analysis revealed that there were no significant differences between either of the 2 groups ($p = 0.13$ in non-diabetic renal diseases; $p = 0.63$ in diabetic nephropathy).

## II. Glu298Asp mutation in exon 7

*Subjects and methods*

We examined 159 patients with ESRD undergoing maintenance HD and 270 genetically unrelated apparently healthy control subjects. Informed consent was obtained from each person enrolled in this study. The underlying causes of ESRD were CGN (n = 68), diabetic nephropathy (n = 48), hypertension (n = 17), polycystic kidney disease (n = 13), lupus nephritis and vasculitis (n = 7), reflux and obstructive nephropathy (n = 5), interstitial nephritis (n = 1).

Genomic DNA was extracted from peripheral lymphocytes with a DNA extraction kit (Wako, Osaka, Japan). Polymerase chain reaction - restriction fragment length polymorphism analysis (PCR-RFLP) were employed to identify Glu298Asp mutation. We originally designed a set of primers: forward 5-GAAACGGTCGCTTCGACGTGCT and reverse 5-CCA-

*Table 2.* Frequencies of ecNOS genotype among patients and controls and results of multiple logistic regression analysis

| | Control | All patients | Non -DM | | DM |
|---|---|---|---|---|---|
| | | | Total | CGN | |
| Genotype | | | | | |
| *bb* | 201 (81.0%) | 226 (74.8%) | 168 (72.7%) | 94 (70.7%) | 58 (81.7%) |
| *ba* | 47 (19.0%) | 71 (23.5%) | 58 (25.1%) | 35 (26.3%) | 13 (18.3%) |
| *aa* | 0 (0.0%) | 5 (1.7%) | 5 (2.2%) | 4 (3.0%) | 0 (0.0%) |
| *aa+ba* | 47 (19.0%) | 76 (25.2%) | 64 (27.3%) | 39 (29.3%) | 13 (18.3%) |
| Total | 248 | 302 | 231 | 133 | 71 |
| Beta | | 0.518 | 0.599 | 0.662 | 0.102 |
| S.E. | | 0.224 | 0.231 | 0.258 | 0.425 |
| Odds ratio (95% CI) | | 1.67 (1.08–2.60) | 1.82 (1.16–2.86) | 1.94 (1.17–3.21) | 1.11 (0.48–2.54) |
| p value* | | 0.021 | 0.0096 | 0.0102 | 0.8107 |
| Allele frequency | | | | | |
| *a* | 0.095 | 0.134 | 0.147 | 0.162 | 0.092 |
| *b* | 0.905 | 0.866 | 0.853 | 0.838 | 0.908 |

DM – diabetic nephropathy; S.E. – standard error; Beta – beta coefficient; C.I. – confidence index; *one degree of freedom.

*Table 3.* ecNOS genotype frequencies according to dialysis periods in non-DM and DM group

| | Non-DM (229 cases) | | DM (71 cases) | |
|---|---|---|---|---|
| | Within 10 years | Over 10 years | Within 5 years | Over 5 years |
| Genotypes | | | | |
| *bb* | 105 (76.1%) | 61 (67.0%) | 31 (83.8%) | 27 (79.4%) |
| *ba* | 32 (23.2%) | 26 (28.6%) | 6 (16.2%) | 7 (20.6%) |
| *aa* | 1 (0.7%) | 4 (4.4%) | 0 (0.0%) | 0 (0.0%) |
| *ba+aa* | 33 (23.9%) | 30 (33.0%) | 6 (16.2%) | 7 (20.6%) |
| Total | 138 | 91 | 37 | 34 |
| $x^2$ | 2.254 | | 0.226 | |
| p value | 0.133 | | 0.634 | |

DM – diabetic nephropathy.

CCCAGTCAATCCCTTTG producing 140-bp amplified fragments. The PCR amplifications were performed 35 cycles under following conditions; denaturation at 94°C for 30 sec, annealing at 56°C for 30 sec, extension at 72°C for 1 min. PCR products were then digested by *Mbo*I, which produces two fragments of 50 and 90 bp in length from alleles with glutamic acid (G) at codon 298 (wild type, G), and *Ban*II, which produces two fragments of 55 and 85 bp in length from alleles with aspartic acid (T) at the same codon (mutant). The digested products were analyzed on 3% agarose gel electrophoresis (Nippon gene, Osaka, Japan).

### *Results*

The frequencies of homozygous or heterozygous Glu298Asp mutation (T/T + T/G) were 7.4% (20/270) in the control group, 22.0% (35/159) in all the patients, 22.5% (25/111) in the non-diabetic group, 23.5% (16/68) in the CGN group and 20.8% (10/48) in the diabetic group. The distribution of Glu298Asp genotypes in controls was in the Hardy-Weinberg equilibrium.

The multiple logistic regression analysis revealed that the frequencies of Glu298Asp mutation in all the patients, in the non-diabetic group, in the CGN group, and in the diabetic group were significantly higher than those in the control subjects ($p = 0.0001$; $p = 0.0019$; $p = 0.0021$; $p = 0.0117$, respectively). Odds ratios ranged from 2.76–3.22 (Table 4).

The patients enrolled in this study had been under renal replacement therapy for 9.2 (mean) years in the non-diabetic group and for 4.2 (mean) years in the diabetic group. To evaluate the survival bias in analyzing the relationship between Glu298Asp mutation and the progression to ESRD, we divided the patients into 2 groups according to the duration of hemodialysis; among the diabetics, patients who had experienced more than 5 years of HD belonged to the long term survival group while patients with HD less than 5 years belonged to the short term group. A similar division into 2 groups was made among the non-diabetic patients; the long-term survival group (HD 10 years or more) and the short-term

*Table 4.* Frequencies eNOS genotypes among patients and controls and results of multiple logistic regression analysis

| | Control | All patients | Non-DM | | DM |
|---|---|---|---|---|---|
| | | | Total | CGN | |
| GG | 250 (92.6%) | 124 (78.0%) | 86 (77.5%) | 52 (77.3%) | 38 (79.2%) |
| TG | 18 (6.7%) | 33 (20.8%) | 25 (22.5%) | 16 (22.7%) | 8 (16.7%) |
| TT | 2 (0.7%) | 2 (1.2%) | 0 (0.0%) | 0 (0.0%) | 2 (4.1%) |
| TT + TG | 20 (7.4%) | 35 (21.5%) | 25 (22.5%) | 16 (22.7%) | 10 (20.8%) |
| Total | 270 | 159 | 111 | 68 | 48 |
| Beta | | 1.163 | 1.016 | 1.107 | 1.17 |
| S.E. | | 0.302 | 0.327 | 0.36 | 0.464 |
| Odds ratio | | 3.2 | 2.76 | 3.02 | 3.22 |
| (95% C.I.) | | (1.77–5.79) | (1.46–5.24) | (1.49–6.12) | (1.30–8.00) |
| p value* | | 0.0001 | 0.0019 | 0.0021 | 0.0117 |
| Allele frequency | | | | | |
| T (Asp) | 0.041 | 0.113 | 0.109 | 0.118 | 0.125 |
| G (Glu) | 0.959 | 0.887 | 0.891 | 0.882 | 0.875 |

DM – diabetic nephropathy; S.E. – standard error; Beta – beta coefficient; C.I. – confidence index; *one degree of freedom.

group (HD less than 10 years). The 2 groups were compared with regard to the frequencies of Glu298Asp mutation by the chi-square test. Statistical analysis revealed that there were no significant differences between either of the 2 groups (Table 5).

## Discussion

In this study the frequency of *a* allele in eNOS intron 4 was significantly higher in patients from the non-diabetic group and the CGN group compared with healthy subjects indicating that the eNOS 4*a* allele is a risk factor for ESRD in non-diabetic renal diseases. Our data are consistent with the results reported by Yokoyama *et al.* [11]. Both Yokoyama's and our study suggest that eNOS gene polymorphism in intron 4 might have some relevance to the progression of ESRD in non-diabetic renal diseases. Tsukada *et al.* [12] recently reported a strong association between the *a* allele of the eNOS gene and the plasma NOx (nitrate and nitrite) levels. The mean plasma NOx level of the subjects who were homozygous for the *a* allele was nearly 20% lower than in the subjects with the *b* allele. They concluded, therefore, that the eNOS gene locus might be responsible for variations in the genetic control of plasma NOx. In our study, the frequencies of eNOS gene polymorphism showed no significant difference between the diabetic group and healthy subjects. The mechanisms behind diabetic nephropathy have not yet been fully elucidated and there have been several hypotheses proposed including impaired vasodilatory function. In fact, an abnormal renal vasomotor tone is present in the early stage of diabetes mellitus [13]. NO, a potent vascular regulator, has been extensively studied as a factor in the pathogenesis and progression of diabetic nephropathy. Several *in vitro* studies have concluded that diabetic animals have a decreased production and/or responsiveness to NO, and this decreased NO activity in diabetes may have relevance to the renal hemodynamic changes associated with this disease [14, 15]. On the other hand, measurements made *in vivo* indicate that there is an increase in endogenous NO production in diabetic rats [16–18]. Over production of NO may play a significant role in the genesis of diabetic hyperfiltration and hyperperfusion [16–19] which are related to glomerular hypertrophy with messangial expansion and increased intracapillary pressure [20]. As for the role of the endothelium and NO in the control of renal hemodynamics in diabetic nephropathy, it still remains disputable. This polymorphism is reported to be related to coronary heart diseases [6, 7], and it might affect other vascular diseases which are lethal. The subjects of the study have been dialysed for relatively long periods. Therefore, we examined the survival bias. The frequencies of eNOS intron 4 gene polymorphism in the long-term survival groups did not differ from those in the short-term group for both in non-diabetes and diabetes. It appears, therefore that neither death nor survival are factors in estimating the role of gene polymorphism in disease progression.

In the second part of this study clarified that the Glu298Asp mutation of the eNOS gene is heavily accumulated in patients with ESRD, whose underlying causes of chronic renal failure include not only non-diabetic renal diseases but also diabetic nephropathy. The results demonstrate that patients with renal diseases including diabetic nephropathy who have Glu298Asp mutation are at higher risk for progression to ESRD. This is the first report that refers to the significance of Glu298Asp mutation in patients with renal diseases. We also evaluated the effects of the Glu298Asp mutation on mortality. We could not find any statistical difference in the frequencies of the Glu298Asp mutation between short-term survival groups and long-term survival groups both in diabetic and non-diabetic patients. Our results therefore indicate that Glu298Asp mutation is not a lethal factor but a risk factor for ESRD, though its association with relatively lethal diseases such as myocardial infarction has been reported.

The molecular mechanism by which eNOS gene polymorphism acts to affect the progression of ESRD is not known, and it is also unclear whether these polymorphisms are causative variants or markers of other functional variants. The polymorphisms in intron may change the transcriptional activity and those in exon may alter the 3 dimensional structure of the enzyme, and may affect NO synthesis though we did not measure the enzyme activity. However, the fact that the distribution of the *a* allele in eNOS intron 4 showed a

*Table 5.* Frequencies of Glu 298 Asp mutation according to dialysis periods in non-DM and DM groups

| | Non-DM (111 cases) | | DM (48 cases) | |
|---|---|---|---|---|
| | Within 10 years | Over 10 years | Within 5 years | Over 5 years |
| Genotypes | | | | |
| TT + TG | 16 (21.1%) | 9 (22.5%) | 7 (26.9%) | 3 (14.3%) |
| GG | 60 (78.9%) | 31 (77.5%) | 19 (73.1%) | 18 (85.7%) |
| $x^2$ | 0.491 | | 1.108 | |
| p value | 0.483 | | 0.292 | |

DM – diabetic nephropathy.

significantly higher incidence in the patients with ESRD, and the report that the plasma NOx levels of the subjects with *a* allele of the VNTR in intron 4 of the eNOS gene were significantly lower than those without the *a* allele [12] suggest that eNOS gene polymorphism in intron 4 is a useful parameter for studying the relationship between NO and the progression of renal diseases. Though a decrease in serum NOx according to the eNOS gene polymorphisms is suggestive, it is still controversial because serum NOx level can vary according to renal function or dietary intake of nitrite or nitrate [21].

In conclusion, among the frequencies of eNOS intron 4 gene polymorphism, the *a* allele showed a significantly higher incidence in the cases with ESRD except those caused by diabetic nephropathy. It would appear that eNOS gene polymorphism in intron 4 affects the progression of renal failure in non-diabetic renal diseases but the same conclusion could not be drawn in diabetic nephropathy. In addition, Glu298Asp mutations of the eNOS gene significantly accumulate in patients with ESRD not only caused by non-diabetic diseases but by diabetic nephropathy. These evidences may indicate that the *a* allele of the VNTR in intron 4 and the Glu298Asp mutation in exon 7 of eNOS gene are genetic risk factors. However, additional studies, such as *in vitro* measurement of eNOS activity using cultured endothelium derived from polymorphic individuals, are required to assess the importance of these polymorphisms as predictive factors in patients with renal diseases.

## Acknowledgements

We are deeply grateful to Dr. Burton D. Cohen for his valuable criticism in preparing the manuscript. This study was supported in part by the Scientific Research Funds of the Ministry of Education, Science and Culture of Japan.

## References

1. Zate R, De Nucci G: Effects of acute nitric oxide inhibition on rat glomerular microcirculation. Am J Physiol 261: F360–F363, 1991
2. Tolins JP, Palmer RMJ, Moncada S, Raij L: Role of endothelium-derived relaxing factor in regulation of renal hemodynamic responses. Am J Physiol 258: H655–H662, 1990
3. Miyahara K, Kawamoto T, Sase K *et al.*: Cloning and structural characterization of the human endothelial nitric-oxide-synthase gene. Eur J Biochem 233: 719–726, 1994
4. Nadaud S, Bonnardeaux A, Lathrop M, Soubrier F. Gene structure polymorphism and mapping of the human endothelial nitric oxide synthase gene. Biochem Biophys Res Commun 198: 1027–1033, 1994
5. Bonnardeaux A, Nadaud S, Charru A, Jeunemaitre X, Corvol P, Soubrier F: Lack of evidence for linkage of the endothelial cell nitric oxide synthase gene to essential hypertension. Circulation 91: 96–102, 1995
6. Wang XL, Sim AS, Badenhop RF, Mccredie RM, Wilcken DEL: A smoking-dependent risk of coronary artery disease associated with a polymorphism of the endothelial nitric oxide synthase gene. Nature Med 2: 41–45, 1996
7. Ichihara S, Yamada Y, Fujimura T, Nakashima N, Yokota M: Association of a polymorphism of the endothelial constitutive nitric oxide synthase gene with myocardial infarction in the Japanese population. Am J Cardiol 81: 83–86, 1998
8. Miyamoto Y, Saito Y, Kajiyama N, Yoshimura M, Shimasaki Y, Nakayama M, Kamitani S, Harada M, Ishikawa M, Kuwahara K, Ogawa E, Hamanaka I, Takahashi N, Kaneshige T, Teraoka H, Akamizu T, Azuma N, Yoshimasa Y, Yoshimasa T, Itoh H, Masuda I, Yasue H, Nakao K: Endothelial nitric oxide synthase gene is positively associated with essential hypertension. Hypertension 32: 3–8, 1998
9. Shimasaki Y, Yasue H, Yoshimura M, Nakayama M, Kugiyama K: Association of the missense Glu298Asp variant of the endothelial nitric oxide synthase gene with myocardial infarction. J Am Coll Cardiol 31: 1506–1510, 1998
10. Yoshimura M, Yasue H, Nakayama M, Shimasaki Y, Sumida H, Sugiyama S, Kugiyama K, Ogawa H, Ogawa Y, Saito Y, Miyamoto Y, Nakao K: A missense Glu298Asp variant in the endothelial nitric oxide synthase gene is associated with coronary spasm in the Japanese. Hum Genet 103: 65–69, 1998
11. Yokoyama K, Tsukada T, Matsuoka H, Hara S, Yamada A, Kawaguchi Y: High accumulation of endothelial nitric oxide synthase (ecNOS): A gene polymorphism in patients with end-stage renal disease. Nephron 79: 360–361, 1998
12. Tsukada T, Yokoyama K, Arai T *et al.*: Evidence of association of the ecNOS gene polymorphism with plasma NO metabolite levels in humans. Biochem Biophys Res Commun 245: 190–193, 1998
13. Miller JA, Floras JS, Zinman B, Skorecki KL, Logan AG: Abnormalities in the renal and vascular responses to LBNP in humans with early diabetes. Am J Physiol 266: R442–R450, 1994
14. Wang YX, Brooks DP, Edwards RM: Attenuated glomerular cGMP production and renal vasodilation in streptozotocin-induced diabetic rats. Am J Physiol 264: R952–R956, 1993
15. Taylor PD, McCarthy AL, Thomas CR, Poston L: Endothelium- dependent relaxation and noradrenaline sensitivity in mesenteric resistance arteries of streptozotocin-induced diabetic rats. Br J Pharmacol 107: 393–399, 1992
16. Tolins JP, Shultz PJ, Raij L, Brown DM, Mauer SM: Abnormal renal hemodynamic response to reduced renal perfusion pressure in diabetic rats: Role of NO. Am J Physiol 265: F886–F895, 1993
17. Bank N, Aynedjian HS: Role of EDRF (nitric oxide) in diabetic renal hyperfiltration. Kidney Int 43: 1306–1312, 1993
18. Komers R, Allen TJ, Cooper ME: Role of endothelium-derived nitric oxide in the pathogenesis of the renal hemodynamic changes of experimental diabetes. Diabetes 43: 1190–1197, 1994
19. Mattar AL, Fujihara CK, Ribeiro MO, De Nucci G, Zatz R: Renal effects of acute and chronic nitric oxide inhibition in experimental diabetes. Nephron 74: 136–143, 1996
20. Steffes MW, Østerby R, Chavers B, Mauer SM: Mesangial expansion as a central mechanism for loss of kidney function in diabetic patients. Diabetes 38:1077–1081, 1989
21. Florin TH, Neale G, Cummings JH: The effect of dietary nitrate on nitrate and nitrite excretion in man. Br J Nutr 64: 387–397, 1990

*Molecular and Cellular Biochemistry* **244:** 119–124, 2003.


# Importance of renal mitochondria in the reduction of TEMPOL, a nitroxide radical

Atsushi Ueda,[1] Sohji Nagase,[1] Hidekatsu Yokoyama,[2] Mika Tada,[2] Hiroyuki Noda,[2] Hiroaki Ohya,[2] Hitoshi Kamada,[2] Aki Hirayama,[1] and Akio Koyama[1]

[1]*Department of Internal Medicine, Institute of Clinical Medicine, University of Tsukuba, Ibaraki;* [2]*Institute for Life Support Technology and Regional Joint Research Project of Yamagata Prefecture, Yamagata Public Corporation for the Development of Industry, Yamagata, Japan*

## Abstract

Spin probing methods using an electron spin resonance (ESR) spectrometer are used extensively and bring us a lot of information about *in vivo* redox mechanisms. However, the *in vivo* reducing mechanisms of exogenous nitroxide radicals, which serve as typical spin probing reagents are not clear. To clarify this, we examined the sequential kinetics of a spin probe, 4-hydroxy 2,2,6,6-tetramethylpiperidine-N-oxyl (TEMPOL) in the *in vivo* organs, tissue homogenates and subcellular fractions of kidney and liver using an *in vivo* and X-band ESR spectrometers. As a parameter of reducing activity, we calculated the half-life of TEMPOL from the decay curve of ESR signal intensity. The half-life of TEMPOL in the whole organs and homogenates of the kidney was significantly shorter than that of the liver, this indicates that the kidney has more reducing activity against TEMPOL as compared to the liver. Subcellular fractional studies revealed that this reducing activity of the kidney mainly exists in the mitochondria. Contrarily, in addition to reduction in the mitochondria, TEMPOL in the liver was reduced by the microsome and cytosol. (Mol Cell Biochem **244**: 119–124, 2003)

*Key words*: free radical, TEMPOL, reducing activity, kidney, mitochondria

## Introduction

Free radicals and reactive oxygen species are recognized as important mediators of cellular events. They participate in many physiological functions, such as neurotransmission, inflammation and host defenses. *In vivo* detection of these free radicals is very difficult because of their small quantity and short life span. Therefore, to understand the dynamics of reactive oxygen species in biological systems it is useful to study the dynamics of exogenous nitroxide radicals which are typical spin probing reagents and can serve as mimics of *in vivo* free radicals. For this purpose, spin probing methods using an electron spin resonance (ESR) spectrometer are used extensively and bring us a lot of information about redox mechanisms. There are several studies on the reduction of nitroxide radicals in microsomes [1–4], mitochondria [5–7], and non-enzymatic processes [8–11]. These studies clarified the reducing mechanisms of a nitroxide radical in a certain cell or organelle [1–12], the differences between various types of spin probing reagents [3, 5, 6, 8, 10–13] and the effects of oxygen concentration in the measuring sample [3, 4, 7, 11, 12]. However there are few reports concerning organ to organ differences in reducing activity [10, 13], and the *in vivo* reducing mechanisms of nitroxide radicals in the liver [14, 15]. These reports showed the different results and the dynamics of the exogenous nitroxide radicals in the organ are still not clear. It is generally considered that the main reduction site *in vivo* is the liver. Though, there are some evidences that the kidney plays an important role in the biological reduction [16–19].

*Address for offprints*: S. Nagase, Department of Internal Medicine, Institute of Clinical Medicine, University of Tsukuba, 1-1-1 Ten-nodai, Tsukuba, Ibaraki 305-8575, Japan (E-mail: sohji-n@md.tsukuba.ac.jp)

Using newly developed surface-coil-type resonator (SCR) for *in vivo* ESR studies [20], we succeeded in observing sequential changes of signal intensity from the exogenous nitroxide radical, 4-hydroxy 2,2,6,6-tetramethylpiperidine-N-oxyl (TEMPOL) in the kidney, liver, rectum, stomach and skin [21]. This SCR method is useful for the direct evaluation of the *in vivo* redox state in localized organs.

In this study, the *in vivo* reducing activity against TEMPOL in the kidney and liver was investigated by using the SCR method. Thereafter, the mechanisms for reduction in the homogenates and organelles in each organ were examined and the features of redox functions in the kidney and liver were clarified. TEMPOL can penetrate readily into aqueous and lipid-rich areas, and is expected to reflects fully the cellular redox state [3, 5, 7, 11, 14, 21–23]. It is well known that the reduction rate of a nitroxide in biological systems is affected by the oxygen concentration [3, 4, 7, 11, 12]. Therefore, *in vitro* measurements in this study were performed under anaerobic conditions which mimic physiological intracellular oxygen concentration.

## Material and methods

### *Chemicals*

TEMPOL was obtained from Sigma Chemical Co. (St. Louis, MO, USA). Sucrose was purchased from Wako Pure Chemical Industries (Osaka, Japan). Tris-HCl was purchased from Kanto Chemical Co. (Tokyo, Japan) and EDTA was from Dojindo Laboratory (Kumamoto, Japan). All other chemicals and reagents were of analytical grade.

### *Animals*

Male Wistar rats (250–300 g body wt) were provided with standard laboratory chow and water. The rats were divided into the kidney (n = 10) and liver (n = 10) groups. The rats were anesthetized with intraperitoneal administration of 50 mg/kg body weight of sodium pentobarbital after one night of starvation.

### In vivo *ESR spectrometer*

A 700 MHz microwave ESR spectrometer equipped with an SCR constructed at our laboratory has been already described in detail [21, 24–26]. It consisted of an SCR, a main electromagnet, a pair of field scan coils, a pair of field modulation coils, power supplies, a personal computer, and a 700 MHz microwave circuit for homodyne detection. The SCR consisted of a single-turn coil (10 mm in inner diameter) and transmission lines formed by flexible coaxial cables [20, 21]. An SCR was connected to the microwave circuit through a three-stub tuner and driven at a frequency of approximately 720 MHz.

### *Half-life of TEMPOL* in vivo

Five rats in each group were subjected to *in vivo* ESR study under anesthesia. For the kidney group, the left kidney of the rat was exposed by an incision from the back. In the liver group, the rat's liver was exposed by an abdominal incision. Each animal was restrained in the static magnetic field and an SCR was attached to the kidney or liver. The animals in each group received a TEMPOL solution (2 ml/kg body wt) via the tail vein that had been prepared by dissolving it in a phosphate buffer solution at a concentration of 200 mM. After injection, ESR measurements were started and spectral data were collected by the personal computer. The measurements were repeated every 4 sec from 20–52 sec after the injection of TEMPOL. One spectrum was obtained from an average of 3 accumulations of 1 sec scans. The spectroscopy setting were as follows: microwave frequency, 720 MHz; microwave power, 52 mW; center field, 26 mT; field modulation width, 0.2 mT at 100 kHz; scan speed, 10 mT/sec; scan width, 10 mT; time constant, 1 msec. The peak height of the lowest components of the triplet spectra was defined as signal intensity. The half-life of TEMPOL was calculated from the temporal change in the signal intensity.

### *Half-life of TEMPOL in the homogenate*

Five rats in each group subjected to *in vitro* ESR study under anesthesia. The kidneys and livers were removed after perfusion with ice-cooled heparinized saline and 1 g of each organ was immediately homogenized in 9 volumes of the homogenate buffer consisting of 0.25 M sucrose (pH 7.4), 3 mM Tris-HCl and 0.1 mM EDTA at 0°C using a Potter-Elvehjem Teflon® homogenizer. The homogenates were incubated at 37°C for 3 min and deoxidized by bubbling 100% nitrogen gas for 5 min. The oxygen concentration of the incubation medium was less than 10 μM confirmed by a Clark oxygen electrode (UC-12, Central Kagaku Co, Japan). TEMPOL was dissolved to 1 mM in phosphate buffered solution (pH 7.4) and was deoxidized in the same way. The TEMPOL solution was added to the homogenate and mixed quickly. Thus the reaction mixture consisted of 450 μl of the homogenate and 50 μl of the TEMPOL solution (final concentration; 100 μM). The ESR signal was measured using a conventional X-band ESR spectrometer (FR-30, JEOL, Japan). The reaction mixture was put into a flat quartz cell, which was then placed at the center of the cavity of the ESR spectrometer. The ESR measurements were repeated every 20 sec from 2–7 min after the mixing of the homogenate and TEMPOL solutions at 25°C. The spectroscopy settings were as follows; microwave frequency, 9.432 GHz; microwave power, 4 mW; center field, 333 mT; field modulation width, 0.1 mT at 100 kHz; scan speed, 0.5 mT/sec; scan width, 10 mT; time constant, 1 msec. The half-life of TEMPOL was calculated from the change of signal intensities in the same manner as that used in the *in vivo* study.

### *Half-life of TEMPOL in the various organelles*

The various organelles were separated from the homogenates by differential velocity centrifugation at 4°C in the following way. The nuclear fraction was obtained by centrifugation of the homogenate at 600 g for 10 min and the supernatant from this step was defined as S1. Then the mitochondrial fraction was obtained by centrifugation of S1 at 9000 g for 10 min and the supernatant from this step was defined as S2. The microsomal fraction was obtained by centrifugation of S2 at 105,000 g for 60 min and the supernatant was defined as S3. This fraction was considered cytosol. All organelles were weighed and freshly separated. These supernatants (S1 ~ S3) of the kidney or liver were incubated and deoxidized, mixed to the TEMPOL solution, and subjected to the ESR measurement in the same way as described above. The oxygen concentration of S1 was less than 10 μM and that of S2 and S3 was less than 20 μM.

### *Statistical analysis*

Statistical significance was estimated by the Student's *t*-test, and the p values less than 0.05 were considered significant.

## Results

### *Half-life of TEMPOL* in vivo

Figure 1 shows typical semilogarithmic plots of signal intensity in the kidney and liver after the injection of TEMPOL. Good linearity on the semilogarithmic plot with high reproductivity was observed (correlation coefficient in the kidney = –0.998 ± 0.003; in the liver = –0.996 ± 0.006. Values are mean ± S.D.) which means that the signal intensity decays exponentially and it follows first-order kinetics. As shown in Fig. 2 the half-life of the ESR signal of TEMPOL in the kidney and liver groups was 14.5 ± 1.0 and 30.5 ± 1.5 sec, respectively. The half life in the kidney was significantly shorter than that in the liver ($p < 0.001$).

### *Half-life of TEMPOL in the homogenate*

Figure 3 shows typical semilogarithmic plots of signal intensity after the mixing of the homogenate and TEMPOL in the kidney and liver. Good linearity on semilogarithmic plot was observed and this indicates that it follows first-order kinetics. As shown in Fig. 4 the half-life of the ESR signal of TEMPOL in the kidney and liver groups was 148 ± 14 and 278 ± 65 sec, respectively. The half-life in the kidney was significantly shorter than that in the liver ($p < 0.001$).

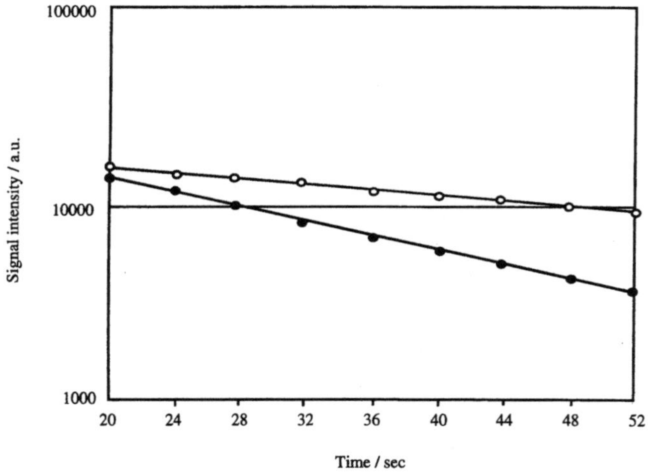


*Fig. 1.* Typical semilogarithmic plot of signal intensity after the injection of TEMPOL in the kidney and liver groups. Open circle, liver; closed circle, kidney.

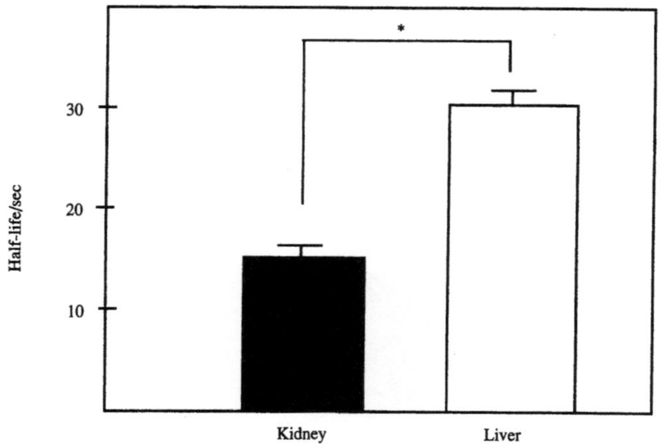


*Fig. 2.* The half-life of TEMPOL injected via the tail vein of rats in the kidney and liver groups. Each group consists of 5 rats and values are expressed as mean ± S.D. Student's *t*-test, *p < 0.001.

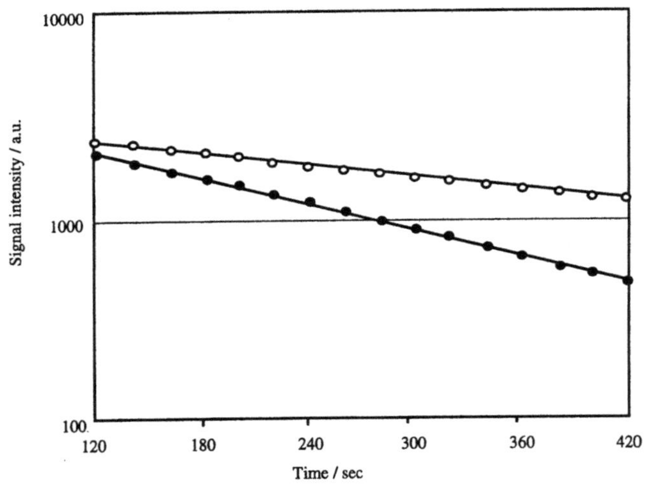


*Fig. 3.* Typical semilogarithmic plot of signal intensity after the mixing of the homogenate and TEMPOL in the kidney and liver groups. Open circle, liver; closed circle, kidney.

### *Half-life of TEMPOL in the various organelles*

Table 1 shows the weights of the various organelles of the kidney and liver. The weights of the nuclear fractions and the microsomal fractions of the liver were heavy as compared to those of the kidney. However the weights of mitochondrial fractions of the kidney were heavy as compared to those of the liver. Figure 5 shows the half-life of TEMPOL in the reaction mixtures containing various fractions of the kidney. The half-life of TEMPOL of the homogenates was almost the same as that of the sample without a nuclear fraction (S1). However, the half-life in the supernatant after the sedimentation of mitochondria (S2) was significantly longer than that of S1 ($p < 0.001$). There was not a significant difference between S2 and S3. Figure 6 shows the half-life of TEMPOL in the reaction mixtures containing various fractions of the liver. There was no significant difference between the half-life in the homogenate and S1. However, the half-life in the supernatant after the sedimentation of mitochondria (S2) was significantly longer than that of S1 ($p < 0.001$), and the half-life in S3 was long compared to that in S2 ($p < 0.05$).

*Table 1.* The weights of various organelles of the kidney and liver

| | Weight (mg/wet organ) | |
|---|---|---|
| Organelles | Kidney | Liver |
| Nucleal fraction | 268 ± 47 | 364 ± 41** |
| Mitochondrial fraction | 148 ± 47 | 106 ± 13* |
| Microsomal fraction | 48 ± 10 | 119 ± 16** |

The organelles were separated from the kidney and liver. Values are expressed as the mean ± S.D. Student's *t*-test, *$p < 0.05$; ** $p < 0.01$.

## Discussion

Nitroxide radicals are widely used as spin probes. The decrease of ESR signal intensity from exogenous nitroxide radicals is assumed to be caused by a one-electron reduction of the nitroxide radical to the corresponding hydroxylamine [3, 6, 12]. This reducing process reflects the redox status in a biological system and several disease conditions such as cancer and inflammation. There are several reports concerning the reduction of nitroxide radicals in subcellular fractions,

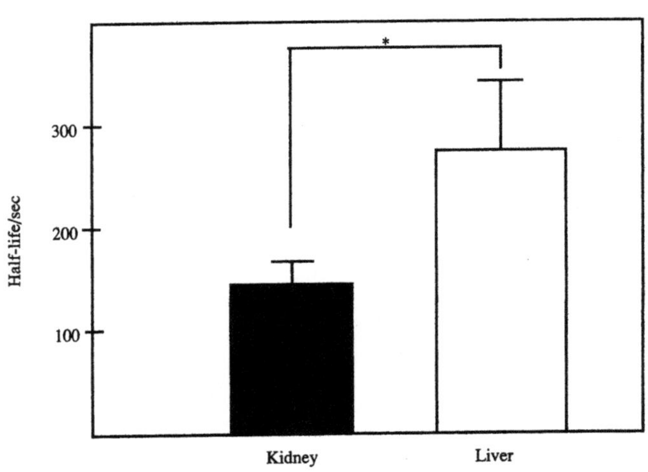


*Fig. 4.* The half-life of TEMPOL after the mixing of the homogenate and TEMPOL in the kidney and liver. Each group consists of 5 rats and values are expressed as mean ± S.D. Student's *t*-test, *$p < 0.001$.

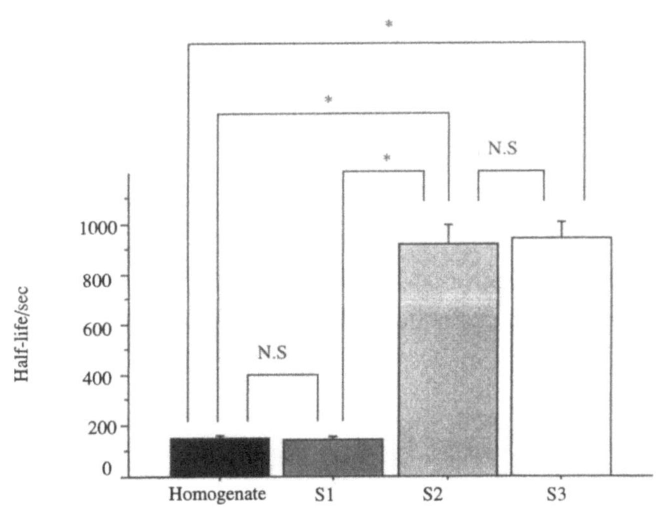


*Fig. 5.* The half-life of TEMPOL mixed the various organelles in the kidney. The supernatant without the nuclear fraction from the homogenate was defined as S1. The supernatant without the mitochondrial fraction from S1 was defined as S2. The supernatant without the microsomal fraction from S2 was defined as S3. Each group consists of 5 rats and values are expressed as mean ± S.D. Student's *t*-test, *$p < 0.001$.

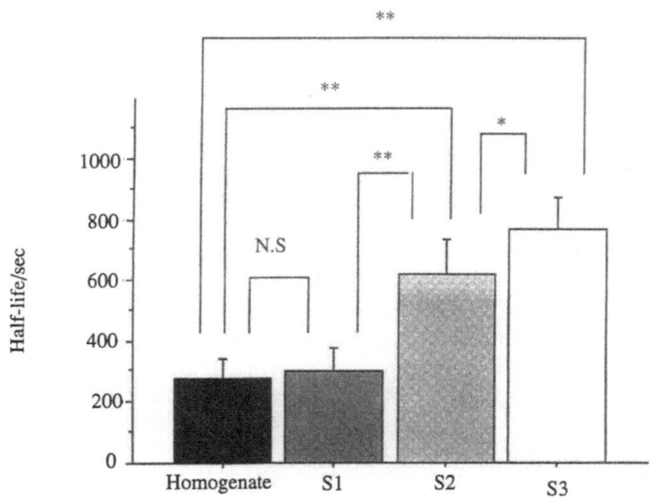


*Fig. 6.* The half-life of TEMPOL mixed the various organelles in the liver. The supernatant without the nuclear fraction from the homogenate was defined as S1. The supernatant without the mitochondrial fraction from S1 was defined as S2. The supernatant without the microsomal fraction from S2 was defined as S3. Each group consists of 5 rats and values are expressed as mean ± S.D. Student's *t*-test, *$p < 0.05$, **$p < 0.001$.

cells, tissues and *in vivo* [1–15, 21–23, 27]. In these reports, the half-life and the rate constant of exogenous nitroxide radicals were used as parameters of reducing activity. It is generally considered that redox reactions in the body mainly occur in the liver, and most of the previous experiments have been performed using the tissue and cells of the liver.

We examined the reducing activity against TEMPOL *in vivo* and in homogenates of both kidney and liver. The results indicate that the half-life of TEMPOL in the kidney is significantly shorter than that in the liver in both experiments. No urinary excretion of TEMPOL throughout the measuring period was found in this study, so it is considered that the diminution of the ESR signal of TEMPOL is due to reduction in the cell. The *in vivo* half-life of TEMPOL in the liver was twice as long as that in the kidney, and this result corresponds to the results of the experiments using homogenates. These findings indicate that the experimental conditions in the homogenates reflect *in vivo* conditions correctly.

Couet *et al.* have shown that the reducing activity against two different nitroxide radicals in the homogenates of the liver or kidney is greater than that in brain, lung, heart, muscle or plasma [13], however, the differences in the reducing activity between the liver and kidney were not clear [8, 10, 13]. In subsequent experiments using the homogenates of the liver and kidney, they conclude that nitroxide radicals are reduced mainly by the sulfhydryl groups on proteins [8], while Eriksson *et al.* demonstrate that this reducing activity is due to ascorbate in the cytosol [10]. Togashi *et al.* have performed *in vivo* ESR imaging of the liver and reported that the reducing mechanisms of nitroxide radicals in the liver are involved in various enzymatic and nonenzymatic reactions [15].

In our experiments using the various subcellular fractions, there is a significant difference in the half-life between S1 and S2. It is considered that the difference reflects the reducing activity of the mitochondria. These results indicate that the main reduction site of TEMPOL in the kidney and liver is the mitochondria. In the same way this reducing activity of TEMPOL exists in the microsome of the liver to some degree and even in the cytosol. However the nuclear fraction is not involved in this phenomenon. These *in vivo* and *in vitro* experiments using homogenates demonstrate that the kidney has great reducing activity against TEMPOL as compared to the liver, and this activity exists primarily in the mitochondria.

The greater reducing activity against TEMPOL in the kidney homogenates is possibly brought about by its relatively higher content of mitochondria as compared to the liver. The kidney contains many more mitochondria than the liver, because the kidney requires enormous energy to reabsorb solutes and water and this energy is supplied by the mitochondria. On the other hand, in the liver, TEMPOL is mainly reduced by the mitochondria, but also by the microsome and cytosol. This seems to be a consequence of the rich content of microsomes which is involved in the detoxification of exogenous substances including nitroxide radicals in the liver. Previous reports proved that nitroxide radicals were reduced by ascorbate in the cytosol, and liver contains much more ascorbate in the cytosol as compared to the kidney [10, 11]. These findings lend support our results.

There are several reports concerning the reducing mechanisms of a nitroxide radicals by the mitochondria. It is considered TEMPOL is reduced to a hydroxylamine during the intracellular process of accepting and passing electrons in the respiratory chain of the kidney. Quintanilha *et al.* using mitochondria, mitoplasts and inverted submitochondria in the liver, reported that an analogue of cetyltrimethylammonium bromide as a spin probing reagent is reduced during the process of the conversion of non-heme iron to ubiquinone or ubiquinone to cytochrome b in the respiratory chain [5]. Chen *et al.* using TB cells reported that 5-doxylstearate was reduced at the same site [6].

Our results have shown that the kidney and its mitochondria are important in the reduction of TEMPOL and demonstrate that each organ has a different mechanism for its reducing process against nitroxide radicals. Moreover, a spin probing method using the ESR spectrometer is thought to be advantageous in the evaluation of the redox state and may serve a useful application in various fields, especially those of medical science.

# Acknowledgements

The authors acknowledge the helpful comments of Dr. Burton D. Cohen.


# References


1. Rosen GM, Rauckman EJ: Formation and reduction of a nitroxide radicals by liver microsomes. Biochem Pharmacol 26: 675–678, 1977
2. Iannone A, Bini A, Swartz HM, Tomasi A, Vannini V: Metabolism in rat liver microsomes of the nitroxide spin probe tempol. Biochem Pharmacol 38: 2581–2586, 1989
3. Iannone A, Tomasi A, Vannini V, Swartz HM: Metabolism of nitroxide spin labels in subcellular fraction of rat liver I. Reduction by microsomes. Biochim Biophys Acta 1034: 285–289, 1990
4. Miura Y, Utsumi H, Kashiwagi M, Hamada A: Effects of oxygen on the membrane structure and the metabolism of lipophilic nitroxide in rat liver microsomes. J Biochem 108: 516–518, 1990
5. Quintanilha AT, Packer L: Surface localization of sites of reduction of nitroxide spin-labeled molecules in mitochondria. Proc Natl Acad Sci USA 74: 570–574, 1977
6. Chen K, Morse PD II, Swartz HM: Kinetics of enzyme-mediated reduction of lipid soluble nitroxide spin labels by living cells. Biochim Biophys Acta 943: 477–484, 1988
7. Chen K, Glockner JF, Morse PD II, Swartz HM: Effects of oxygen on the metabolism of nitroxide spin labels in cells. Biochemistry 28: 2496–2501, 1989
8. Couet WR, Brasch RC, Sosnovsky G, Tozer TN: Factors affecting nitroxide reduction in ascorbate solution and tissue homogenates. Magn Reson Imag 3: 83–88, 1985
9. Giotta GJ, Wang HH: Reduction of nitroxide free radicals in biological materials. Biochem Biophys Res Commun 46: 1576–1580, 1972
10. Eriksson UG, Brasch RC, Tozer TN: Non-enzymatic bioreduction in rat liver and kidney of nitroxyl spin labels, potential contrast agents in magnetic resonance imaging. Drug Metab Disp 15: 155–160, 1987
11. Iannone A, Tomasi A, Vannini V, Swartz HM, Metabolism of nitroxide spin labels in subcellular fraction of rat liver II. Reduction in the cytosol. Biochim Biophys Acta 1034: 290–293, 1990
12. Swartz HM, Sentjurc M, Morsell PD: Cellular metabolism of water-soluble nitroxides: Effect on rate of reduction of cell/nitroxide ratio, oxygen concentrations and permeability of nitroxides. Biochim Biophys Acta 888: 82–90, 1986
13. Couet WR, Eriksson UG, Tozer TN, Tuck LD, Wesbey GE, Nitecki D, Brasch RC: Pharmacokinetics and metabolic fate of two nitroxides potentially useful as contrast agents for magnetic resonance imaging. Pharm Res 1: 203–209, 1984
14. Inaba K, Nakashima T, Shima T, Mitsuyoshi H, Sakamoto Y, Okanoue T, Kashima K, Hashiba M, Nishikawa H, Watari H: Hepatic damage influences the decay of nitroxide radicals in mice – an *in vivo* ESR study. Free Radic Res 27: 37–43, 1997
15. Togashi H, Shinzawa H, Ogata T, Matsuo T, Ohno S, Saito K, Yamada N, Yokoyama H, Noda H, Oikawa K, Kamada H, Takahashi T: Spatiotemporal measurement of free radical elimination in the abdomen using an *in vivo* ESR-CT imaging system. Free Radic Biol Med 25: 1–8, 1998
16. Fillit H, Elion E, Sullivan J, Sherman R, Zabriskie JB: Thiobarbituric acid reactive material in uremic blood. Nephron 29: 40–43, 1981
17. Dasgupta A, Hussain S, Ahmad S: Increased lipid peroxidation in patients on maintenance hemodialysis. Nephron 60: 56–59, 1992
18. Gotoh M, Nagase S, Aoyagi K, Hirayama A, Takemura K, Ueda A, Tomoda C, Kikuchi H, Koyama A: Thiobarbituric acid reactive substances are increased in the subcutaneous fat tissue of patients with end-stage renal disease. Nephrol Dial Transplant 12: 713–717, 1997
19. Nagase S, Aoyagi K, Hirayama A, Gotoh M, Ueda A, Tomida C, Kamezaki T, Nagai Y, Kikuchi H, Koyama A: Favorable effects of hemodialysis on decreased serum antioxidant activity in hemodialysis patients demonstrated by electron spin resonance. J Am Soc Nephrol 8: 1157–1163, 1997
20. Hirata H, Iwai H, Ono M: Analysis of a flexible surface-coil-type resonator for magnetic resonance measurements. Rev Sci Instrum 66 : 4529–4534, 1995
21. Tada M, Yokoyama H, Toyoda Y, Ohya H, Ito T, Ogata T: Surface-coil-type resonators for *in vivo* temporal ESR measurements in different organs of nitroxide-treated rats. Appl Magn Reson 18: 575–582, 2000
22. Utsumi H, Muto E, Hamada A: *In vivo* ESR measurement of free radicals in whole mice. Biochem Biophys Res Commun 172: 1342–1348, 1990
23. Takeshita K, Hamada A, Utsumi H: Mechanisms related to reduction of radical in mouse lung using an L-band ESR spectrometer. Free Radic Biol Med 26: 951–960, 1999
24. Ishida S, Matsumoto S, Yokoyama H, Mori N, Kumashiro H, Tsuchihashi N, Ogata T, Yamada M, Ono M, Kitajima T, Kamada H, Yoshida H: An ESR-CT imaging of the head of a living rat receiving an administration of a nitroxide radical. Magn Reson Imag 10: 21–27, 1992
25. Lin Y, Yokoyama H, Ishida S, Tsuchihashi N, Ogata T: *In vivo* electron spin resonance analysis of nitroxide radicals injected into a rat by a flexible surface-coil-type resonator as an endoscope- or a stethoscope-like device. Magn Reson Mater Phy 2: 99–104, 1997
26. Yokoyama H, Ogata T, Tsuchihashi N, Hiramatsu M, Mori N: A spatiotemporal study on the distribution of intraperitoneally injected nitroxide radical in the rat head using an *in vivo* ESR imaging system. Magn Reson Imag 10: 21–27, 1992
27. Takeshita K, Utsumi H, Hamada A: ESR measurement of radical clearance in lung of whole mouse. Biochem Biophys Res Commun 177: 874–880, 1991

*Molecular and Cellular Biochemistry* **244:** 125–128, 2003.


# Effect of caffeine on metabolism of L-arginine in the brain

Jelenka Nikolic, Gordana Bjelakovic and Ivana Stojanovic
*Institute of Biochemistry, Faculty of Medicine, University of Nis, Nis, Yugoslavia*

## Abstract

Methylxanthines are widely consumed because of their stimulating effect primarily on the central nervous system. Their diuretic and respiratory stimulant action is used in clinical medicine. L-Arginine metabolism in the brain is very important for normal brain function. In addition to brain protein synthesis, arginine is a substrate for the production of urea, creatine, nitric oxide, agmatine, glutamic acid, ornithine, proline and polyamines. As known, many of these compounds are very important in brain function. There is no information relating to effects of caffeine on arginine metabolism in the brain, however, there is a lot of new information about arginine metabolism and caffeine action on the central nervous system. So, we have hypothesized the existence of a relationship that may be of interest in understanding mechanisms of caffeine effects on the central nervous system that may have utility in the clinical applications.

In our experiment protocol we used male Wistar rats weighing about 200 g. Caffeine was added to the drinking water in gradually increasing amounts, from 2 g/l over the first 3 days, to 4 g/l over the last 7 days. A control group was given drinking water without caffeine. The level of lipid peroxidation, arginase and diamine oxidase (DAO) activity in the brain was measured. The results of our study show that arginase and diamine oxidase were decreased in animals treated with caffeine. The level of lipid peroxidation (MDA) was decreased also.

The inhibitory effect of caffeine on arginase activity indicates that caffeine provides more arginine for consumption in other metabolic pathways. Considering the central stimulant effects of caffeine and the decreased lipid peroxidation level, it can be assumed that moderate short-term consumption of caffeine may be beneficial for brain function. (Mol Cell Biochem **244**: 125–128, 2003)

*Key words*: caffeine, arginase, brain, lipid peroxidation, polyamines, diamine oxidase

## Introduction

Methylxanthines: caffeine, theophylline and theobromine are present in several food products such as coffee, tea, cocoa, chocolate, etc. They are widely consumed primarily because of their stimulating effect on the central nervous system. Xanthine drugs are used in clinical medicine as diuretics, analgetics, in the treatment in brain disorders such as vascular headaches, Parkinson's disease [1–3].

L-Arginine metabolism in the brain is very important for normal brain function. The physiological significance of arginine includes protein synthesis, production of urea, agmatine, nitric oxide (NO), proline, glutamate and polyamines [4]. As it is well known, all of these components have very important contributions to brain function. Metabolism of arginine depends on the activity of arginase, argininosuccinate synthetase, arginine decarboxylase and nitric oxide synthetase. Changes in activities of these enzymes will have effects on the metabolic fate of L-arginine in the cell.

We have studied the influence of caffeine on metabolism of L-arginine by measuring arginase activity. As it is known, arginase is an enzyme that leads to the degradation of L-arginine to urea and ornithine, which is the main function of this liver enzyme. In extrahepatic tissues arginase is more important in the synthesis of other products such as NO, proline, polyamines, creatine-P, glutamic acid [5, 6]. The effects of caffeine on the central nervous system is very much dependent on dose. The cortical stimulation produced by small

*Address for offprints*: J. Nikolic, Institute of Biochemistry, Faculty of Medicine, Brace Taskovica 81, 18000, Nis, Yugoslavia (E-mail: nikdany@bankerinter.net)

amounts of caffeine results in mental alertness, decreased fatigue and decreased drowsiness. Large doses of caffeine may produce irritability, insomnia, tremor and headache. Rats ingested high doses of caffeine reproduce the self-destructive behavior that have obtained in Lesch- Nyhan syndrome [7, 8].

The aim of this study was to investigate the possible effect of moderate doses of caffeine during short-term consumption on arginine metabolism, considering the importance of its metabolic products on brain functions.

## Materials and methods

Male Wistar rats weighing about 200 g were used in each experiment. Caffeine was added to the drinking water in gradually increasing amounts: 2 g/l for 3 days and then 4 g/l for the next 7 days. A control group was given drinking water without caffeine. Rats were killed by decapitation and the brains were quickly removed and frozen.

Brain arginase activity was measured in whole brain homogenate on the basis of formed ornithine, according to the method of Porembska and Kedra [9]. Lipid peroxidation levels (MDA) were determined utilizing thiobarbituric acid [10]. Polyamine oxidase activity (PAO) was measured according to the method of Bashrach and Reches, using spermidine as substrate [11]. Proteins in tissue homogenates were estimated according to Lowry *et al.* [12]. Blood urea, creatinine and uric acid levels were measured by standard biochemical analyses.

Statistical significance between groups was determined using Student's *t*-test.

## Results

Blood levels of urea, creatinine and uric acid were increased in the group of animals treated with caffeine with respect to the control group (Table 1). Urea and creatinine were elevated but not significantly. Elevation of uric acid in caffeine treated group of animals was significant, $p < 0.001$.

The influence of caffeine on brain arginase is illustrated in Fig. 1. Arginase was significantly reduced ($p < 0.01$) compared to control. Diamine oxidase activity was depressed also in caffeine treated rats ($p < 0.01$, Fig. 2).

Malon dialdehyde, a measure of the level of lipid peroxidation, was decreased with respect to the control group ($p < 0.01$, Fig. 3).

*Table 1.* Caffeine effects on blood levels of urea (mmol/l), creatinine (μmol/l) and uric acid (μmol/l)

| | Urea | Creatinine | Uric acid |
|---|---|---|---|
| Control | 7.35 ± 0.07 | 44.70 ± 5.90 | 67.82 ± 10.7 |
| Caffeine | 9.05 ± 0.08 | 54.51 ± 9.22 | 115.6 ± 21.2*** |

Urea, creatinine and uric acid were determinated in plasma using heparin as a anticoagulant. Results are mean ± S.D. ***p < 0.001.

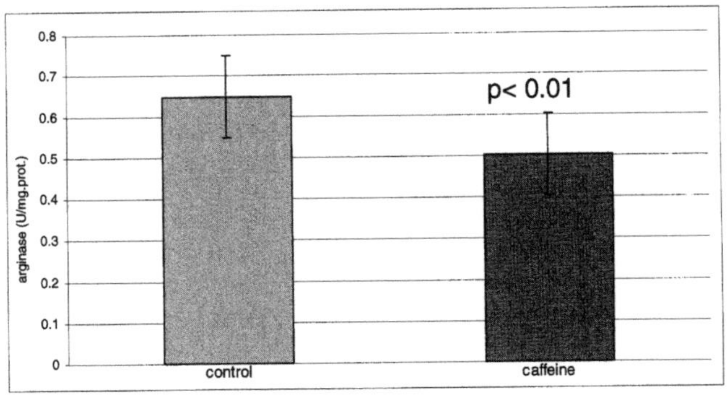


*Fig. 1.* Arginase activity in the brain of caffeine treated rats. Results are expressed as mean ± S.D.

## Discussion

The results of our study show that short-term treatment of animals with small doses of caffeine decreases brain arginase activity. There are two distinct isoenzymes of arginase. Type I arginase is highly expressed in liver as an enzyme of the urea cycle. Type II arginase is expressed in brain, kidney, mammary gland, small intestine and macrophages. The existing differences in regulation of arginase isoenzymes in response to diet, hormones and cytokines show that arginase may be a regulator of the metabolic fate of arginine [4, 13].

Methylxanthines (i.e. caffeine, theophylline and theobromine) belong to a chemical group of purine bases that include important endogenous substances such as adenine, guanine, hypoxanthine and uric acid. This chemical similarity of

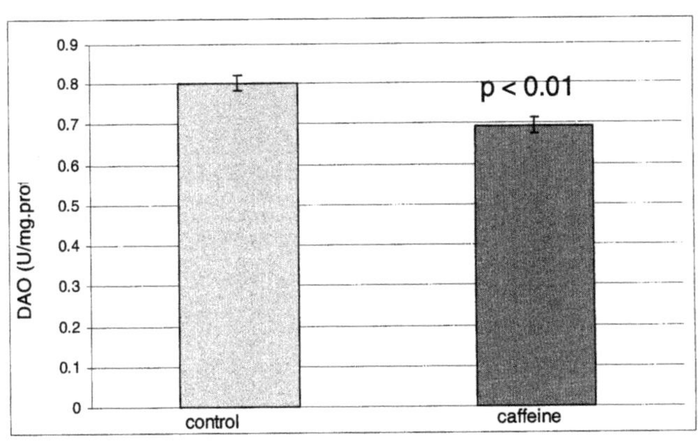


*Fig. 2.* DAO activity in caffeine treated rats. Enzymes activities are expressed as mean ± S.D.

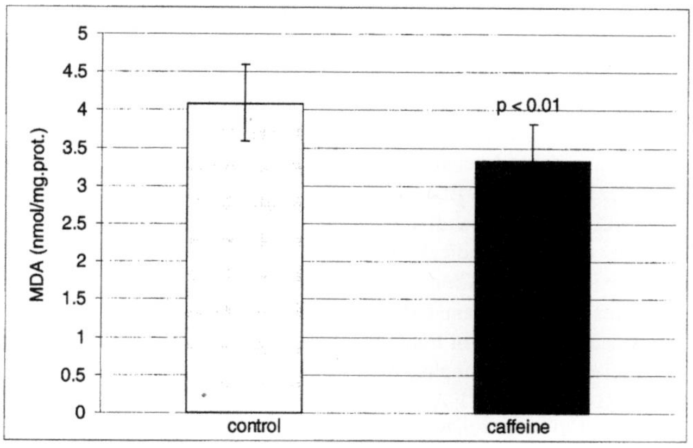


*Fig. 3.* MDA levels in the brain of caffeine treated rats. Results are expressed as mean ± S.D.

xanthines to purine molecules is important with respect to their interaction with the important biochemical regulator, cAMP. Depressed arginase activity in the brain after consumption of caffeine may be a result of caffeine's effect on the level of 3′5′-cyclic adenosine monophosphate (cAMP). A number of metabolic reactions are controlled via cAMP levels. The cellular level of cyclic AMP depends of the activity adenylate cyclase and phosphodiesterase. Xanthines inhibit phosphodiesterase and breakdown cAMP. Cyclic AMP was found to be increased after caffeine treatment [14, 15].

Caffeine stimulates adrenocortical and adrenomedular hormone secretion [16]. Cyclic AMP is a secondary messenger molecule of many hormones, such as ACTH, catecholamines, glucagon, thyroxine and insulin. Glucocorticoids, catecholamines and glucagon increase adenylate cyclase activity. Increased levels of cAMP may affect tissue metabolism in different ways.

Adenosine may be involved in modulation of brain arginase activity. The central stimulant effect of caffeine is linked to the blockade of adenosine receptors [17]. Caffeine removes adenosine from its receptors and increases free adenosine level. Adenosine, adenine, inosine and uric acid are competitive inhibitors of arginase [18]. Changes in the aminoacid pool in the brain may also affect arginase activity. Valine, leucine, isoleucine and ornithine all have inhibitory effects on arginase activity [19].

The important metabolic pathway for arginine is the synthesis of polyamines. Ornithine, a product of arginase activity, is a precursor for synthesis of polyamines. Ornithine decarboxylase catalyses the first step in the biosynthesis of polyamines and its activity is controlled by cAMP [20]. Decarboxylation of ornithine leads to synthesis of putrescine, which is the precursor of spermidine and spermine. Ornithine decarboxylase, the limiting enzyme in polyamine synthesis is not increased in the brain after caffeine treatment [21]. Results of our study show that caffeine leads to a decrease in polyamine catabolism by depression of diamine oxidase activity.

Considering that the activity of arginase and NO synthetase have different regulation it is likely that depressed arginase activity is followed by an increase in arginine levels that can be utilized in the production of NO. Nitric oxide has vasodilatatory and antioxidative effects, and acts as important modulator of brain function.

The results of our study show that caffeine decreases the level of lipid peroxidation in the brain. The lipid peroxidation levels may be influenced by the caffeine itself, although arginine, adenosine and nitro oxide are also antioxidants [22–24]. The relationship between caffeine and L-arginine metabolism indicates a new aspect of caffeine action. It is of interest because both caffeine and arginine have important functions in different physiological and pathological conditions such as vascular tone, neurotransmission, immune response, tumor biology, intoxication, inflammation, etc. [25–28].

# Conclusions

The results show that caffeine changes the metabolism of L-arginine in the brain. Arginase, an enzyme that hydrolyses L-arginine to ornithine and urea, is decreased after caffeine treatment.

Catabolism of polyamines is depressed as a result of decreased activity diamine oxidase. Caffeine consumption decreases the lipid peroxidation level in the brain.

# References

1. Curatolo PW, Robertson D: The health consequences of caffeine. Ann Intern Med 98: 641–653, 1983
2. Barbier A: Caffeine protects against Parkinson's disease? Trends Pharmacol Sci 22: 500, 2001
3. Kaiser SM, Quinn RJ: Adenosine receptors as potential therapeutic targets. Drug Discovery Today 4: 542–551, 1999
4. Wu G, Morris SM Jr: Arginine metabolism: Nitric oxide and beyond. Biochem J 336: 1–17, 1998
5. Herzfeld A, Raper S: The heterogenity of arginase in rat tissues. Biochem J 153: 469–478, 1976
6. Wiesinger H: Arginine metabolism and the synthesis of nitric oxide in the nervous system. Prog Neurobiol 64: 365–391, 2001
7. Farrer I, Costell M, Grisolia S: Lesch-Nyhan syndrome-like behavior in rats from caffeine ingestion. FEBS Lett 141: 275–278, 1982
8. Minana D, Portoles M, Jorda A, Grisolia S: Lesch-Nyhan syndrome, Caffeine model: Increase of purine and pyrimidine enzymes in rat brain. J Neurochem 43: 1556–1560,1984
9. Porembska Z, Kedra M: Early diagnosis of myocardial infarction by arginase determination. Clin Chim Acta 60: 555–561, 1975
10. Stroev EA, Makarova VG: In: Laboratory Manual in Biochemistry. Mir, Moscov, 1989, pp 251–255

11. Bashrach U, Reches B: Enzymic assay for spermine and spermidine. Anal Biochem 17: 38–48, 1966
12. Lowry OHR, Rosebrough NJ, Farr AL, Randall RJ: Protein measurement with Folin phenol reagent. J Biol Chem 193: 265–275, 1951
13. Mimic-Oka J, Cupic Z, Japundzic I: Effect of adrenal function on level of hepatic and extrahepatic arginase. Experientia 27: 1477–1478, 1971
14. Fredholm BB: On the mechanism of action of theophylline and caffeine. Acta Med Scand 217: 149–153, 1985
15. Nahorski SR, Rogers KJ: Inhibition of 3′,5′- nucleotide phosphodiesterse and the stimulation of cerebral cyclic AMP formation by biogenic amines *in vitro* and *in vivo*. Neuropharmacology 15: 609–612, 1976
16. Bellet S, Kostis J, Roman L, DeCastro O: Effect of coffee ingestion on adrenocortical secretion in young men and dogs. Metabolism 18: 1007–1012, 1969
17. Fredholm BB: Adenosine actions and adenosine receptors after one-week treatment with caffeine. Acta Physiol Scand 115: 283–286, 1982
18. Rosenfeld JL, Duta SD, Chedda GB, Tritsch GL: Purine and pyrimidine inhibitors of arginase. Biochem Biophys Acta 410: 164–166, 1975
19. Reddy PUM, Ramana Rao JV: Inhibition of arginase in sheep brain homogenates by some L-amino acids. Experientia 37: 814, 1981
20. Byus CV, Russell DH: Ornithine decarboxylase activity: Control by cyclic nucleotides. Science 187: 650–652, 1975
21. Colombatto S, Fasulo L, Mondardini A, Nmalabaila A, Grillo MA: Effect of caffeine on ornithine metabolism in rat brain, liver and kidney. Ital J Biochem 38: 75–82, 1989
22. Krycevskaya AA, Shugaley VS, Tsvetnenko EZ: Brain and liver arginase and polyamines in the mechanism of arginine protective effect under hypoxia. Bilt Exp Biol Med 4: 445–447, 1981
23. Lee C: Antioxidant ability of caffeine and its metabolites based on the study of oxygen radical absorbing capacity and inhibition of LDL peroxidation. Clin Chim Acta 295: 141–154, 2000
24. Devasagaya TP, Kamat JP, Mohan H, Kesavan PC: Caffeine as an oxidant: Inhibition of lipid peroxidation induced by reactive oxygen species. Biochim Biophys Acta 1282: 63–70, 1996
25. Wu G, Meininger CJ: Arginine nutrition and cardiovascular function. J Nutr 130: 2626–2629, 2000
26. Nikolic J: Alcoholic intoxication. In: G.A. Qureshi *et al.* (eds). Neurochemical Markers of Degenerative Nervous Diseases and Drug Addiction. VSP, 1998, pp 193–221
27. De Mendoca A, Sebastiao AM, Ribeiro JA: Adenosine: Does it have a neuroprotective role after all? Brain Res Rev 33: 258–274, 2001
28. Essayan DM: Cyclic nucleotide phosphodiesterase (PDE) inhibitors and immunomodulation. Biochem Pharmacol 57: 965–973, 1999

*Molecular and Cellular Biochemistry* **244:** 129–133, 2003.


# Role of nitric oxide synthase activity in experimental ischemic acute renal failure in rats

Miki Komurai,[1] Yasuko Ishii,[1] Fumiaki Matsuoka,[1] Katsuhide Toyama,[1] Masayuki Ominato,[2] Takeo Sato,[1] Teruhiko Maeba,[1] Kenjiro Kimura[1] and Shigeru Owada[2]

[1]*St. Marianna Univerity School of Medicine, Department of Internal Medicine, Division of Nephrology and Hypertension, Kawasaki;* [2]*Akabane Central Hospital, Department of Internal Medicine, Tokyo, Japan*

## Abstract

To determine the role of nitric oxide (NO) in acute renal failure (ARF), we have studied the time course change activities to activity of nitric oxide synthase (NOS) isoform activities, both calcium dependent and independent NOS, in experimental ischemic ARF. We have also analyzed change activities to activity of the NOS activities in both renal cortex and medulla. Male SD rats (n = 5) were inducted to ARF by ischemia-reperfusion injury and divided into the following groups; Control group (sham operation), Day 0 group, (measurement performed on that day of operation), Day 1 group, (measurement performed one day after induction of ARF), Day 3 group and Day 7 group. Measurement of NOS activity was based on the following principles; NO is synthesized from arginine by nitric oxide synthase (NOS) and NO is converted to $NO_2^-/NO_3^-$ (NOx) by oxidation. Detection of the final metabolite of NO, NOx was done using flow injection method (Griess reaction). The results were, (1) calcium dependent NOS activity in the cortex and medulla decreased, however it increased in the recovery period in the renal cortex (Cortex; Control, 0.941 ± 0.765, D0, 0.382 ± 0.271, D1, 0.118 ± 0.353, D3, 2.030 ± 0.235, D7, 3.588 ± 2.706, Medulla; Control, 1.469 ± 0.531, D0, 0.766 ± 0.156, D1, 0.828 ± 0.187, D3, 2.078 ± 0.094, D7, 1.289 ± 0.313 μmol NOx produced/mg protein/30 min). (2) On the other hand, iNOS activity increased in the early phase of ARF, both in the cortex and medulla, but returned to control values during the recovery phase in cortex and was maintained at higher levels in the medulla (Cortex; Control, 0.333 ± 0.250, D0, 0.583 ± 0.428, D1, 1.167 ± 0.262, D3, 0.250 ± 0.077, D7, 0.452 ± 0.292, Medulla; Control, 0.139 ± 0.169, D0, 0.279 ± 0.070, D1, 1.140 ± 0.226, D3, 0.452 ± 0.048, D7, 0.625 ± 0.048 μmol NOx produced/mg protein/30 min). These findings suggest that the role of NOS in ARF are different for the different NOS isoforms and have anatomic heterogeneity. (Mol Cell Biochem **244**: 129–133, 2003)

*Key words*: nitric oxide synthase, acute renal failure

## Introduction

There has been a growing body of evidence for the roles of reactive oxygen species (ROS) and reactive nitrogen species (nitric oxide; NO) in the pathogenesis of a variety of renal diseases including acute renal failure (ARF) [1–4]. It is well known that inhibitors of NO synthase (NOS) exacerbate renal blood flow (RBF) or glomerular filtration rate (GRF) and that these effects are protected by arginine administration [5, 6]. On the other hand, blocking NO production through inducible NOS (iNOS) inhibition using anti-sense oligonucleotide or specific inhibitors, reduces tubular dysfunction or injury [7, 8]. These dissimilar effects of NO depend upon the site or rate of NO production, timing of the activation of NOS isoforms, metabolites of NO (for example peroxynitrite formation) and their reactivity with ROS or metals. Therefore, time course changes of NOS activity will provide us with some important information needed to clarify the pathophysiology in ARF.

*Address for offprints*: T. Maeba, St. Marianna University School of Medicine, Department of Internal Medicine, Division of Nephrology and Hypertension, 2-16-1, Sugao, Miyamae-ku, Kawasaki, 216-8511, Japan (E-mail: t2maeba@marianna-u.ac.jp)

This study was undertaken to elucidate these time course changes of NOS isoform activities, both calcium dependent and independent NOS, in ARF. We have analyzed the NOS activities phase in both renal cortex and medulla from ischemic ARF, and the results were, (1) calcium dependent NOS activity in the cortex and medulla decreased; however it increased in the recovery period in the renal cortex. (2) On the other hand, iNOS activity increased in early phase of ARF, both in the cortex and medulla but returned to control values during the recovery phase in cortex and was maintained at higher levels in the medulla. These findings suggest that the roles of NOS in ARF are different for the different NOS isoforms and have anatomic heterogeneity.

## Materials and methods

*Animals*

Male Sprague-Dawley rats (10-weeks-old) weighing around 250 g were used. Rats were anesthetized with an intraperitoneal injection of pentobarbital (40 mg/100 g) and both kidneys were exposed posteriorly by back incision. After resection of the right kidney, ARF was induced in the rats by ischemia (45 min, lt. renal artery clamp) – reperfusion injury. They were divided into 5 groups to evaluate the time course change in NOS activity.

The groups were as follows (n = 5); Control group (sham operation), Day 0 group, (measurement performed on that day of operation, that is one hour after reperfusion), Day 1 group, (measurement performed one day after induction of ARF), Day 3 group and Day 7 group. After induction of ARF, in each rat was housed individually in a metabolic cage and urine was collected daily. The rats had free access to food and water.

All experiments were performed according to the 'Guiding Principles for the Care and Use of Laboratory Animals' of the Japanese Pharmacological Society.

*Chemicals*

NADPH, tetrahydrobiopterin(BH4), leupeptin, pepstatin A were purchased from Sigma (St. Louis, MO, USA), Calmodulin, L-arginine, dithiothreitol (DTT) were from WAKO (Osaka, Japan) and other chemicals were reagent grade.

*Measurement of nitric oxide synthase activity (NOS)*

*Preparation of enzyme solution*
Figure 1 shows the preparation method for the enzyme solution. While under anesthesia, blood was drawn through a 23 gauge plastic catheter which was placed into the abdominal aorta. The kidney was perfused with 10 ml of ice cold homogenize buffer A solution. The excised kidney was immediately dissected on ice to isolate the cortex and medulla. This was frozen using liquid nitrogen. After adding 5 volumes of homogenization buffer A to frozen sample, the kidney was homogenized with an ultrasonic homogenizer for 30 sec. The homogenized solution was filtered through a nylon mesh and centrifuged at 50 × g for 7 min. The same volume of homogenization buffer B was placed on the supernatant and 3 centrifugation steps were performed. Then 24,000 × g supernatant was passed through Dowex 50W-X8(Na) column to remove intrinsic arginine. This solution was used as a crude enzyme sample. Components of the homogenize buffer A and B are shown in Table 2.

*Measurement of NOS activity Table 1*
NO is synthesized from arginine by nitric oxide synthase (NOS) and NO is converted to $NO_2^-/NO_3^-$ (NOx) by oxidation. Yokoi *et al.* [9] developed the method for detecting NOS activity without using an isotope. The principle is based on the detection of NOx, the final metabolite of NO, using a flow injection method (Griess reaction). The measurement of NOx was performed using the flow injection system (TCI NOx 1000 m; Tokyo Kasei Kogyou Co., Tokyo) [10]. The sensitivity is above 0.5 μM and CV is within 5% in this apparatus.

First, we checked whether this reaction was depended on protein concentration and reaction time or not. There were 3 protein concentrations used for the enzyme solutions at; 15, 30 and 60 min for time course experiments. The amount of NOx production was increased linearly in proportion to both the protein concentration and the time course (data not shown). Enzyme activity was expressed as the amount of μmol NOx produced/mg protein/30 min (specific activity).

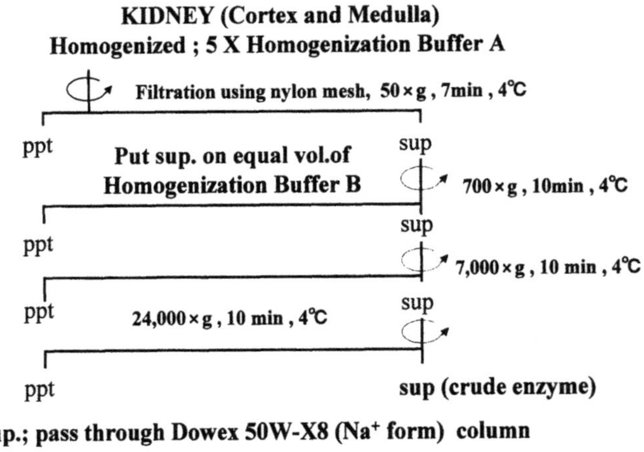


*Fig. 1.* Sample preparation for NOS activity; excised kidney was immediately weighed, dissected and frozen. All procedures were done under cooling condition.

*Table 1.* Method for NOS activity determination

| | Blank | Total NOS | iNOS |
|---|---|---|---|
| 50 mM Tris-HCl (pH 7.4) | + | + | + |
| 1 mM $CaCl_2$ | + | + | – |
| 1 mM NADPH | – | + | + |
| 5 µg/ml Calmodulin | + | + | – |
| 1 µM $BH_4$ | + | + | – |
| 1 mM EGTA | – | – | + |
| 1 mM L-arginine | – | + | + |

Pre-incubate at 37°C, for 20 min. Add enzyme solution (6–8 mg/ml protein). Incubate at 37°C, for 30 min. Stop reaction add NaOH and $ZnSO_4$ (0.09 N and 1.5%). Centrifuge at 3,000 rpm, RT. NOx concentration was measured by Greiss method.

*Table 2.* Components of homogenization buffer

| |
|---|
| 50 mM Tris-HCl pH 7.4 |
| 0.1 mM EDTA-2Na |
| 0.1 mM EGTA |
| 1 mM PMSF |
| 1 mM DTT |
| 10 µg/ml Leupeptin |
| 1 µM Pepstatin |
| Buffer A: 0.25 M sucrose |
| Buffer B: 0.34 M sucrose |

*Measurements of laboratory data*

Creatinine measurement in the urine and plasma, blood urea nitrogen and protein determination were performed using standard laboratory kits (Creatinine WAKO Test, BUN WAKO Test and Bio Rad Protein Assay Kit). Creatinine clearance was calculated from plasma creatinine content and 24 h urine creatinine concentration, and expressed as ml/min/00 g of body weight.

*Statistical analysis*

Data are expressed as the means ± S.D. Data were analyzed by one-way analysis of variance in combination with Fisher's protected least significant difference and p values of less than 0.05 were regarded as significant.

# Results

*Body weight and renal function*

There were significant differences of body weight found in control and ARF groups. Plasma creatinine (Cr) and blood urea nitrogen (BUN) were significantly increased and 24 h creatinine clearance (Ccr) was decreased in the ARF group (Table 3). The maximum decline of renal function was around 1 day after induction of ARF and Ccr increased but did not return to control values 7 days after induction of ARF.

*Kidney NOS activities*

*Renal cortex (Fig. 2)*

Calcium dependent NOS, mainly e-NOS, activity significantly decreased from 1 h after reperfusion and showed minimal values 1 day after induction of ARF. However, e-NOS activities increased significantly compared to control value from day 3 to day 7 after induction of ARF. Conversely, calcium independent NOS, i-NOS, activity significantly increased after 1 h of reperfusion and reached the maximum value 1 day after induction of ARF. Three and 7 days after induction of ARF, they had returned to control values (e-NOS activity; Control, 0.941 ± 0.765, D0, 0.382 ± 0.271, D1, 0.118 ± 0.353, D3, 2.030 ± 0.235, D7, 3.588 ± 2.706, i-NOS activity; Control, 0.333 ± 0.250, D0, 0.583 ± 0.428, D1, 1.167 ± 0.262, D3, 0.250 ± 0.077, D7, 0.452 ± 0.292).

*Table 3.* Laboratory data

| | BW (g) | s-Cr (mg/dl) | BUN (mg/dl) | Ccr (ml/min·100 g BW) |
|---|---|---|---|---|
| Control | 260 ± 17 | 0.40 ± 0.09 | 17.6 ± 3.6 | 4.42 ± 1.40 |
| ARF Day 0 | 243 ± 20* | 0.72 ± 0.13 | 24.5 ± 4.2*** | – |
| Day 1 | 219 ± 21*** | 1.42 ± 0.77*** | 62.5 ± 35.0*** | 0.84 ± 0.66*** |
| Day 3 | 233 ± 18*** | 1.38 ± 1.39** | 68.9 ± 54.7 | 1.78 ± 0.92*** |
| Day 7 | 225 ± 23*** | 0.69 ± 0.34 | 43.3 ± 30.2 | 2.84 ± 1.32** |

Values are mean ± S.D. *$p < 0.05$; **$p < 0.01$; ***$p < 0.001$ vs. control.

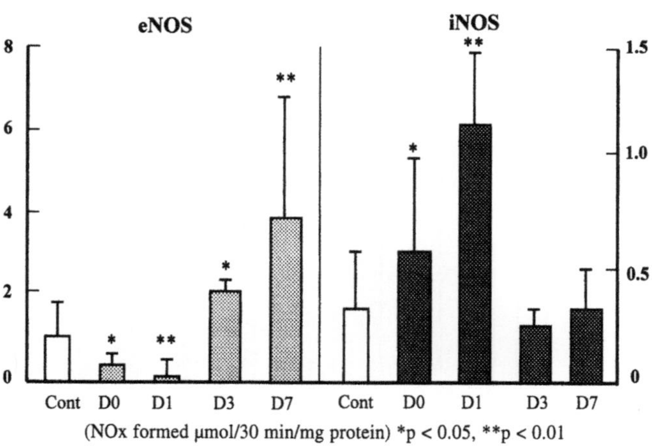


*Fig. 2.* NOS activity in ARF rat kidney cortex; eNOS and iNOS activities were different in time course of ARF.

*Renal medulla (Fig. 3)*
e-NOS activity in medulla decreased significantly after the early phase of ARF similar to the cortex and this inhibitory effect continued for 1 day after induction of ARF, but there was no significant increase in eNOS activity compared to the control value during the following days. Distinct from eNOS activity in cortex, eNOS in the medulla did not exceed control the values. i-NOS activity, however, increased 1 day after induction of ARF, but did not return to control values, even 7 days after induction of ARF (e-NOS activity; Control, 1.469 ± 0.531, D0, 0.766 ± 0.156, D1, 0.828 ± 0.187, D3, 2.078 ± 0.094, D7, 1.289 ± 0.313, i-NOS activity; Control, 0.139 ± 0.169, D0, 0.279 ± 0.070, D1, 1.140 ± 0.226, D3, 0.452 ± 0.048, D7, 0.625 ± 0.048).

## Discussion

The kidney is one of the most active organs metabolically. In fact renal blood flow is up to 20% of cardiac out-put and the oxygen consumption rate is extremely high. Therefore reduction of blood flow to the kidney leads to functional and morphological changes. Acute renal failure is defined as an abrupt decline of renal function caused by many intrinsic or extrinsic factors. One of these causes is reduction of blood supply to the kidney.

In the clinical setting, the mortality rate for ARF patients is still over 50% in spite of the progress for dialysis therapy. The reason for this is based on the fact that the affected organ is not only the kidney but also lung and liver, i.e. pulmonary distress syndrome and/or liver dysfunction, are involved. To elucidate the mechanisms of ARF is an important issue for the development of strategies for treating this syndrome.

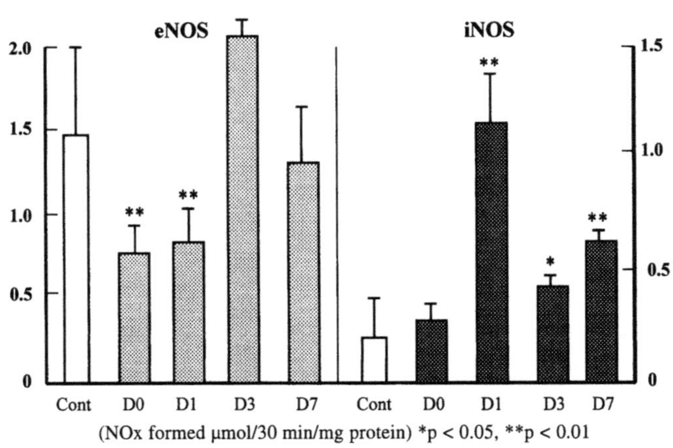


*Fig. 3.* NOS activity in ARF rat kidney medulla; eNOS and iNOS activities were different in time course of ARF.

There has been evidence accumulating for a role of reactive oxygen species (ROS) and reactive nitrogen species (i.e. nitric oxide; NO) in the pathogenesis of ARF. However, there are controversial reports of pathogenetic roles of ROS and NO in experimental ischemic ARF. ROS and NO are produced in ARF, and many antioxidants were used to treat in ARF model [11], but the effects of these compounds on ARF are not sufficient at present for therapies [12] because the organ dysfunction is caused by ROS which is the result of multiple reactions. In the intact kidney, ROS is the result of complex reactions. During normal kidney function, superoxide anion radical is scavenged by superoxide dismutase (SOD) and forms hydrogen-peroxide. Then hydrogen-peroxide is metabolized to water by catalase or peroxidases. When there is over production of superoxide anion radical and/or dysfunction of SOD, catalase and peroxidases, hydroxyl radical, one of the more reactive radicals, is formed from the superoxide anion radical and/or hydrogen-peroxide by reaction with transition metals. SOD, catalase, iron chelators and some kinds of radical scavengers have been used in experimental models of ARF in animals [13, 14], but these antioxidants effects did not completely normalize the condition. NO is also formed in ischemia-reperfusion injury and this small molecule is converted to peroxynitrite by reacting with the superoxide anion radical. Peroxynitrite is a strong oxidative substance and produces hydroxyl radicals [15]. NO is produced from each of the isoforms of NOS, NO has an important physiological roles in the regulation of blood flow and Na balance via the Na-K ATPase in the kidney [16, 17]. On the other hand, NO has a cytotoxic effect on renal tubular cells [18]. Such biphasic effects of NO create an extremely complex pathophysiological situation in the kidney. Therefore, NO may have distinctly different roles in different renal diseases.

These dissimilar effects of NO depend upon the concentration of NO, the integrity of NO protective cell types, the timing of activation of NOS isoforms, metabolites of NO (for example peroxynitrite formation) and reactivity with ROS or metals. There are no published reports concerning the time course changes of NOS activity in experimental models of ARF as far as we have surveyed. Therefore, time course changes of NOS activity will provide us with some important information to clarify the pathophysiology of ARF.

This study was undertaken to elucidate some of the time course changes of the NOS isoforms, both calcium dependent (e) and independent (i) NOS, in ARF. We have analyzed the NOS activities in both renal cortex and medulla from ischemic ARF, and the results were, (1) eNOS activity in the cortex and medulla decreased, however it increased significantly during the recovery period in the renal cortex. (2) On the other hand, iNOS activity increased in the early phase of ARF both in the cortex and medulla but in the cortex it returned to control values during the recovery phase. However, iNOS in the medulla is still significantly higher than that of the controls during the recovery phase. These findings suggest that the roles of NOS in ARF are different for different types of NOS and also depend on the time course of ARF.

In conclusion, it is not an effective protocol to manage only NO production using inhibitors or stimulators on experimental models of ARF because of the correlation of ROS and NO is a multifactorial system. Further studies will be needed to evaluate the combination of effects for the antioxidants and NO regulators on ARF. Such studies could provide us some important information regarding the future management strategies of human ARF.

## References


1. Shah SV: Role of reactive oxygen metabolism in experimental glomerular disease. Kidney Int 35: 1093–1106, 1989
2. Diamond JR: The role of reactive oxygen species in animal model of glomerular disease. Am J Kidney Dis 19: 292–300, 1992
3. Kone BC: Nitric oxide in renal health and disease. Am J Kidney Dis 30: 311–333, 1997
4. Lieberthal W: Biology of ischemic and toxic renal tubular cell injury: Role of nitric oxide and the inflammatory response. Curr Opin Nephrol Hypertens 7: 289–295, 1998
5. Schramm L, Heidbreder E, Schmitt A, Kartenbender K, Zimmermann J, Ling H, Heidland A: Role of L-arginine derived NO in ischemic acute renal failure in the rat. Renal Failure 16: 555–569, 1994
6. Jerkic M, Varagic J, Javovic D, Radujkovic-Kuburovic G, Nastic-Miric D, Adanja-Grujic G, Markovic-Lipkovski J, Dimitrijevic J, Miloradovic Z, Vojvodic SB: L-arginine reduces tubular cell injury in acute post-ischemic renal failure. Nephrol Dial Transplant 14: 1398–1407, 1999
7. Peresleni T, Noiri E, Bahou W, Goligorsky M: Antisense oligodeoxynucleotides to inducible NO synthase rescue epithelial cells from oxidative stress injury. Am J Physiol 270: F971–F977, 1996
8. Noiri E, Peresleni T, Miller F, Goligorsky MSL: *In-vivo* targeting of inducible NO synthase with oligodeoxynucleotides protects rat kidney against ischemia. J Clin Invest 97: 2377–2383, 1996
9. Yokoi I, Habu H, Kabuto H, Mori A: Analysis of nitrite, nitrate, and nitric oxide synthase activity in brain tissue by automated flow injection technique. Meth Enzymol 268: 152–159, 1996
10. Higuchi K, Motomizu S: Flow-injection spectrophotometric determination of nitrite and nitrate in biological samples. Anal Sci 15: 129–134, 1999
11. Leach M, Frank S, Olbrich A, Pfeilschiter J, Thiemermann C: Decline in the expression of copper/zinc superoxide dismutase in the kidney of rats with endotoxic shock: Effects of the superoxide anion radical scavenger, tempol, on organ injury. Br J Pharm 125: 817–825, 1998
12. Zager RA: Hypoperfusion-induced acute renal failure in the rat: An evaluation of oxidant tissue injury. Circ Res 62: 430–435, 1988
13. Zager RA, Fuerstenberg SM, Baehr PH, Myerson D, Torok-Storb B: An evaluation of antioxidant effects on recovery from postischemic acute renal failure. J Am Soc Nephrol 4: 1588–1597, 1994
14. Fischereder M, Trick W, Nath KA: Therapeutic strategies in the prevention of acute renal failure. Semin Nephrol 14: 41–52, 1994
15. Beckman JS, Beckman TW, Chen J, Marshall PA, Freeman BA: Apparent hydroxyl radical production by peroxynitrite – implications for endothelial injury from nitric oxide and superoxide. Proc Acad Sci USA 87: 1620–1624, 1990
16. McKee M, Scavone C, Nathanson JA: Nitric oxide, cGMP and hormone regulation of active sodium transport. Proc Natl Acad Sci USA 91: 12056–12060, 1994
17. Guzman NJ, Fang MZ, Tang SS, Ingelginger JR, Garg LC: Autocrine inhibition of Na/K-ATPase by nitric oxide in mouse proximal tubule epithelial cells. J Clin Invest 95: 2083–2088, 1995
18. Yaqoob M, Edelstein C, Weider E, Alkhunaizi A, Gengaro P, Nemenoff R, Schrier R: Nitric oxide kinetics during hypoxia in proximal tubules: Effects of acidosis and glycine. Kidney Int 49: 1314–1319, 1996

*Molecular and Cellular Biochemistry* **244:** 135–138, 2003.


# Alteration of energy production by the heart in CRF patients undergoing peritoneal dialysis

Goichi Ogimoto,[1] Tsutomu Sakurada,[1] Keiko Imamura,[2] Shingo Kuboshima,[1] Teruhiko Maeba,[1] Kenjiro Kimura[1] and Shigeru Owada[3]

[1]*Division of Nephrology and Hypertension, Department of Medicine;* [2]*Department of Radiology, St. Marianna University School of Medicine, Kawasaki;* [3]*Department of Internal Medicine, Akabane Central Hospital, Tokyo, Japan*

## Abstract

Cardiovascular disease is commonly observed in patients with chronic renal failure and this is a leading cause of death in patients with end-stage renal disease undergoing maintenance dialysis. Myocardial energy production is a very crucial aspect of cardiac function. Therefore, to evaluate energy metabolism of myocardial muscle in peritoneal dialysis (PD) patients, we carried out the following study using Magnetic resonance spectroscopy (MRS).

Fourteen chronic renal failure patients and eight healthy volunteers were enrolled. The ratio of the phosphocreatine peak to the beta-phosphate to ATP peak (PCr/β-ATP) was calculated from their MR spectra obtained by $^{31}$P-MR spectroscopy (Gyroscan S15, Philips). To determine the correlation between cardiac function and energy status, the left atrial diameter, the left ventricular (LV) end-diastolic diameter, the ejection fraction, the fraction of shortening and the LV mass index were measured by echocardiography. Peripheral blood sampling was also performed for creatinine, blood urea nitrogen, hematocrit, hemoglobin, $\beta_2$-microglobuline, intact parathyroid hormone.

PCr/β-ATP was significantly lower in PD (1.03 ± 0.15 vs. 1.40 ± 0.18: $p = 0.0002$), although all patients showed normal systolic function. No correlation was found between PCr/β-ATP and cardiac function or hematological or biochemical markers. A negative correlation was present between PCr/β-ATP and dialysis duration ($r = 0.57$, $p < 0.05$).

Altered energy status of the myocardium in PD should be considered even if the patients did not show any systolic dysfunction. $^{31}$P-MRS is a useful tool to evaluate the energy status of the myocardium. (Mol Cell Biochem **244**: 135–138, 2003)

*Key words*: energy metabolism, chronic renal failure, myocardium, magnetic resonance spectroscopy

## Introduction

Cardiovascular disease is frequently recognized in patients with chronic renal failure and is a leading cause of death in patients with end-stage renal diseases [1]. The incidence of heart failure as a cause of death was 23% in the Japanese dialysis population [2] at the end of 2000, despite the appearance of recombinant human erythropoietin which has made renal anemia less common. Recent advances in dialysis therapy resulted in an increased longevity for uremic patients but dialysis related complications such as atherosclerosis [3], amyloidosis [4] and muscle weakness [5, 6] affect the quality of life [1].

Cardiac dysfunction has been reported frequently in chronic renal failure, and several pathologies have been hypothesized as follows; carnitine deficiency [7], anemia, arteriovenous fistula, chronic volume overload [8], hypertension, reduced vascular compliance [9] and uremic toxins [1, 10].

However, little is known regarding the role of biochemical alterations on cardiac energy status in dialysis patients. Magnetic resonance spectroscopy (MRS) is an excellent way to evaluate energy status non-invasively [11]. Therefore, to evaluate energy metabolism in myocardial muscle of dialysis patients, we carried out this study, specifically on peritoneal dialysis (PD) subjects, using MRS.

*Address for offprints*: G. Ogimoto, St. Marianna University School of Medicine, Department of Medicine, Division of Nephrology and Hypertension, 2-16-1 Sugao Miyamae-ku, Kawasaki, 216-8511, Japan (E-mail: g2ogi@marianna-u.ac.jp)

## Materials and methods

### *Subjects*

Fourteen patients undergoing PD who provided informed consent to participate in this study were examined. There were 5 men and 9 women. The mean age was 49.5 ± 11.7 years. The mean duration of dialysis at entry to this study was 4.6 ± 3.5 years. The etiology of ESRD were glomerulonephritis in 10 patients, diabetic nephropathy in 2, Alport syndrome in 1 and systemic lupus erythematosis in 1. No patients had any history of heart failure or ischemic heart disease. The control group consisted of 8 healthy volunteers (3 males and 5 females). The mean age of the volunteers was 33.4 ± 8.5 years.

### *Spectroscopy*

A $^{31}$P-NMR signal was acquired using a surface coil with a14 cm diameter that was turned to 25.89 MHz for phosphorus. Conventional proton NMR images were obtained to confirm the location of the volume of interest (VOI).

VOI selected extended from the anterior myocardial wall to the ventricular septum. Data were collected from the myocardium at rest. Measurements were carried out as follows: Image selected *in vivo* spectroscopy (ISIS), repetition time (TR) 1200–1700 msec (actual TR 1400–2600 msec), echo time 136 msec, 300 msec delay from R wave for pulse. Sampling point: 1024 point, sampling number: 1024 times. To obtain MR spectra, a baseline correction was made after Fourier transformation. There were four well defined peaks identified on a representative spectrum reading from the left as follows: phosphocreatine (PCr), gamma- phosphate peak to ATP ($\gamma$-ATP), alpha-phosphate peak to ATP ($\alpha$-ATP) and beta-phosphate peak to ATP ($\beta$-ATP). Because of the short repetition time in this study, both phosphocreatine and ATP could not be fully saturated. To exclude the affect of short TR, we have computed the estimated signal intensities of PCr and $\beta$-ATP signals by a saturation correction according to the $T_1$ values of PCr of 4.28, and $\beta$-ATP of 2.99 that were reported by van Dobbenburgh [12].

### *Echocardiography and blood sampling*

To determine the correlation between cardiac function and energy status, the left atrial diameter, the left ventricular (LV) end-diastolic diameter, ejection fraction, fraction of shortening and LV mass index were measured in the PD patients by echocardiography. Serum was obtained from both groups for serum creatinine (Cr), urea nitrogen (BUN), $\beta_2$-microglobuline ($\beta_2$-MG), intact parathyroid hormone (intact-PTH). Hematocrit and hemoglobin were also measured in both groups.

### *Statistical analysis*

Values are presented as the mean ± S.D. Comparison of data from controls and patient's was performed by a Mann-Whitney U test. Correlation between the PCr to ATP ratios and echocardiograms or blood data were analyzed by the Spearman rank test. Statistical significance was defined as $p < 0.05$.

## Results

Hematological and biochemical parameters are listed on Table 1. In the PD patients group, Cr, BUN and $\beta_2$-MG were significantly increased over controls reflecting their renal insufficiency. Hct and Hgb levels in the CRF group are lower than controls.

The echocardiographic data from the PD patients showed prominent left ventricular (LV) hypertrophy (LV mass index; 210.1 g/m$^2$) and minor dilatation of the LV (LV diameter; 50.8 mm). However, the ejection fraction (73%) and fractional shortening (31%) were maintained within normal limits.

In Fig. 1, a representative MR spectrum obtained from a PD patient is shown. Reading from the left, the peaks of PCr, $\gamma$-ATP, $\alpha$-ATP, $\beta$-ATP are identified. The ratio of PCr to $\beta$-ATP (PCr/ATP) was derived from the integrated areas of under the peaks resonance of PCr and $\beta$-ATP.

PCr/ATP was markedly reduced in PD patients (PD 1.03 ± 0.15 vs. cont 1.40 ± 0.18, p = 0.0002) which suggests an alteration of energy status in chronic renal failure.

*Table 1.* Characteristics of biochemical parameters in blood obtained from peritoneal dialysis patients (PD) and control group (cont).

| | n | Cr | BUN | Hct | Hgb | $\beta_2$-MG | Intact-PTH |
|---|---|---|---|---|---|---|---|
| | | mg/dl | | % | g/dl | mg/l | pg/ml |
| PD | 11 | 10.7 ± 2.1 | 57.3 ± 14.0 | 25.3 ± 3.6 | 8.4 ± 1.3 | 34.0 ± 8.2 | 216.5 ± 253.9 |
| Cont | 5 | 0.8 ± 0.1 | 12.4 ± 2.9 | 36.7 ± 3.0 | 12.4 ± 1.3 | 1.2 ± 0.1 | |

Cr – creatinine; BUN – blood urea nitrogen; Hct – hematocrit; Hb – hemoglobin; $\beta_2$-MG – $\beta_2$-microglobulin; intact-PTH – intact-parathyroid hormone.

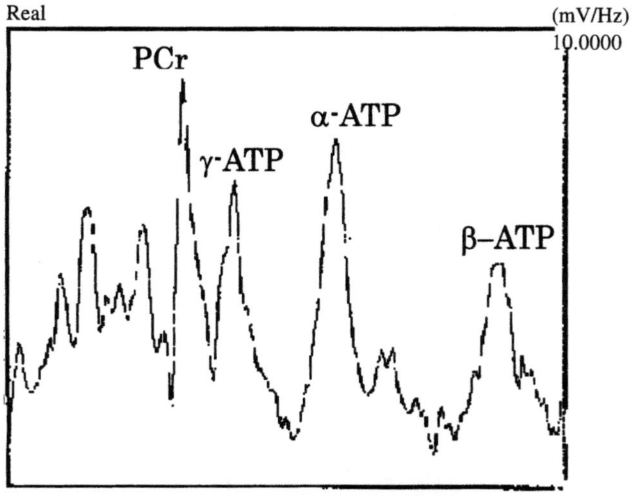


??Unable to read text around figure - please supply new figure??

*Fig. 1.* Representative MR spectrum obtained from a PD subject. Four well-defined peaks identified on this figure. Phosphocreatine (PCr), gamma-phosphate peak to ATP (γ-ATP), alpha-phosphate peak to ATP (α-ATP) and beta-phosphate peak to ATP (β-ATP).

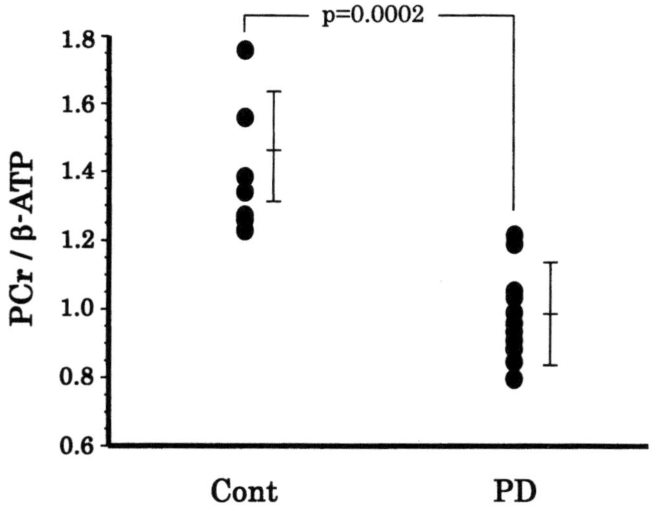


*Fig. 2.* The myocardial PCr/β-ATP obtained from control group (Cont) and peritoneal dialysis patients (PD) by $^{31}$phosphorus magnetic resonance spectroscopy. PCr – phosphocreatine; β-ATP – beta-phosphate peak to ATP.

Moreover, comparison of PCr/ATP to the duration of dialysis showed a statistically significant negative correlation ($p < 0.05$; Fig. 2). However, no correlation was found between PCr/ATP and the hematological or biochemical data in this study.

In Table 2, a case is presented that showed a dramatic amelioration in myocardial energy status by correction of renal anemia. This patient's severe renal anemia (Hgb 6.6 g/dl) had been documented at the beginning of erythropoietin (EPO) therapy of 6000 U per week. Following EPO therapy, the patient's hemoglobin level increased dramatically to around 12 g/dl where it remained 6 months. After treatment, the patient's PCr/ATP reached nearly normal despite an initial PCr/ATP which was 0.91 at the beginning of the therapy.

## Discussion

Left ventricular hypertrophy is the most frequent cardiac alteration in end-stage renal disease (ESRD) [13]. While LV hypertrophy exist in ESRD patients, systolic function has been found mostly normal [14] and this is confirmed by our data. Assessment of left ventricular metabolism and function is very important for patients on maintenance dialysis because

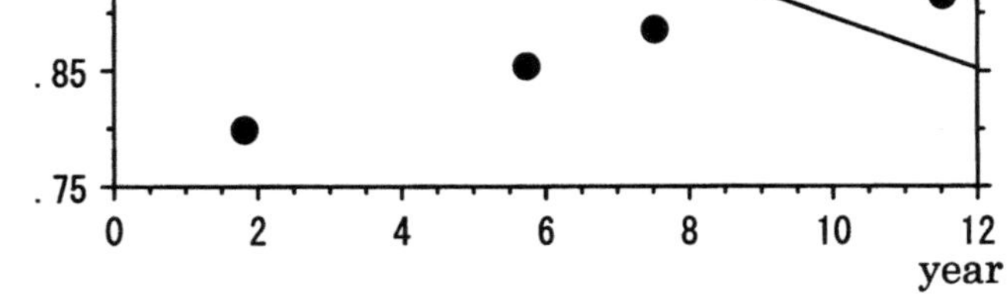


*Fig. 3.* Negative correlation between PCr/β-ATP and the duration of dialysis therapy in the patients undergoing peritoneal dialysis. PCr – phosphocreatine; β-ATP – beta-phosphate peak to ATP.

*Table 2.* A case showed dramatic amelioration in myocardial energy status by correction of renal anemia. PCr/β-ATP increased to almost normal range without any change in dialysis quantity of the patient

| | PCr/β-ATP | Cr | BUN | Hct | Hgb | $\beta_2$-MG | Intact-PTH |
|---|---|---|---|---|---|---|---|
| | | mg/dl | | % | g/dl | mg/l | pg/ml |
| 0 month | 0.91 | 13.3 | 37 | 21 | 6.6 | 48/9 | 182.8 |
| 10 month | 1.33 | 11.8 | 39 | 38 | 12.0 | 44.4 | 246.4 |

PCr – phophocreatine; β-ATP – beta-phospate peak to ATP; Cr – creatinine; BUN – blood urea nitrogen; Hct – hematocrit; Hgb – hemoglobin; $\beta_2$-Mg – $\beta_2$-microglobulin; intact-PTH – intact-parathyroid hormone.

congestive heart failure occurs quite often leading to a tragic outcome.

MRS provides intracellular information non-invasively from living tissue. Additionally, we are able to learn the content of high-energy phosphate in living tissue using $^{31}$P-MRS. PCr/ATP is useful as an excellent marker of the energy status of the heart [15, 16]. Weiss *et al.* [16] reported a decrease in the PCr/ATP during handgrip exercises in patients with coronary heart disease that reflects a transient imbalance between oxygen supply and demand in the myocardium.

In several reports on left ventricular hypertrophy in patients without heart failure [17, 18], there was no significant decrease in the myocardial PCr/β-ATP ratio. Hardy *et al.* [19] found an altered PCr/β-ATP ratio in the patients with dilated cardiomyopathy. All of the patients studied had symptoms of congestive heart failure. Although there were no symptoms of heart failure in our cases, almost all the patients had myocardial hypertrophy and a reduced myocardial PCr/β-ATP ratio. Additionally, no correlation was found between the ratio of PCr/β-ATP ratio and hematological or echocardiographic parameters on this study.

The case presented in Table 2 showed a dramatic amelioration of the ratio of PCr/β-ATP through correction of her renal anemia using erythropoietin. Erythropoietin has been shown to promote angiogenesis and stimulate endothelial and vascular smooth muscle cell proliferation [20]. Thompson *et al.* [21] reported a bioenergetic effect of erythropoietin in skeletal muscle by correction of renal anemia but they also mentioned other possibilities such as a deficiency in carnitine and an increase in peripheral vascular resistance as causes of abnormal energy production. Carnitine is a vitamin-like compound synthesized mainly in mammalian liver, brain and kidney starting from lysine and methionine [22]. It plays an important role in fatty acid transport into mitochondria before its oxidation to $H_2O$ and $CO_2$.

Therefore we must consider the direct effect of increased oxygen supply and angiogenesis in evaluating the effect of erythropoietin therapy in this case. Additionally, the negative correlation between the PD duration and the PCr/β-ATP ratio ($p < 0.05$, $r = 0.57$) might support a role for carnitine deficiency and/or diminished fiber density of the myocardium in the gradual deterioration of energy production in long-term dialysis patients.

MR spectroscopy is useful to estimate energy status directly and provide precise information about improvement in energy status in the target organ.

# References


1. Horl WH, Riegel W: Cardiac depressant factors in renal disease. Circulation 87(supple IV): IV77–IV82, 1993
2. Akiba T (ed).: An Overview of Regular Dialysis Treatment in Japan, 2001, p 89
3. Kawaguchi T, Nishizawa Y, Konishi T *et al.*: High-resolution B-mode ultrasonography in evaluation of atherosclerosis in uremia. Kidney Int 48: 820–826, 1995
4. Gejo F, Yamada T, Odani S, Nakagawa Y *et al.*: A new form of amyloid protein associated with chronic hemodialysis was identified as β2-microglobuli. Biophys Res Commun 129: 701–706, 1985
5. Isaacs H: Electromyographic study of muscle weakness in chronic renal failure. S A Med J: 683–688, 1969
6. Brautbar N: Skeletal myopathy in uremia: Abnormal energy metabolism. Kid Int 24: S81–S86, 1983
7. Van Es A, Henny FC, Kooista MP, Lobatto S, Scholtte HR: Amelioration of cardiac function by L-carnitine administration in patients on hemodialysis. Contrib Nephrol 92: 28–35, 1992
8. Kramer W, Wizemann V, Lämmlein, Thormann J, Kindler M, Schlepper M, Schütterle G: Cardiac dysfunction in patients on maintenance hemodialysis. Contrib Nephrol 52: 110–124, 1986
9. Kooman JP, Wijnen JAG, Draaijer P, van Bortel L, Gladziwa U, Struyker-Bboudier HAJ, Peltenburg HG, van Hooff JP, Leunissen KML: Compliance and reactivity of the peripheral venous system in patients treated with chronic intermittent hemodialysis. Kidney Int 41: 1041–1048, 1992
10. Ritz E, Rambausek M, Mall G, Ruffman K, Mandelbaum A: Cardiac changes to uremia and their possible relation to cardiovascular instability on dialysis. Contrib Nephrol 78: 221–229, 1990
11. Ogimoto G: Energy metabolism of skeletal muscle on chronic renal failure. St. Marianna Med J 20: 540–549, 1992
12. Van Dobbenburgh JO, Lekkerkerk C, van Echteld JA: Saturation correlation in human cardiac $^{31}$P MR spectroscopy at 1.5T. *NMR* Biomed 7: 218–224, 1994
13. London GM, Guefin AP, Pannier B, Safar ME, Dat M, Metivier F: Cardiac and arterial interactions in end-stage renal disease. Kidney Int 50: 600–608, 1996
14. Hüting J, Kramer W, Reitinger J, Kühn K, Wizemann V, Schütterle G: Cardiac structure and function in continuous ambulatory peritoneal dialysis: Influence of blood purification and hypercirculation. Am Heart J 119: 344–352, 1990
15. Tagami T, Sakuma H, Matsumura K, Takeda K, Mori S, Takeuchi T, Nakano T: Evaluation of altered myocardial high energy phosphate metabolism in patients on maintenance dialysis using phosphorus-31 magnetic resonance spectroscopy. Invest Radiol 33: 171–176, 1998
16. Weiss RG, Bottomley PA, Hardy CJ, Gerstenblith G: Regional myocardial metabolism of high-energy phosphates during isometric exercise in patients with coronary artery disease. N Engl J Med 323: 1593–1600, 1990
17. Schaefer S, Gober JR, Schwartz GG, Twieg DB, Weiner MW, Massie B: *In vivo* phosphorus-31 spectroscopic imaging in patients with global myocardial disease. Am J Cardiol 65: 1154–1161, 1990
18. Masuda Y, Tateno Y, Ikehira H *et al.*: High-energy phosphate metabolism of the myocardium in normal subjects and patients with various cardiomyopathies: The study using ECG gated MR spectroscopy with a localization technique. Jpn Circ J 56: 620–626, 1992
19. Hardy CJ, Weiss RG, Bottomley PA, Gerstenblith G: Altered myocardial high-energy phosphate metabolites in patients with dilated cardiomyopathy. Am Heart J 122: 795–801, 1991
20. Vaziri ND: Cardiovascular effects of erythropoietin and anemia correction. Curr Opin Nephrol Hypertens 10: 633–637, 2001
21. Thompson CH, Kemp GJ, Taylor DJ, Radda GK: Bioenergetic effects of erythropoietin in skeletal muscle. Nephron 74: 239–240, 1996
22. Bremer J: Carnitine: Metabolism and functions. Physiol Rev 63: 1420–1480, 1983

*Molecular and Cellular Biochemistry* **244:** 139–141, 2003.


# Oral L-arginine can reverse digital necrosis in Raynaud's phenomenon

Christopher M. Rembold and Carlos R. Ayers
*Cardiovascular Division, Departments of Internal Medicine and Physiology, University of Virginia Health System, Charlottesville, VA, USA*

## Abstract

Raynaud's phenomenon is characterized by transient reduction in blood supply through the small arteries in the hands and feet. Severe Raynaud's phenomenon can cause digital necrosis. It has been hypothesized that nitric oxide may have a role in Raynaud's phenomenon. We report two cases in which oral L-arginine reversed digital necrosis in Raynaud's phenomenon and two additional cases in which the symptoms of severe Raynaud's phenomenon were improved with oral L-arginine. These reports suggest that a defect in nitric oxide synthesis or metabolism is present in Raynaud's phenomenon. They also suggest a potential role for oral L-arginine therapy in Raynaud's phenomenon, especially in Raynaud's phenomenon with digital necrosis. (Mol Cell Biochem **244**: 139–141, 2003)

*Key words*: L-arginine, nitric oxide, Raynaud's phenomenon, digital necrosis

## Introduction

Raynaud's phenomenon is characterized by transient reduction in blood supply through the small arteries in the hands and feet [1]. Typically, exposure to cold induces digital artery spasm associated with pain and a characteristic change in skin color from cyanotic to white to red. Most cases of Raynaud's phenomenon are not associated with other conditions, i.e. they are considered primary Raynaud's phenomenon [2]. People with collagen vascular diseases such as rheumatoid arthritis and scleroderma have a higher incidence and more severe form of Raynaud's phenomenon. Severe Raynaud's phenomenon can cause digital necrosis. Treatment for digital necrosis typically involves dihydropyridine type calcium channel blockers and avoidance of cold. Unfortunately, treatment of digital necrosis is frequently not successful.

Nitric oxide is an endogenous vasodilator that relaxes smooth muscle by myoplasmic calcium dependent and independent mechanisms [3, 4]. Inadequate nitric oxide production can cause vascular spasm. Nitric oxide is synthesized from the amino acid L-arginine [5]. It was hypothesized that a deficiency of nitric oxide could be responsible for Raynaud's phenomenon. Chronic oral administration of L-arginine had no effect on digital blood flow in patients with primary Raynaud's phenomenon [6] or with scleroderma [7]. Intra-arterial administration of L-arginine had no effect on digital blood flow in patients with primary Raynaud's phenomenon [8]. However, in scleroderma patients, intra-arterial administration of L-arginine had reduced the number of fingers that exhibited Raynaud's phenomenon in response to cold exposure (9).

We hypothesized that nitric oxide had a role in severe Raynaud's phenomenon with digital necrosis. We report two cases in which oral L-arginine reversed digital necrosis in Raynaud's phenomenon and two additional cases in which the symptoms of severe Raynaud's phenomenon were improved with oral L-arginine.

## Case reports

### *Case #1*

MM is a 63 year old woman with a 4 year history of rheumatoid arthritis treated with plaquenil. She first developed Raynaud's phenomenon in 12/1999 and digital necrosis developed

*Address for offprints*: C.M. Rembold, Box 800146, Cardiovascular Division, University of Virginia Health System, Charlottesville, CA 22908-0146, USA (E-mail: crembold@virginia.edu)

in 2/2000. She described severe burning pain that was not controlled with Percocet 1 q 4 h. Topical tetracaine relieved the pain for approximately 15 min. She was initially treated with nifedipine 60 mg and prednisone 40 mg daily for 6 weeks without benefit. She presented to the University of Virginia on 3/23/2000. There were necrotic regions on the ends of the 2nd, 3rd, and 4th digits of the right hand (a picture is not available). Pus was draining from several of these necrotic regions. All digits were red and various stages of cracked skin. She had swan neck deformities of several of her digits. She had no signs of scleroderma. Amlodipine 10 mg qd and L-arginine 2 g bid were begun. The nifedipine was discontinued and the prednisone tapered over 2 weeks. On 4/6/2000 (day 14), the patient mistakenly stopped the L-arginine. Examination on 4/14/2000 (day 22) revealed less discharge but no significant healing (see Fig. 1, left). The L-arginine was restarted and a 5 day course of azithromycin 500 g daily was given. She was also prescribed topical nitroglycerin paste, but this caused no relief and was discontinued after three days. Examination on 4/28/2000 (day 36) revealed reduced discharge with partial healing. The L-arginine was increased to 2 g tid and hydrochlorothiazide 25 mg qd prn was added to control edema. Examination on 5/25/2000 (day 63) revealed almost complete healing (Fig. 1, right). Her burning pain was relieved. She no longer required Percocet or topical tetracaine.

*Case #2*

MC is a 61 year old woman with a 20 year history of primary Raynaud's phenomenon. She had a sympathectomy in 1990 without relief. She presented to the University of Virginia on 6/28/1996. There was a necrotic region with draining pus on the end of the 2nd digit of the right hand. All of the digits had a red rash and various stages of cracked skin. She had no signs of scleroderma. Amlodipine 5 mg and amoxacillin 500 mg tid for 2 weeks were begun. On 8/9/1996 (day 41), the right 2nd digit was improved but the left 4th digit had a new necrotic lesion. The amlodipine was increased to 10 mg daily and another course of amoxacillin 500 mg tid for 2 weeks was prescribed. The LDL cholesterol was 160 and the HDL cholesterol was 34, so pravastatin 20 mg and gemfibrozil 600 mg bid were added. The patient soon stopped pravastatin secondary to constipation. On 10/8/1996 (day 100), the left 4th digit still had a draining necrotic lesion and L-arginine 500 mg qid was added. On 2/4/1997 (day 216), the necrotic lesions were totally healed and her Raynaud's phenomenon was under much better control, despite the cold weather. Her Raynaud's phenomenon remained quiescent until she stopped taking L-arginine in 7/1998 (day 723). On 10/26/1998 (day 838), her hands showed no necrotic lesions, but she reported increasing symptoms of Raynaud's phenomenon. Her L-arginine was restarted. She was contacted by phone on 5/30/2000 (day 1412). She continues on L-arginine 2 g daily. When she is unable to obtain L-arginine from the store, the pain from her Raynaud's phenomenon develops in approximately 2 weeks and resolves when L-arginine is restarted. She reports no necrotic lesions.

*Case #3*

SO is a 54 year old woman with a 15 year history of primary Raynaud's phenomenon. She presented to the University of Virginia on 2/27/1992. Her toes were cyanotic and there were focal regions on the verge of breaking down. She had no signs of rheumatoid arthritis or scleroderma. Felodipine 5 mg was begun and then increased to 10 mg daily. On 5/22/1992 (day

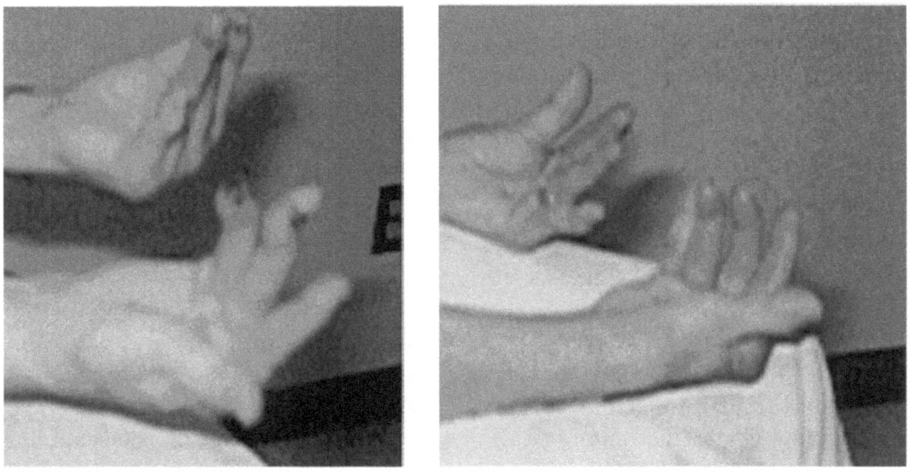

*Fig. 1.* Photographs of subject #1 hands 22 days (left) after starting 4 g of L-arginine daily (left) and 26 days after increasing L-arginine to 6 g daily (total 66 days of L-arginine).

82), her toes were improved and felodipine was increased to 20. Her LDL was 97 and HDL 81. On 1/2/1996 (day 1385), the felodipine was changed to amlodipine 10 mg daily. On 9/27/1996 (day 1650), since her Raynaud's phenomenon was not well controlled, L-arginine 500 mg qid was added. On 1/24/1997 (day 1767), she reported that her Raynaud's phenomenon had remarkably improved since starting the L-arginine. She was switched from amlodipine back to felodipine 10 mg bid. Her Raynaud's phenomenon remained quiescent until she stopped taking L-arginine in the summer of 1998. On 4/20/1998 (day 2213), she had more symptoms of Raynaud's phenomenon. She agreed to take L-arginine year round. She doubled the dose of L-arginine for 2 weeks in 10/1999 for an exacerbation of her Raynaud's phenomenon. On 4/25/2000 (day 2938) she reported having a good winter.

*Case #4*

CM is a 63 year old woman with a 23 year history of primary Raynaud's phenomenon primarily involving the hands and chronic phlebitis of the calfs. She presented to the University of Virginia on 10/5/1987. She had no signs of rheumatoid arthritis or scleroderma. She was first treated with diltiazem 60 qid with some improvement. She was changed from diltiazem to Felodipine 10 mg daily on 4/10/1992 with little change. On 10/7/1996, L-arginine 500 mg qid was added. On 1/24/1997, she reported that her symptoms of Raynaud's phenomenon had improved. On 10/3/1997, after stopping L-arginine but still taking felodipine, she reported worsening of her Raynaud's phenomenon. The L-arginine was restarted. On 10/23/1998 she reported an improvement on the L-arginine. By the summer of 1999, she had stopped the L-arginine again and complained of bloating. Her felodipine was reduced to 5 mg and L-arginine 1 g bid was restarted. On both 4/30/1999 and 4/28/2000 she reported minimal Raynaud's phenomenon symptoms. On examination, there was some digital cyanosis.

## Discussion

In 1929, an intrinsic defect in arterial blood supply was shown to be present in Raynaud's phenomenon [1]. We hypothesized that reduced function of the endogenous vasodilator nitric oxide may be the cause of reduced blood flow in severe Raynaud's phenomenon. We further hypothesized that oral supplementation with the amino acid L-arginine, the substrate for nitric oxide synthetase [5], may improve Raynaud's phenomenon. We report two cases in which oral L-arginine reversed digital necrosis in Raynaud's phenomenon and two additional cases in which the symptoms of Raynaud's phenomenon were improved with oral L-arginine. These case reports suggest that a defect in nitric oxide synthesis or metabolism is present in Raynaud's phenomenon. They also suggest a potential role for oral L-arginine therapy in Raynaud's phenomenon, especially in Raynaud's phenomenon with digital necrosis. We suggest that a randomized trial be initiated to test the efficacy of oral L-arginine in Raynaud's phenomenon with digital necrosis.

## References


1. Lewis T: Experiments relating to the peripheral mechanisms involved in the spasmodic arrest of the circulation in the fingers, a variety of Raynaud's disease. Heart 15: 7–99, 1929
2. Belch JJ: Raynaud's phenomenon. Cardiovasc Res 33: 25–30, 1997
3. Chen XL, Rembold CM: Nitroglycerin relaxes rat tail artery primarily by lowering $Ca^{2+}$ sensitivity and partially by repolarization. Am J Physiol Heart Circ Physiol 271: H962–H968, 1996
4. Rembold CM, Foster B, Strauss JD, Wingard CJ, Van Eyk JE: cGMP mediated phosphorylation of heat shock protein 20 may cause smooth muscle relaxation without myosin light chain dephosphorylation. J Physiol (Lond) 524: 865–878, 2000
5. Murad F: Regulation of cytosolic guanylyl cyclase by nitric oxide: the NO-cyclic GMP signal transduction system. Adv Pharmacol 26: 19–33, 1994
6. Khan F, Litchfield SJ, McLaren M, Veale DJ, Littleford RC, Belch JJ: Oral L-arginine supplementation and cutaneous vascular responses in patients with primary Raynaud's phenomenon. Arth Rheum 40: 352–357, 1997
7. Khan F, Belch JJ: Skin blood flow in patients with systemic sclerosis and Raynaud's phenomenon: Effects of oral L-arginine supplementation. J Rheumatol 26: 2389–2394, 1999
8. Singh S, De Trafford JC, Baskerville PA, Martin JF: Response of digital arteries to endothelium dependent and independent vasodilators in patients with Raynaud's phenomenon. Eur J Clin Invest 25: 182–185, 1995
9. Freedman RR, Girgis R, Mayes MD: Acute effect of nitric oxide on Raynaud's phenomenon in scleroderma (letter). Lancet 28: 354–739, 1999

*Molecular and Cellular Biochemistry* **244:** 143–150, 2003.


# Creatine deficiency syndromes

Andreas Schulze

*Division of Metabolic and Endocrine Diseases, University Children's Hospital, Heidelberg, Germany*

## Abstract

Since the first description of a creatine deficiency syndrome, the guanidinoacetate methyltransferase (GAMT) deficiency, in 1994, the two further suspected creatine deficiency syndromes – the creatine transporter (CrT1) defect and the arginine:glycine amidinotransferase (AGAT) deficiency were disclosed.

GAMT and AGAT deficiency have autosomal-recessive traits, whereas the CrT1 defect is a X-linked disorder. All patients reveal developmental delay/regression, mental retardation, and severe disturbance of their expressive and cognitive speech. The common feature of all creatine deficiency syndromes is the severe depletion of creatine/phosphocreatine in the brain. Only the GAMT deficiency is in addition characterized by accumulation of guanidinoacetic acid in brain and body fluids. Guanidinoacetic acid seems to be responsible for intractable seizures and the movement disorder, both exclusively found in GAMT deficiency. Treatment with oral creatine supplementation is in part successful in GAMT and AGAT deficiency, whereas in CrT1 defect it is not able to replenish creatine in the brain. Treatment of combined arginine restriction and ornithine substitution in GAMT deficiency is capable to decrease guanidinoacetic acid permanently and improves the clinical outcome. The lack of the creatine/phosphocreatine signal in the patient's brain by means of *in vivo* proton magnetic resonance spectroscopy is the common finding and the diagnostic clue in all three diseases. In AGAT deficiency guanidinoacetic acid is decreased, whereas creatine in blood was found to be normal. On the other hand the CrT1 defect is characterized by an increased concentration of creatine in blood and urine whereas guanidinoacetic acid concentration is normal.

The increasing number of patients detected very recently suffering from a creatine deficiency syndrome and the unfavorable outcome highlights the need of further attempts in early recognition of affected individuals and in optimizing its treatment. The study of creatine deficiency syndromes and their comparative consideration contributes to the better understanding of the pathophysiological role of creatine and other guanidino compounds in man. (Mol Cell Biochem **244**: 143–150, 2003)

*Key words*: creatine deficiency, inborn errors of metabolism, GAMT deficiency, AGAT deficiency, creatine transporter deficiency, guanidinoacetic acid

## Introduction

The creatine (Cr)/phosphocreatine (PCr) system plays an important role in energy storage and transmission. Beside the mitochondrial and cytosolic creatine kinase (CK) system as a shuttle of high-energy phosphates, synthesis and transport of Cr are integral parts of cellular energy metabolism. Major achievements made over the last couple of years have been attempted to a better understanding of the mechanism of cellular energy homeostasis and their pathophysiological consequences in man. Studies in several animal models (e.g. CK knockout mouse) and on cellular models as well as investigations in patients with inborn errors of energy metabolism (e.g. mitochondrial cytopathies) were undertaken in order to clarify the role of the Cr/PCr and CK systems. Three main proteins are the basis of Cr metabolism, namely arginine:glycine amidinotransferase (AGAT, EC 2.1.4.1), *S*-adenosyl-L-methionine:N-guanidinoacetate methyltransferase (GAMT, EC 2.1.1.2), and the Cr transporter (CrT). In the recent years inborn errors of metabolism have been identified for all three proteins - AGAT deficiency (AGAT-D) [1, 2], GAMT deficiency (GAMT-D [MIM 601240]) [3], and CrT1 defect (CrT1-D [MIM 300036]) [4, 5]. A common feature in all these disorders is the complete lack of Cr/PCr in the brain measured by *in vivo* magnetic resonance spectroscopy (MRS), so that they are subsumed as Creatine Deficiency Syndromes

*Address for offprints*: A. Schulze, Division of Metabolic and Endocrine Diseases, University Children's Hospital, Im Neuenheimer Feld 150, 69120 Heidelberg, Germany (E-mail: andreas_schulze@med.uni-heidelberg.de)

(CDS). The clinical presentation of all CDS patients is characterized by developmental delay/arrest, mental retardation, and disturbance of active and comprehensible speech. In GAMT-D guanidinoacetic acid (GAA) is increased in brain, blood, cerebrospinal fluid (CSF), and urine, and seizures refractory to anti-epileptic drugs complicate this disease.

In order to more fully comprehend the clinical presentation, treatment and outcome in the different CDS, it seems important to provide a short introduction into the basics of Cr metabolism in humans (Fig. 1) (for a review see [6]). Cr is either taken up from the food by intestinal absorption, and/or it is synthesized endogenously, primarily in kidney, pancreas, and liver. AGAT catalyzes the reversible transamidination of the guanidino group from arginine to glycine yielding GAA and ornithine. GAMT subsequently catalyzes *S*-adenosyl-L-methionine-dependent methylation of GAA to yield Cr and *S*-adenosyl-L-homocysteine [7]. Cr is then transported through the blood and is taken up into Cr-requiring tissues against a large concentration gradient (plasma [Cr] ~ 50 µM; intracellular [Cr + PCr] up to 40 mM). Uptake into the tissues is afforded by a $Na^+$- and $Cl^-$-dependent Cr transporter [8]. Cr and PCr are nonenzymatically converted at an almost constant rate (~ 1.7%/day) into creatinine (Crn) which passively diffuses out of the cells and is excreted by the kidneys into the urine. The urinary Crn excretion therefore represents a convenient indicator of the total Cr stores in the body. A 70-kg man contains ~ 120 g Cr, of which > 90% are found in muscle tissue.

The increasing number of recently new recognized patients with CDS opens possibility of comprehensive view on the clinical symptoms, metabolic patterns and treatment outcome

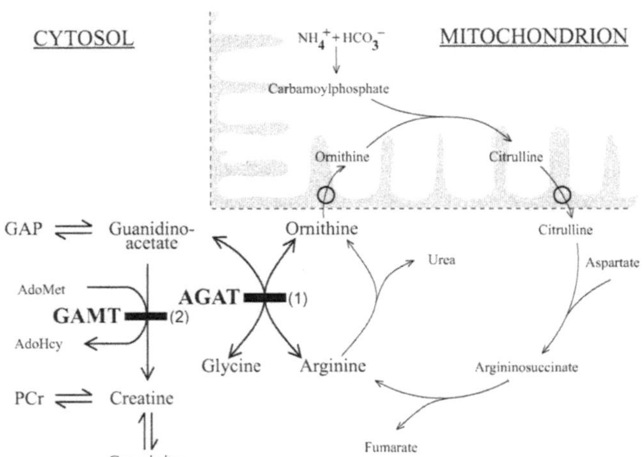


*Fig. 1.* The metabolic pathway of creatine/phosphocreatine. AdoHcy – S-adenosylhomocysteine; AdoMet – S-adenosylmethionine; AGAT – arginine:glycine amidinotransferase; GAMT – guanidinoacetate methyltransferase; GAP – guanidinoacetophosphate; PCr – phosphocreatine. Sides of metabolic bloc in the two steps of creatine biosynthesis are displayed as filled box and numbered as (1), AGAT deficiency and (2), GAMT deficiency.

in the different disorders. Thus, the study of CDS patients will improve the knowledge of the crucial role of Cr and other guanidino compounds in healthy and disease.

## Guanidinoacetate methyltransferase deficiency

GAMT-D is the first detected deficiency in creatine biosynthesis in man (for review see [9]).

### *Enzyme defect*

This disorder is caused by deficiency of *S*-adenosyl-L-methionine:N-guanidinoacetate methyltransferase. This enzyme is expressed in liver, kidney, and pancreas [7, 10] and with lower extend also in brain [11], lymphocytes, fibroblasts [12], and other tissues. GAMT catalyzes the *S*-adenosyl-L-methionine-dependent methylation of GAA to yield Cr and *S*-adenosyl-L-homocysteine (Fig. 1). In consequence, GAMT-D leads to a lack of Cr and an accumulation of GAA.

### *Inheritance and molecular defect*

GAMT-D is inherited as an autosomal recessive trait. The GAMT gene mapped to chromosome 19p 13.3 [13]. Four GAMT-D alleles have been characterized so far, 327G → A/309ins13/c.491insG/IVS5-3C → G [12, 14].

### *Patients*

The first patient with GAMT-D was described in 1994 [3]. The clue to its detection was the finding of absent Cr/PCr signal in the brain by *in vivo* proton MRS. Immediately after this first description further patients could be disclosed through verification of missed Cr/PCr signal in the brain by *in vivo* proton MRS [15–22].

### *Clinical manifestation*

The clinical presentation of GAMT-D is heterogeneous. However, in generally it is characterized by developmental delay attracting attention at 6–12 months of age and/or developmental arrest in the second year of life, muscular hypotonia, dyskinetic involuntary movements, no active or comprehensible speech development, severe mental retardation, and seizures in part not controllable by anti-epileptic drugs. In older patients, autism with self-injurious behavior comes along. In patients with a severe phenotype severe extrapyramidal movement disorder and therapy-refractory epilepsy are predominant, whereas milder affected patients only show developmental delay and mild epilepsy. In some patients abnormalities of brain magnetic resonance imaging (MRI) were described, which consisted in myelination delay or increased signal intensity in T2 weighted images of the globus pallidus.

*Diagnostic findings*

The pathognomonic laboratory findings in GAMT-D consist of decreased concentration of Cr and Crn and accumulation of GAA. Cr and Crn are lowered in blood and CSF and their excretion in urine is decreased. It is essential to mention, that Crn measurement in blood by the still widely used Jaffé method may lead to misdiagnose the decreased Crn. Verhoeven *et al.* [23] reported that Crn in plasma from 2 GAMT-D patients appeared normal when measured by the Jaffé method but was decreased when measured enzymatically or by HPLC. In urine, the Jaffé method and the enzymatic method gave similar results, indicating that in urine no false elevations of Crn can be expected [23]. For the metabolic urine screening of inborn errors of metabolism the determination of several compounds, e.g. amino acids, organic acids, uric acid, purines, pyrimidines, were usually expressed per mol Crn, because of the relative constancy of the latter (see above). In case of pathological decreased Crn excretion, as in GAMT-D, these compounds might appear to be elevated if expressed per mol Crn. This finding was in fact the clue in the diagnosis of two GAMT-D patients [19, 23]. When generalized elevation of such compounds are found, the possibility of GAMT-D should be considered. GAA, the precursor of Cr, is elevated distinctly in GAMT-D. Increased concentration are found in blood, CSF, and urine. For GAA measurement several methods are used. A simple qualitative test in urine applicable in every metabolic laboratory uses the Sakaguchi reaction [24]. One GAMT-D patient was initially diagnozed by this method [15]. For quantitative results more sophisticated methods, e.g. cation exchange chromatography with pre/post-column derivatization [25] or stable isotope dilution gas chromatography-mass spectrometry [26] can be applied.

The *in vivo* MRS proved as an useful tool in detection of GAMT-D. Cr depletion in brain measured by means of proton MRS is reflected by the absence of the Cr/PCr resonance (Fig. 2). By *in vivo* phosphorus MRS of the brain one can detect the decrease of PCr with concomitant appearance of a usually not detectable resonance assigned to be guanidinoacetophosphate (Fig. 2). Whereas GAMT-D patients reveal a complete lack of the Cr/PCr signal in the brain, the decrease of Cr in their muscle seems not to be so pronounced but even verifiable by MRS (Fig. 3) [15, 20].

For confirmatory diagnosis determination of GAMT activity in liver, fibroblasts, or lymphoblasts is available [27].

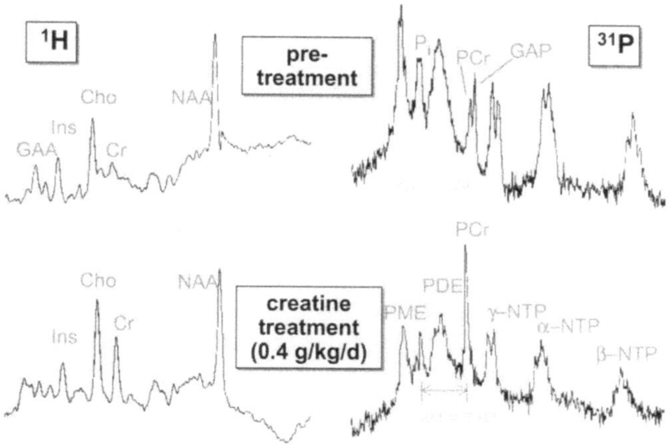


*Fig. 2. In vivo* proton and phosphorus magnetic resonance spectroscopy of the brain in a patient with guanidinoacetate methyltransferase deficiency [18]. GAA – guanidinoacetic acid; Ins – inositols; Cho – choline-containing compounds; Cr – total creatine (mainly creatine and phosphocreatine); NAA – N-acetyl-L-aspartate; Pi – inorganic phosphate; PCr – phosphocreatine; GAP – guanidinoacetophosphate; PME – phosphomonoester; PDE – phosphodiester; NTP – nucleoside 5′-triphosphate.

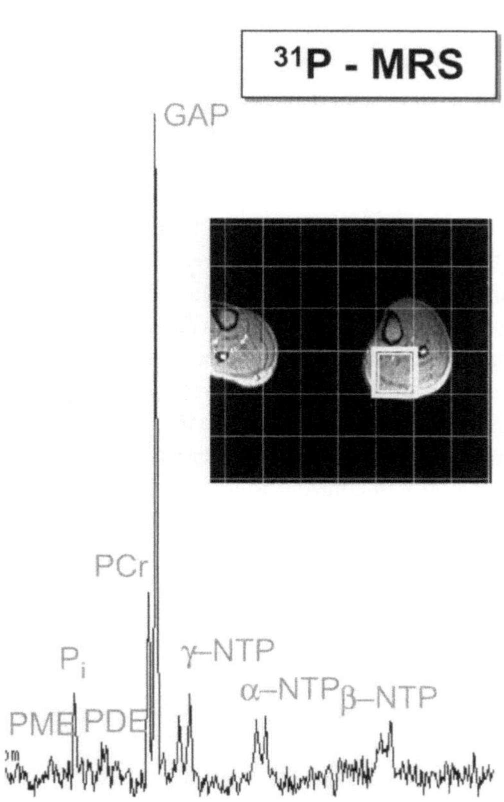


*Fig. 3. In vivo* phosphorus magnetic resonance spectroscopy of the calve muscle in a patient with guanidinoacetate methyltransferase deficiency [20]. PME – phosphomonoester; Pi – inorganic phosphate; PDE – phosphodiester; PCr – phosphocreatine; GAP – guanidinoacetophosphate; NTP – nucleoside 5′-triphosphate.

*Treatment*

In GAMT-D treatment approaches attempt to restore depleted creatine in the brain by supplementation of creatine in pharmacological doses. Oral supplementation with 0.35–2.0 g/kg/day of Cr slowly increased the Cr/PCr concentration in the brain. However, even after several months, Cr/PCr in these patients' brain remained significantly below the normal range [9]. Even if Cr replacement also causes decreasing GAA formation, GAA concentration remains largely elevated in CSF, serum and urine [28] which may explain the persistence of some of the clinical symptoms, especially epilepsy. Dietary arginine restriction (15 mg/kg/day) in combination with ornithine supplementation (100 mg/kg/day) led to a substantial and permanent decrease of GAA in body fluids of one patient. Biochemical effects were accompanied by a marked clinical improvement. Distinctly reduced epileptogenic activities in electroencephalography accompanied by almost completely disappearance of seizures demonstrates the positive effect of GAA reduction, indicating for the first time that GAA may exert an important epileptogenic potential in man [28].

*Outcome*

All patients benefited from Cr supplementation, although to different degrees; however, none has returned to a normal developmental level, and all patients still lack active speech. In the first months after initiating Cr treatment an improvement affecting dyskinesia, muscular hypotonia, seizures, alertness, social contact, and behavior is ascertainable. However, thereafter delay in further clinical improvement becomes evident. Especially in patients with the severe phenotype their clinical circumstances may deteriorate and therapy-refractory seizures reoccur. This clinical course is timely correlated with the observed changes of Cr/PCr in the brain. Cr/PCr concentration increases during the first months of treatment reaching only ~ 50% of controls after some years of treatment [9]. Beside the inability to fully correct the Cr/PCr deficit in brain the neurotoxic action of GAA can contribute to the unsatisfying clinical course. The impact of a combined approach of Cr supplementation and GAA lowering still has to be elucidated.

## X-linked creatine transporter deficiency

The CrT1-D is the second CDS which was disclosed [4, 5].

*Protein defect*

CrT1-D is caused by a defective CrT1. The CrT1 gene is expressed in most tissues, with highest levels in skeletal muscle and kidney and somewhat lower levels in colon, brain, heart, testis, and prostate [29, 30]. *In situ* hybridization studies in adult rat brain revealed the presence of the CrT1 in neurons and oligodendrocytes, but not in astrocytes [11]. A second Cr transporter, CrT2, is expressed in testis only. The Cr transporters are members of solute-carrier family 6 (neurotransmitter transporters). Cr, mainly synthesized in the liver, is transported through the blood and is taken up into Cr-requiring tissues against a large concentration gradient. Uptake into the tissues is afforded by a $Na^+$- and $Cl^-$-dependent Cr transporter.

*Inheritance and molecular defect*

CrT1-D is an X-linked disorder. The CrT1 gene, now called 'SLC6A8' (MIM 300036), has been mapped to chromosome Xq28 [31]. One CrT1-D allele, 1539C → T, causing a hemizygous nonsense mutation has been reported so far [5].

*Patients*

The first patient, a 6 years old boy, has been described recently [4]. Furthermore, two brothers of an independent family were detected in the same institution (Neurological Division at Children's Hospital Medical Center in Cincinnati, USA) [32, 33]. Very recently in three brothers of an other family the CrT1-D could been disclosed [33].

*Clinical manifestation*

The index patient initially presents with mild mental retardation, mild epilepsy, but with severe delay both in speech and in expressive-language function. In addition, mild central hypotonia was observed, but gross and fine motor functions were normal. In three female relatives of the index patient, mild biochemical abnormalities and learning disabilities were reported [4]. In all three male patients from Cincinnati common findings were developmental delay, mild epilepsy, but expressive dysphasia. None of the patients has muscle or heart problems. Increasing behavior problems and development of brain atrophy in adolescence points to a slowly progressive disorder [32].

*Diagnostic findings*

In CrT1-D the Cr concentration both in plasma and urine is elevated. Crn in plasma was found to be normal. In contrast to GAMT-D the GAA concentration in plasma and urine is normal. *In vivo* proton MRS revealed the complete lack of the Cr/PCr signal in the brain. Confirmatory diagnosis can be made by a recently developed investigation of Cr uptake in fibroblasts, by which it is possible to discriminate between patients, carriers, and controls [5].

*Treatment*

Treatment of the index patient with oral Cr monohydrate (0.34 g/kg/day) for 3 months resulted in increased Cr concentration in CSF and urine. However, the follow-up proton MRS performed 4 months later demonstrated a similar absence of Cr/PCr as seen before treatment. Due to the fact that the clinical symptoms even did not improve, Cr treatment does not appear to be useful. The oral substitution of Cr was discontinued.

*Outcome*

Since there is no effective treatment, the course of the slowly progressive neurological syndrome with mental retardation, sever speech impairment and progressive atrophy of the brain can not be influenced so far.

## Arginine:glycine amidinotransferase deficiency

AGAT-D, at first reported as 'reversible brain Cr deficiency' in two sisters [1], is the third CDS which could be disclosed [2].

*Enzyme defect*

AGAT-D is caused by deficiency of L-arginine:glycine amidinotransferase (EC 2.1.4.1), which catalyzes the first and rate-limiting step in creatine biosynthesis, which is the reversible transamidination of the guanidino group from arginine to glycine to yield GAA and ornithine (Fig. 1). Immunoreactive enzyme was proven with highest content in the proximal tubules of the kidney, but also in hepatocytes and in alpha cells of the pancreas of the rat [10]. *In situ* hybridization studies in adult rat brain revealed an ubiquitous neuronal and glial expression of AGAT [11].

*Inheritance and molecular defect*

AGAT-D is an autosomal recessive disorder. The AGAT gene has been mapped to chromosome 15q11.2. One AGAT-D allele, 9279G $\rightarrow$ A, homozygously causing a truncated protein has been reported so far [2].

*Patients*

Two sisters of unrelated Italian parents have been detected so far [1].

*Clinical manifestation*

The two sisters, 4 and 6 years of age, suffered from mild mental retardation and severe language delay. They started walking unaided at 24 months and started speaking the first words at 30 months. Beside one uncomplicated febrile seizure in one girl, they had no further seizures. Their brain magnetic resonance imaging were normal. Both of them never expressed muscular or other neurological symptoms.

*Diagnostic findings*

Cr in serum as well as the amino acids arginine and glycine was reported to be normal. The GAA concentration in serum was slightly decreased, whereas the excretion of GAA in urine was found to be extremely low. *In vivo* proton MRS revealed the total absence of Cr/PCr in the brain, thus suspecting a CDS. The underlying enzyme defect could be established by undetectable AGAT activity, as investigated radiochemically in fibroblasts and lymphoblasts [2].

*Treatment*

Cr supplementation at a rate of 400 mg/kg/day increased Cr/PCr in the brain to 40 and 80% of controls within 3 and 9 months, respectively. After 16 months of treatment a nearly complete replenishment of the Cr/PCr signal has been reported.

*Outcome*

Cr supplementation led to a rapid progress in the acquisition of visual perceptual and fine motor skills, together with a slower rate of general cognitive development in the younger sister, but not in the elderly. Language abilities also improved, but more slowly than nonverbal skills.

## Comparative consideration of creatine deficiency syndromes

The summarized comparison of the clinical features in the different CDS is shown in Table 1.

*Table 1.* Comparative findings in creatine deficiency syndromes

| | GAMT deficiency | CrT1 deficiency | AGAT deficiency |
|---|---|---|---|
| Patients | 10<br>(7 published, incl. 1 in abstr./3 unpublished) *1 | 6<br>(3 published, incl. 2 in abstr./3 unpublished)*2 | 2<br>(2 published)*3 |
| Gender | 1 female/8 male | 3 male | 2 female |
| Origin | Kurdish/German/Welsh/2 Turkish/Italian | Caucasian | 2 Italian |
| Consanguinity | 2 consang./4 unrelated | ? | None |
| Age at onset | Developm. delay median 5 mo (3–7 mo)<br>seizures median 2½ yrs (10 mo–4 yrs) | 7 mo in 1/3 | ? |
| Age at diagnosis | Median 3 yrs 8 mo (19 mo–26 yrs) | 6, 16, 20 yrs | 4 yrs 4 mo/6 yrs 5 mo |
| Developmental delay/arrest | 8/8 | mild 3/3 | 2/2 |
| Hypotonia | 7/8 | 1/3 | 0/2 |
| Dyskinesia | 4/8 | 0/3 | 0/2 |
| Reflexes, increased | 2/7 | 0/3 | 0/2 |
| Seizures | 7/8<br>untractable seizures: 5/8 | mild 3/3 | None (only one febrile seizure) |
| Mental retardation | Severe 7/8, mild 1/8 | mild 3/3 | Severe 2/2 |
| Autism/self-injurious behaviour | 7/7 | 1/3 | 0/2 |
| Active speech | None 7/8, single words 1/8 | Single words,<br>severe expressive dysphasia 3/3 | Delayed 2/2 |
| MRI | Myelination delay 3/8, T2 intens Pallidum 3/8 (1 pat. with both), no MRI changes 3/8 | White matter lesion 2/3<br>Brain atrophia 2/3 | Normal |
| EEG (pathol) | 7/7 | 1/3 | ? |
| Treatment | Creatine 350–1250 (–2000) mg/kg/day 8/8<br>Arginine restrictive diet 3/8<br>Phenylbutyrate 1/8<br>Sodium benzoate 1/8 | None | Creatine 400 mg/kg/day 2/2 |
| Outcome | Unfavourable | | Satisfying |
| Motor development | Improved | | Improved 1/2 |
| Active speech | None | | Improved 1/2 |
| Social contact/behaviour | Improved | No changes | Improved 1/2 |
| Seizures | Improved (after GAA lowering) | | |
| Mental retardation | Severe | | |

*1 references [3, 12, 14–22, 28, 35]; *2 references [4, 5, 32, 33, 36]; *3 references [1, 2].

## Conclusion

The different Cr metabolism defects are highly instructive, thus allowing interesting conclusions, but also raise new challenging questions.

The GAMT-D revealed the most severe phenotype. Extrapyramidal symptoms and intractable seizures were exclusively present in this disease. In addition, autistic, self-injurious behavior seems to be characteristic for GAMT-D. Clinical symptoms only found in GAMT-D might be attributed to the influence of GAA, which is highly increased in this disease. Furthermore, the combined impact of Cr deficiency and GAA accumulation might be responsible for the most severe phenotype.

Developmental delay/arrest, mental retardation, even if in different degrees, and impairment of active and cognitive speech are common findings in all three disorders. This points to the special effects caused by Cr deficiency.

Mildest clinical symptoms were found in AGAT-D. The finding of a normal serum concentration of Cr in the two sisters with AGAT-D indicates that they efficiently take up dietary Cr (which in adults on average contributes ~ 50% to the daily Cr requirements) into blood and most likely also into muscle. The complete lack of the Cr/PCr signal in MRS in their brain despite of normal plasma Cr concentration suggests that the synthesis of Cr is also affected within the brain. Assuming that the normal brain is self-sufficient in terms of Cr biosynthesis, one would expect a more severe phenotype in AGAT-D than in CrT1-D, what in fact could not be established in the patients reported so far. The lack of Cr in the brain of CrT1-D patients and their severe phenotype reveals some doubts about the brain's capacity for de novo Cr biosynthesis. Similar to the rest of the body where the liver is the primary site of Cr biosynthesis, even containing only low levels of Cr itself, there may be a strict separation of Cr-synthesizing and Cr-accumulating cells in the brain as well. If this is the case, the Cr transporter would be required for significant accumulation of Cr in the brain.

The fact that normal concentrations of Cr in serum in AGAT-D or even supranormal concentrations by supplementation with rather high doses of Cr in GAMT-D failed to normalize brain Cr/PCr points to the impact of a limited permeability of the blood-brain barrier for Cr. Beside other findings, this is further corroborated by the observation that in GAMT-D patients ingesting Cr, Cr uptake into the brain is a very slow process and steady-state concentrations still below the normal range were only reached after several months. Why in both AGAT-D patients the Cr replenishment was nearly completely after 16 months of treatment and if there is an compromising impact of GAA on the Cr uptake in the brain has to be established in the future.

The therapeutic outcome in CDS is different. Whereas in CrT1-D there exists no therapeutic option so far, the preliminary results of therapeutic attempts in AGAT-D seems to be promising. Despite of partial improvement by Cr supplementation the outcome in GAMT-D is still unfavorable. Even GAA lowering attempts by arginine restriction and ornithine substitution revealed distinct additional effects on the clinical course, it was not possible to normalize general condition. Further therapeutic attempts must be focussed on alternative modes of Cr application and on additional approaches to lower the GAA concentrations.

The early detection of affected patients and the timely onset of treatment might also improve the outcome especially in GAMT-D and AGAT-D. The detection of GAMT-D already in the newborn period is feasible. In one patient increased GAA was proven in the retrospective analyzed dried blood spot specimen kept from neonatal screening [34]. This points to the usefulness of a neonatal screening of GAMT-D by measurement of GAA. However, as our preliminary results with such an screening by means of tandem mass spectrometry have shown, we did not find any affected child out of more than 150.000 neonates (unpublished observation). Nevertheless, at least in all patients with developmental delay and speech impairment of unknown origin, and particularly if accompanied by an extrapyramidal movement disorder and/or seizures, a CDS should be excluded carefully.

## References


1. Bianchi MC, Tosetti M, Fornai F, Alessandri' MG, Cipriani P, De Vito G, Canapicchi R: Reversible brain creatine deficiency in two sisters with normal blood creatine level. Ann Neurol 47: 511–513, 2000
2. Item CB, Stockler-Ipsiroglu S, Stromberger C, Muhl A, Alessandri MG, Bianchi MC, Tosetti M, Fornai F, Cioni G: Arginine:glycine amidinotransferase deficiency: The third inborn error of creatine metabolism in humans. Am J Hum Genet 69: 1127–1133, 2001
3. Stöckler S, Holzbach U, Hanefeld F, Marquardt I, Helms G, Requart M, Hanicke W, Frahm J: Creatine deficiency in the brain: A new, treatable inborn error of metabolism. Pediatr Res 36: 409–413, 1994
4. Cecil KM, Salomons GS, Ball WS Jr, Wong B, Chuck G, Verhoeven NM, Jakobs C, Degrauw TJ: Irreversible brain creatine deficiency with elevated serum and urine creatine: A creatine transporter defect? Ann Neurol 49: 401–404, 2001
5. Salomons GS, van Dooren SJ, Verhoeven NM, Cecil KM, Ball WS, Degrauw TJ, Jakobs C: X-Linked creatine-transporter gene (SLC6A8) defect: A new creatine-deficiency syndrome. Am J Hum Genet 68: 1497–1500, 2001
6. Wyss M, Kaddurah-Daouk R: Creatine and creatinine metabolism. Physiol Rev 80: 1107–1213, 2000
7. Walker JB: Creatine: Biosynthesis, regulation, and function. Adv Enzymol 50: 177–242, 1979
8. Guimbal C, Kilimann MW: A Na(+)-dependent creatine transporter in rabbit brain, muscle, heart, and kidney. cDNA cloning and functional expression. J Biol Chem 268: 8418–8421, 1993
9. von Figura K, Hanefeld F, Isbrandt D, Stöckler-Ipsirogly S: Guanidinoacetate methyltransferase deficiency. In: C.R. Scriver, A. Beaudet, W.S. Sly, D. Valle (eds). The Metabolic and Molecular Basis of Inherited Disease. McGraw Hill, New York, 2001, pp 1897–1908
10. McGuire DM, Gross MD, Elde RP, van Pilsum JF: Localization of L-arginine-glycine amidinotransferase protein in rat tissues by immunofluorescence microscopy. J Histochem Cytochem 34: 429–435, 1986.
11. Braissant O, Henry H, Loup M, Eilers B, Bachmann C: Endogenous synthesis and transport of creatine in the rat brain: An *in situ* hybridization study. Mol Brain Res 86: 193–201, 2001
12. Stöckler S, Isbrandt D, Hanefeld F, Schmidt B, von Figura K: Guanidinoacetate methyltransferase deficiency: The first inborn error of creatine metabolism in man. Am J Hum Genet 58: 914–922, 1996
13. Chae YJ, Chung CE, Kim BJ, Lee MH, Lee H: The gene encoding guanidinoacetate methyltransferase (GAMT) maps to human chromosome 19 at band p13.3 and to mouse chromosome 10. Genomics 49: 162–164, 1998

14. Carducci C, Leuzzi V, Carducci C, Prudente S, Mercuri L, Antonozzi I: Two new severe mutations causing guanidinoacetate methyltransferase deficiency. Mol Genet Metab 71: 633–638, 2000
15. Ensenauer R, Thiel T, Schwab KO, Lehnert W: Presence of muscle creatine in a patient with guanidinacetate methyltransferase (GAMT) deficiency. J Inherit Metab Dis 23: 212, 2001
16. Ganesan V, Johnson A, Connelly A, Eckhardt S, Surtees RAH: Guanidinoacetate methyltransferase deficiency: New clinical features. Pediatr Neurol 17: 155–157, 1997
17. Leuzzi V, Bianchi MC, Tosetti M, Carducci C, Cerquiglini CA, Cioni G, Antonozzi I: Brain creatine depletion: Guanidinoacetate methyltransferase deficiency (improving with creatine supplementation). Neurology 55: 1407–1409, 2000
18. Schulze A, Hess T, Wevers R, Mayatepek E, Bachert P, Marescau B, Knopp MV, De Deyn PP, Bremer HJ, Rating D: Creatine deficiency syndrome caused by guanidinoacetate methyltransferase deficiency: Diagnostic tools for a new inborn error of metabolism see comments. J Pediatr 131: 626–631, 1997
19. van der Knaap MS, Verhoeven NM, Maaswinkel-Mooij P, Pouwels PJ, Onkenhout W, Peeters EA, Stockler-Ipsiroglu S, Jakobs C: Mental retardation and behavioral problems as presenting signs of a creatine synthesis defect. Ann Neurol 47: 540–543, 2000
20. Schulze A, Bachert P, Schlemmer H, Harting I, Polster T, Salomons GS, Verhoeven NM, Jakobs C, Fowler B, Hoffmann GF, Mayatepek E: Lack of creatine in muscle and brain in an adult with guanidinoacetate methyltransferase deficiency. Ann Neurol (in press)
21. Wilichowski E (personal communication)
22. Zinn A (personal communication)
23. Verhoeven NM, Guerand WS, Struys EA, Bouman AA, van der Knaap MS, Jakobs C: Plasma creatinine assessment in creatine deficiency: A diagnostic pitfall. J Inherit Metab Dis 23: 835–840, 2000
24. Schulze A, Mayatepek E, Rating D, Bremer HJ: Sakaguchi reaction: A useful method for screening guanidinoacetate-methyltransferase deficiency. J Inherit Metab Dis 19: 706, 1996
25. Marescau B, Deshmukh DR, Kockx M, Possemiers I, Qureshi IA, Wiechert P, De Deyn PP: Guanidino compounds in serum, urine, liver, kidney, and brain of man and some ureotelic animals. Metabolism 41: 526–532, 1992
26. Struys EA, Jansen EE, ten Brink HJ, Verhoeven NM, van der Knaap MS, Jakobs C: An accurate stable isotope dilution gas chromatographic-mass spectrometric approach to the diagnosis of guanidinoacetate methyltransferase deficiency. J Pharm Biomed Anal 18: 659–665, 1998
27. Ilas J, Muhl A, Stockler-Ipsiroglu S: Guanidinoacetate methyltransferase (GAMT) deficiency: Non-invasive enzymatic diagnosis of a newly recognized inborn error of metabolism. Clin Chim Acta 290: 179–188, 2000
28. Schulze A, Ebinger F, Rating D, Mayatepek E: Improving treatment of guanidinoacetate methyltransferase deficiency: Reduction of guanidinoacetic acid in body fluids by arginine restriction and ornithine supplementation. Mol Genet Metab 74: 413–419, 2001
29. Nash SR, Giros B, Kingsmore SF, Rochelle JM, Suter ST, Gregor P, Seldin MF, Caron MG: Cloning, pharmacological characterization, and genomic localization of the human creatine transporter. Receptors Channels 2: 165–174, 1994
30. Sora I, Richman J, Santoro G, Wei H, Wang Y, Vanderah T, Horvath R, Nguyen M, Waite S, Roeske WR, l. a: The cloning and expression of a human creatine transporter. Biochem Biophys Res Commun 204: 419–427, 1994
31. Gregor P, Nash SR, Caron MG, Seldin MF, Warren ST: Assignment of the creatine transporter gene (SLC6A8) to human chromosome Xq28 telomeric to G6PD. Genomics 25: 332–333, 1995
32. Degrauw TJ, Cecil KM, Salomons GS, van Dooren SJ, Verhoeven NM, Ball WS, Jakobs C: The clinical syndrome of creatine transporter deficiency. Oral presentation on 6th International Meeting on Guanidino Compounds in Biology and Medicine, Cincinatti, Ohio, USA, 31.08–03.09, 2001
33. Salomons GS, van Dooren SJ, Bunea D, Verhoeven NM, Cecil KM, Ball WS, Degrauw TJ, Jakobs C: X-linked creatine transporter defect: The first two families. Oral presentation on 6th International Meeting on Guanidino Compounds in Biology and Medicine, Cincinatti, Ohio, USA, 31.08.– 03.09, 2001
34. Carducci C, Birarelli M, Santagata P, Leuzzi V, Carducci C, Antonozzi I: Automated high-performance liquid chromatographic method for the determination of guanidinoacetic acid in dried blood spots: A tool for early diagnosis of guanidinoacetate methyltransferase deficiency. J Chromatogr B Biomed Sci Appl 755: 343–348, 2001
35. Ensenauer R. (personal communication)
36. Degrauw TJ, Cecil KM, Ball WS, Wong B, Jakobs C, Verhoeven NM: A new disorder of creatine metabolism; a patient with possible creatine transport defect. J Inherit Metab Dis 23: 211, 2001

*Molecular and Cellular Biochemistry* **244:** 151–157, 2003.


# Human skeletal muscle creatine transporter mRNA and protein expression in healthy, young males and females

Robyn M. Murphy,[1] Rebecca J. Tunstall,[1] Kate A. Mehan,[1] David Cameron-Smith,[1] Michael J. McKenna,[2] Lawrence L. Spriet,[3] Mark Hargreaves[1] and Rodney J. Snow[1]

[1]*School of Health Sciences, Deakin University, Burwood;* [2]*School of Human Movement, Recreation and Performance, Victoria University of Technology, Footscray, Australia;* [3]*Department of Human Biology and Nutritional Sciences, University of Guelph, ON, Canada*

## Abstract

The present study investigated whether there were any differences between males and females in respect to creatine transporter (CreaT) gene expression and/or total creatine (TCr) content in human vastus lateralis muscle. Skeletal muscle obtained from young healthy male (n = 13, age: 23.2 ± 5.0 years) and female subjects (n = 12, age: 21.7 ± 4.3 years) was analyzed for CreaT mRNA, CreaT protein and TCr content. Total CreaT protein content in the muscle was similar ($p > 0.05$) between the sexes. Two bands (~ 55 and 73 kDa) of the CreaT protein were detected in all muscle samples. Both the 55 and the 73 kDa bands were present in similar ($p > 0.05$) amounts in males compared with females. The 73 kDa band was in greater abundance ($p < 0.05$) than the 55 kDa band, irrespective of gender. In addition, CreaT mRNA expression relative to β-actin mRNA and the TCr content (males: 117.8 ± 2.2, females: 125.3 ± 4.3 mmol.$kg^{-1}$ dry mass) were also unaffected ($p > 0.05$) by gender. These data demonstrate that gender does not influence skeletal muscle TCr content and CreaT gene expression in young human subjects. (Mol Cell Biochem **242**: 151–157, 2003)

*Key words*: creatine metabolism, gene expression, high-energy phosphates

## Introduction

Creatine in its free and phosphorylated form plays an important role in skeletal muscle metabolism and function [1]. The total creatine content (TCr = creatine; Cr plus creatine phosphate; CrP content) of a skeletal muscle fibre is dependent upon the rate of Cr entry into the cell, intracellular Cr trapping and the rate of Cr degradation to creatinine and its subsequent loss from the cell. It has been known for many years that skeletal muscle fibres do not possess the enzymes required to synthesise Cr [2], so this process cannot contribute to the intramuscular Cr pool. Given that the degradation of Cr to creatinine occurs by a non-enzymatic reaction at a constant rate of about 2% of the Cr pool per day, the rate of Cr uptake is likely to be the most important process controlling myofibre Cr content [3]. Several studies using cultured cells have demonstrated that cellular Cr uptake is enhanced, at least initially, by elevated extracellular Cr concentrations [4] and the increased expression [5, 6] and/or activity [7] of the Cr transporter (CreaT) protein. Previous research has identified that Cr transport is a saturable, sodium chloride dependent process [4, 6, 8]. In the past decade, researchers [9–12] have established that the CreaT gene is found at two chromosomal locations (Xq28; CreaT1 and 16p11.1; CreaT2). CreaT2 mRNA transcripts are only expressed in testicular cells [10]. Recently, antibodies for the CreaT1 protein have been produced [13, 14] and at least two isoforms (~ 70 and ~ 55 kDa) have been identified in rat [13, 15] and human [16]

*Address for offprints*: R.J. Snow, School of Health Sciences, Deakin University, Burwood, 3125 Australia (E-mail: rsnow@deakin.edu.au)

skeletal muscle. The two isoforms are likely to result from alternative splicing of the CreaT mRNA and/or variation in glycosylation state [14, 17]. At present, the functional significance of these isoforms is unclear.

Surprisingly, to date there have been no studies that have examined the CreaT mRNA content in humans and only one study that has investigated the CreaT protein content in human skeletal muscle [16]. The major focus of that research was on determining the effect of various neuromuscular diseases on the expression of the transporter protein in middle-aged control and diseased subjects (35–59 years old) [16]. There are currently no data available on the expression of the transporter in young adult subjects. Many aspects of muscle metabolism are differentially regulated between gender (for reviews see [18]), however there is no research on the effect of gender on the expression of the CreaT protein, and this is the major focus of this study. Although most of the indirect evidence suggests that skeletal muscle CreaT expression would be similar between males and females, this has not been verified. For example, most [19–22], but not all [23] studies, report that muscle TCr content is similar between the sexes. Furthermore, the magnitude of skeletal muscle Cr loading following a period of oral Cr supplementation is not different in males compared with females [19, 24]. Although the indirect evidence suggests there is unlikely to be any gender effect on CreaT expression in human skeletal muscle, no studies have specifically examined this possibility. The major aim of the present study therefore was to compare the CreaT protein and mRNA content in skeletal muscle of young healthy male and female subjects. We hypothesised that gender would not affect the expression of the transporter in human skeletal muscle.

## Materials and methods

### *Materials*

CreaT antibody was kindly provided by Professor Theo Wallimann, (ETH-Zurich) and anti-muscle myosin heavy chain I (MHC1) antibody was from Chemicon International (Temecula, CA, USA). Anti-rabbit horse-radish peroxidase (HRP) secondary antibody was from Silenus (Boronia, Australia) with AlexaFluor secondary antibodies being Molecular Probes (Eugene, OR, USA). Protein assay and chemiluminescent substrate were obtained from Pierce (Selby Biolab, Australia). Fluoroguard was from BioRad (Hercules, CA, USA). Nitrocellulose membrane and primers were purchased from Geneworks (Adelaide, Australia). FastPrep RNA extraction kit was purchased from BIO 101 (Vista, CA, USA) and the reverse transcription kit was from Promega (Madison, WI, USA). SYBR green buffer was acquired from Applied Biosystems (Foster City, CA, USA). Chemicals, actin antibody and enzymes were purchased from Sigma (Castle Hill, Australia) or Roche Diagnostics (Sydney, Australia).

### *Subjects*

Thirteen males and twelve females (see Table 1 for subject details) volunteered to participate in this study. They were all physically active and healthy but not specifically trained. The experiment was approved by the Deakin University Human Research Ethics Committee. Subjects were fully informed of the procedures and signed informed consent prior to participation. No attempt was made to control for the influence of the menstrual cycle on the variables measured in this study. It should be noted that no research has been conducted on the influence of the menstrual cycle on CreaT protein and mRNA expression in skeletal muscle so it is unclear if failing to control for this variable will significantly affect our findings.

### *Muscle sampling*

A resting muscle sample was obtained under local anaesthesia (Xylocaine 1%) from the vastus lateralis using the percutaneous needle biopsy technique modified to include suction. The sample was quickly frozen in liquid nitrogen ($LN_2$), and stored in $LN_2$ until analysis. Portions of some muscle samples (n = 2, male subjects) were aligned, mounted in embedding matrix (Tissue Tek Oct,) and frozen in iso-pentane cooled by $LN_2$ for subsequent immunohistochemical analysis.

### *Immunoblotting*

Total cellular protein was extracted as previously described [15]. Denatured protein (50 µg; CreaT and 30 µg; actin) isolated from muscle samples (~ 10 mg) was separated on 10% SDS-polyacrylamide gels and semi-dry transferred onto 0.45 µm nitrocellulose membrane. Following 2 h blocking (5% skim milk powder in Tris buffered saline and 0.25% Tween (TBST)), the membranes were exposed overnight at 4°C to either 1 in 2000 dilution C-terminal anti-CreaT antibody [13] or 1 in 200 dilution anti-actin antibody. After several washes in blocking buffer, the membranes were incubated

*Table 1.* Subject characteristics

| | n | Age (years) | Body mass index ($kg.m^{-2}$) | Body mass (kg) |
|---|---|---|---|---|
| Males | 13 | 23.2 ± 5.0 | 22.3 ± 1.9 | 72.0 ± 8.4 |
| Females | 12 | 21.7 ± 4.3 | 22.1 ± 1.9 | 60.1 ± 7.1* |

*$p < 0.05$, females different to males. Means ± S.D.

for 45 min in 1 in 8000 dilution of anti-rabbit HRP conjugated antibody. Following washes in TBST, light sensitive film detected the immuno-reactive bands using chemiluminescent substrate. Band density was analyzed using Kodak 1D software. An internal control of previously extracted human muscle was used in each gel to normalise for variation in signal observed across the membranes. CreaT protein was measured in muscle samples from only 10 males and 11 females due to insufficient sample.

*Immunohistochemistry*

Immunohistochemical analysis was conducted to determine if CreaT protein content differed in type I and II fibres in human skeletal muscle. Muscle samples obtained for this analysis were sectioned (10 µm slices) at –16°C, mounted on microscope slides and stored at –80°C. On thawing, the sections were fixed in 4% paraformaldehyde and after a series of washes (2 × 5 min, phosphate buffered saline; PBS, 1 × 10 min 5% Triton-X 100™, 2 × 5 min PBS), non-specific binding sites were blocked by overnight incubation with 3% bovine serum albumin (BSA) at 4°C. The sections were incubated with the primary antibodies, rabbit anti-CreaT [12] diluted 1 in 2000 in 1% BSA and mouse anti-MHC1 monoclonal antibody, diluted 1 in 200 in 1% BSA, overnight at 4°C. Following PBS washes (2 × 5 min) sections were incubated with the secondary antibodies, anti-rabbit AlexaFluor-488 and anti-mouse AlexaFluor-568, both diluted 1 in 2000 in 1% BSA, for 90 min. Negative control sections followed the above protocol except that they were incubated with 1% BSA for the 2nd overnight incubation period. A drop of Fluoroguard was added to the sections and a coverslip applied. Epifluorescence was viewed with an Olympus AX70 microscope with an UplanFL X40 0.5 NA objective. Digital images were collected (Pulnix TM-6CN monochrome CCD camera; Optimas 5.2 software, Optronics, Goleta, CA, USA). The intensity of the CreaT protein fluorescent signal from type I and II fibres was quantified using ImagePro Express software (Media Cybernetics, Silver Spring, MD, USA).

*Total RNA isolation and reverse transcription*

Total RNA was extracted from 10–25 mg of muscle using a modification of the phenol/chloroform extraction, isopropanol precipitation protocol, using the FastPrep™ Instrument (BIO 101, Vista, CA, USA). Total RNA pellets were resuspended in warm EDTA treated water and total RNA concentration was determined spectrophotometrically at 260 nm. RNA (1 µg) was heated at 65°C for 10 min immediately prior to first strand cDNA being generated using AMV Reverse Transcriptase, in the presence of 1 mM of each dNTP, 20 U Recombinant RNAsin Ribonuclease Inhibitor and 0.5 µg Oligo $(dT)_{15}$ Primer by incubation at 42°C for 60 min followed by 5 min incubations at 99 and 4°C . The cDNA was stored at –80°C until subsequent analysis.

*Real time PCR*

Samples were analysed in triplicate to detect CreaT and β-actin mRNA transcripts using real time RT-PCR (GeneAmp 5700 Sequence Detection System, Applied Biosystems, Foster City, CA, USA) as previously described [15]. β-actin mRNA content was used in the present study as the control to account for any variations due to efficiencies of the reverse transcription and PCR steps. A melting point dissociation curve generated by the instrument was used to confirm that only a single product was present. A threshold cycle ($C_T$) value was obtained for each sample for CreaT and β-actin mRNA transcripts, and a $\Delta C_T$ value was calculated by subtracting the $C_T$ for β-actin from the $C_T$ for CreaT. The relative expression of CreaT was calculated using the expression $2^{-\Delta CT}$. Due to insufficient sample the mRNA data presented are for 11 males and 12 females.

*Enzymatic assays*

Muscle samples were freeze dried for 24 h, weighed and powdered, removing any visible connective tissue. Muscle samples were extracted using 0.5 M perchloric acid and 1 mM ethylenediaminetetraacetic acid, and neutralized with 2.1 M $KHCO_3$ [25]. These extracts were subsequently analysed for CrP and Cr levels using an enzymatic fluorometric technique [26, 27]. Muscle TCr content was calculated from the sum of the muscle CrP and Cr contents. Data are for 13 males with the female metabolite data from 11 subjects due to problems encountered with the analysis of a sample obtained from one female subject.

*Statistics*

The results were analyzed where appropriate by using a two-tailed, unpaired Student's *t*-test. The relationship between muscle TCr content and CreaT protein expression was analysed using a Pearsons correlation coefficient. The level of statistical significance was set at $p < 0.05$. All data are reported as means ± S.E.

# Results

*Muscle CreaT protein and mRNA*

Immunoblotting of the samples with CreaT antibody revealed two bands, of apparent molecular weights of ~ 73 and 55 kDa

(Fig. 1), and a single band (43 kDa) using the actin antibody (data not shown). There were no differences ($p > 0.05$) between either of the CreaT isoforms (Fig. 2) or the total CreaT protein expression across gender (Fig. 2). This remained the case even when the CreaT protein data was normalized to the myofibrillar protein actin (data not shown). The 73 kDa band was in greater abundance ($p < 0.05$) than the 55 kDa band, irrespective of gender. The immunohistochemical results (Figs 3 and 4) indicate that CreaT protein expression occurred to a greater extent ($p < 0.05$) in type I compared with type II fibres. CreaT mRNA expression relative to β-actin was not different ($p > 0.05$) between males and females (Fig. 5).

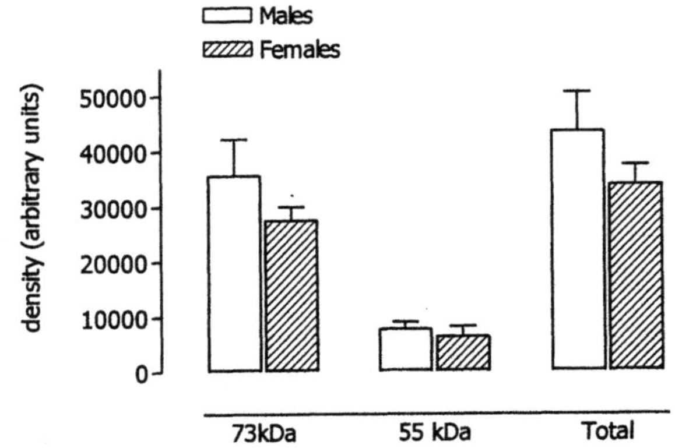


*Fig. 2.* Creatine transporter protein for males (n = 10, open bars) and females (n = 11, hatched bars). Values are means ± S.E.

*Muscle Cr and CrP content*

Intracellular Cr, CrP and TCr content were not different ($p > 0.05$) between males and females (Fig. 6).

*Relationship between TCr content and CreaT protein expression*

A significant inverse relationship was found between the TCr and CreaT protein content for females ($r = -0.77$; $p < 0.01$), whereas for the males this relationship only tended to be significant ($r = -0.61$; $p = 0.059$). When the data from the male and female groups were combined (Fig. 7) the correlation between the variables was $r = -0.60$ ($p < 0.01$). Additionally, when the apparent outlier (TCr, 102.6 mmol $kg^{-1}$ dw; CreaT protein, 93300 arbitary units) was removed from this data set, the correlation, although attenuated, remained statistically significant ($r = -0.54$, $p < 0.05$).

## Discussion

This is the first study to simultaneously report CreaT protein and mRNA expression in skeletal muscle from healthy, young adults. We show for the first time the presence of two CreaT isoforms in male and female subjects. We further demonstrate that the expression of CreaT mRNA and protein in human skeletal muscle was not influenced by gender, in support of our initial hypothesis. We have also provided evidence that CreaT protein expression occurs to a greater extent in type I compared with type II human skeletal muscle fibres. Finally, we have confirmed that the resting intramuscular content of Cr, CrP and TCr, when expressed per dry muscle mass, is similar between young males and females, in support of several studies [19–21].

The factors regulating CreaT mRNA expression are not known, although it is recognized that the regulation of gene

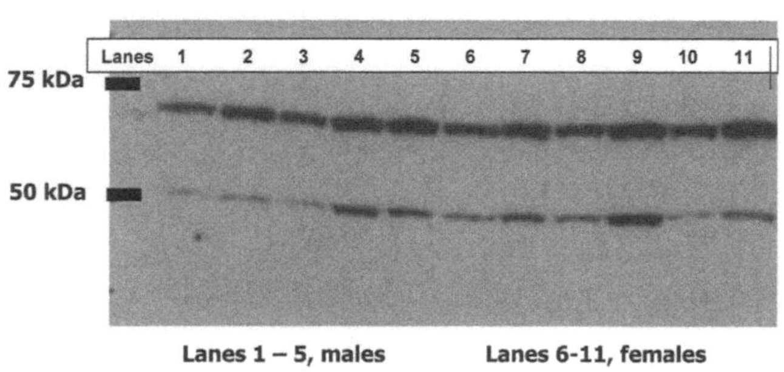


*Fig. 1.* Representative immunoblot of skeletal muscle creatine transporter isoforms in males (n = 5) and females (n = 6). Two isoforms were identified at approximately 73 and 55 kDa.

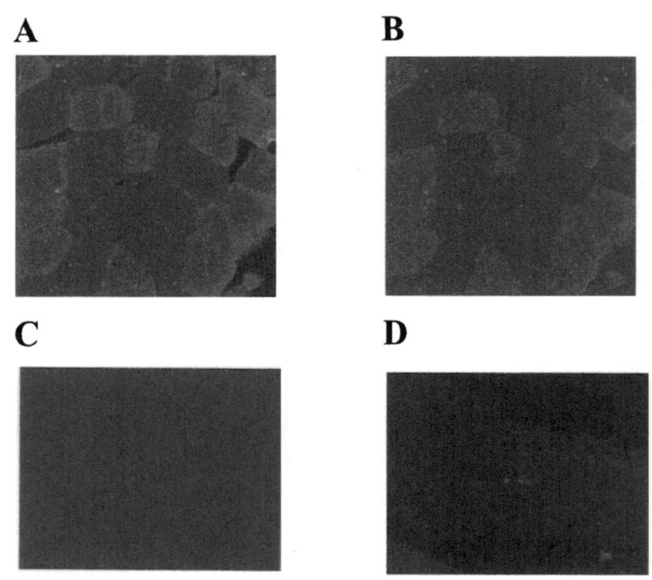


*Fig. 3.* Epifluorescence (×40 amplification) of 10-μm cross-section of human skeletal muscle. (A) detection of CreaT protein using AlexaFluor-488 secondary antibody. (B) detection of MHC1 protein using AlexaFluor-568 antibody. Negative control sections with AlexaFluor-488 (C) and AlexaFluor-568 (D) secondary antibodies.

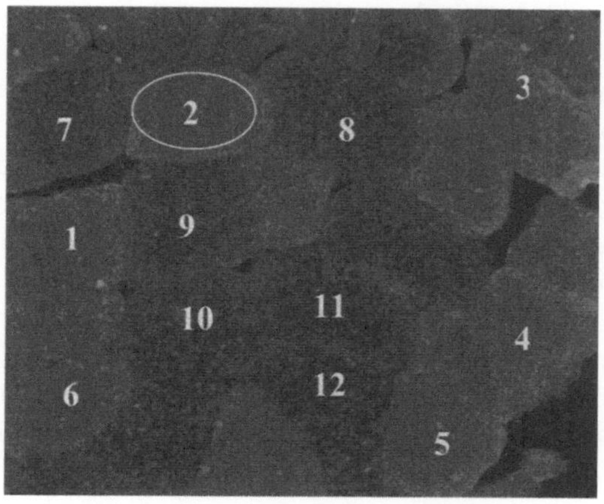


*Fig. 4.* Epifluorescence (×40 amplification) of 10-μm cross-section of human skeletal muscle. Detection of CreaT protein using AlexaFluor-488 secondary antibody fluorescence units from 360 pixels (indicated by ellipse) from fibres (labelled 1–12) were counted using ImagePro Express software. Type I fibres, labelled 1–6, had a greater fluorescence than type II fibres, labelled 7–10 ($p < 0.05$).

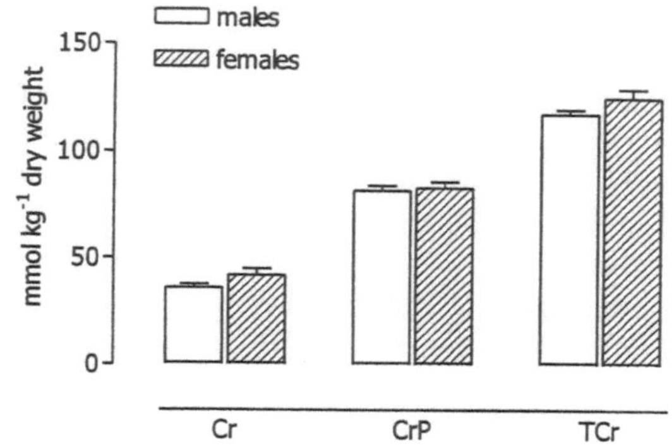


*Fig. 6.* Muscle metabolites (mmol.kg$^{-1}$ dry weight) for males (n = 13, open bars) and females (n = 11, hatched bars). Creatine (Cr), creatine phosphate (CrP) and total creatine (TCr = Cr + CrP). Values are means ± S.E.

expression is generally a complex process, involving many potential levels of control such as the abundance and activity of specific transcription factors within the nuclei, transcription rate and mRNA stability [28]. The present study has demonstrated that skeletal muscle CreaT mRNA expression normalised to β-actin was similar between males and females (see Fig. 5), perhaps indicating that the factors regulating the transporter mRNA expression in this tissue are likely to be similar between the sexes. To date only one study has attempted to examine the regulation of CreaT mRNA expression [29]. These researchers induced an elevated TCr content in rat myocardium by supplementing the diet of these rats with Cr for 16 days and found that the myocardial CreaT mRNA expression increased along with the enhanced TCr content. Perhaps surprisingly, these findings suggest that CreaT gene transcription rates are elevated and/or transporter mRNA degradation rates are attenuated with an increased TCr content. It is not known if skeletal muscle CreaT mRNA content would respond in a similar manner to Cr supplementation, although it has been demonstrated that CreaT protein levels are decreased in rat skeletal muscle following 3–6 months of Cr supplementation.

The present study identified two CreaT isoforms of apparent molecular weight ~ 73 (CreaT73) and 55 (CreaT55) kDa in all muscle samples (see Fig. 1). This observation is consistent with previous research using rat [13, 15] and middle-aged human [16] skeletal muscle. There was no influence of gender on the relative expression of either isoform or on the total expression of the protein (see Fig. 2). The major CreaT species, ~ 70 and 55 kDa, have been localised to the inner mitochondrial membrane of rat cardiomyocytes [30]. At least one of these isoforms is likely to be responsible for the active Cr transport that has been observed into isolated mitochondria obtained from rat cardiac muscle [30]. Additionally, Tran *et al.* [14] and Walzel *et al.* [30] have reported that the

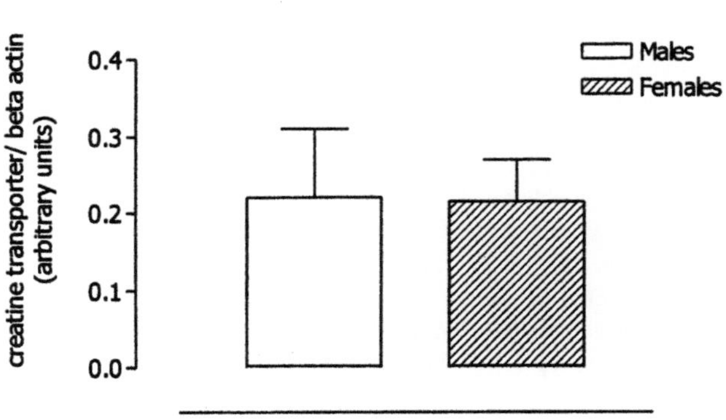


*Fig. 5.* Creatine transporter mRNA normalised to β-actin mRNA for males (n = 11, open bars) and females (n = 12, hatched bars). Values are means ± S.E.

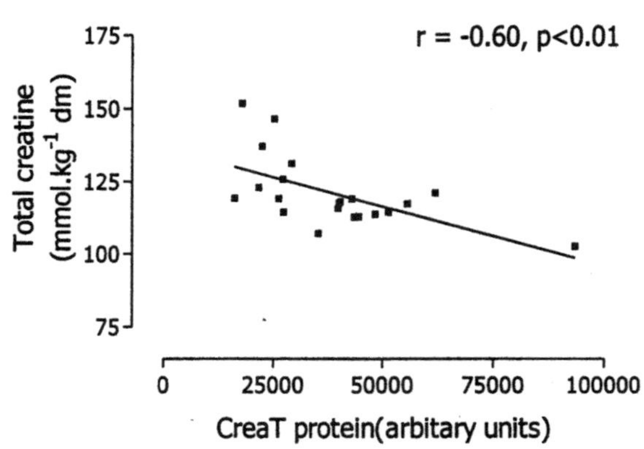


*Fig. 7.* The relationship between creatine transporter protein and total creatine content in human skeletal muscle (n = 21). Linear regression, $y = -0.0004x + 136.72$.

CreaT protein is also located on the plasma membrane in mouse myoblast cells and rat cardiomyocytes, respectively. Preliminary work [30] suggests that the sarcolemmal CreaT isoform is slightly larger (i.e. 58 kDa) than the ~ 55 kDa isoform and is relatively low in abundance compared with the other isoforms. Taken together, these data suggest that the CreaT proteins identified in the current experiment are likely to be the mitochondrial isoforms. The role(s) that these transporters play in mitochondria and cellular function are not known, but have been suggested to allow a mitochondrial pool of Cr and CrP [30].

The predominant localisation of the CreaT to mitochondria [30] probably explains our earlier findings that the CreaT protein content is fibre type dependent in rat skeletal muscle [15]. Interestingly, this relationship is likely to be similar in human skeletal muscle, since the data presented in the current study demonstrate that human fibres expressing myosin heavy chain type I (MHC1) protein (i.e. type I fibres) display greater CreaT content compared with type II fibres (Figs 3 and 4). These data indicate that differences in muscle fibre type distribution may be a factor that needs to be considered in future studies when measuring CreaT protein expression. Whilst conflicting data exist on whether fibre type composition differs between men and women [31, 32], a more recent study has demonstrated that fibre type distribution of the vastus lateralis muscle of 150 young, physically active, but untrained subjects was similar between genders [33]. Consequently, differences in fibre type composition is unlikely to confound the interpretation of our data.

The inverse relationship between the TCr and CreaT protein content when the male and female data are combined (see Fig. 7), or when each gender is treated separately, is possibly explained by the fact that human type I fibres have a lesser TCr but a higher mitochondrial content [34] than type II fibres [20]. Assuming that the current study has measured CreaT isoforms predominantly from mitochondrial origin (see above), then it is perhaps not surprising that those fibres with more mitochondria will have more CreaT protein and also tend to have lower levels of TCr. A further regulatory mechanism of the CreaT is observed in rat skeletal muscle, whereby chronic, high dose supplementation Cr resulted in a decrease in both the 55 and 70 kDa CreaT isoform expression in skeletal muscle [13]. Although not reported by these authors, Cr supplementation would be expected to increase Cr and TCr content in muscle [35]. Hence mitochondrial content and/or intramuscular TCr stores may be involved in the regulation of the CreaT protein expression and provide an explanation for the inverse relationship between CreaT protein and TCr content observed in the present experiment.

Previous studies have consistently reported that the intramuscular TCr content expressed relative to dry muscle mass is similar between males and females [19, 21, 22, 36]. In agreement with this research, the current study also found no gender difference in the resting intramuscular contents of Cr, CrP or TCr, when comparing values expressed as $mmol.kg^{-1}$ dry mass (see Fig. 6) or when these metabolites were normalised to intramuscular ATP content (data not shown).

In conclusion, the present study has demonstrated that CreaT mRNA and protein (CreaT-55 and CreaT-73) are expressed in skeletal muscle in healthy young adults, with a similar expression between male and female subjects. The present experiment has also demonstrated that CreaT protein expression in human skeletal muscle appears to be fibre type dependent. This knowledge provides an important fundamental first step in attempting to understand the regulation of CreaT gene expression in human skeletal muscle.

## Acknowledgements

We would like to acknowledge the subjects in this study who kindly volunteered to undergo the invasive procedures employed in this experiment. We would also like to thank Drs. Andrew Garnham and Peter Braun, both of whom performed the medical procedures described in this study.

## References

1. Wallimann T, Wyss M, Brdiczka D, Nicolay K, Eppenberger HM: Intracellular compartmentation, structure and function of creatine kinase isoenzymes in tissues with high and fluctuating energy demands: The 'phosphocreatine circuit' for cellular energy homeostasis. Biochem J 281: 21–40, 1992
2. Van Pilsum JF, Stephens GC, Taylor D: Distribution of creatine, guanidinoacetate and the enzymes for their biosynthesis in the animal kingdom. Implications for phylogeny. Biochem J 126: 325–345, 1972
3. Hoberman HD, Sims EAH, Peters JH: Creatine and creatinine metabolism in the normal male adult studied with the aid of isotopic nitrogen. J Biol Chem 172: 45–58, 1948
4. Loike JD, Zalutsky DL, Kaback E, Miranda AF, Silverstein SC: Extracellular creatine regulates creatine transport in rat and human muscle cells. Proc Natl Acad Sci 85: 807–811, 1988
5. Guimbal C, Kilimann MW: A Na(+)-dependent creatine transporter in rabbit brain, muscle, heart, and kidney. cDNA cloning and functional expression. J Biol Chem 268: 8418–8421, 1993
6. Dai WX, Vinnakota S, Qian XJ, Kunze DL, Sarkar HK: Molecular characterization of the human CRT-1 creatine transporter expressed in *Xenopus* oocytes. Arch Biochem Biophys 361: 75–84, 1999
7. Odoom JE, Kemp GJ, Radda GK: The regulation of total creatine content in a myoblast cell line. Mol Cell Biochem 158: 179–188, 1996
8. Daly MM, Seifter S: Uptake of creatine by cultured cells. Arch Biochem Biophys 203: 317–324, 1980
9. Eichler EE, Lu F, Shen Y, Antonacci R, Jurecic V, Doggett NA, Moyzis RK, Baldini A, Gibbs RA, Nelson DL: Duplication of a gene-rich cluster between 16p11.1 and Xq28: A novel pericentrometric-directed mechanism for paralogous genome evolution. Hum Mol Gen 5: 899–912, 1996
10. Iyer GS, Krahe R, Goodwin LA, Doggett NA, Siciliano MJ, Funanage VL, Proujansky R: Identification of a testis-expressed creatine transporter gene at 16p11.2 and confirmation of the X-linked locus to Xq28. Genomics 34: 143–146, 1996

11. Sandoval N, Bauer D, Brenner V, Coy JF, Drescher B, Kioschis P, Korn B, Nyakatura G, Poustka A, Reichwald K, Rosenthal A, Platzer M: The genomic organization of a human creatine transporter (CRTR) gene located in Xq28. Genomics 35: 383–385, 1996
12. Gregor P, Nash SR, Caron MG, Seldin MF, Warren ST: Assignment of the creatine transporter gene (SLC6A8) to human chromosome Xq28 telomeric to G6PD. Genomics 25: 332–333, 1995
13. Guerrero-Ontiveros ML, Wallimann T: Creatine supplementation in health and disease. Effects of chronic creatine ingestion *in vivo*: Down-regulation of the expression of creatine transporter isoforms in skeletal muscle. Mol Cell Biochem 184: 427–437, 1998
14. Tran TT, Dai W, Sarkar HK: Cyclosporin A inhibits creatine uptake by altering surface expression of the creatine transporter. J Biol Chem 275: 35708–35714, 2000
15. Murphy R, McConell G, Cameron-Smith D, Watt K, Ackland L, Walzel B, Wallimann T, Snow R: Creatine transporter gene expression and protein localization in rat skeletal muscle. Am J Physiol 280: C415–422, 2001
16. Tarnopolsky MA, Parshad A, Walzel B, Schlattner U, Wallimann T: Creatine transporter and mitochondrial creatine kinase protein content in myopathies. Muscle Nerve 24: 682–688, 2001
17. Snow RJ, Murphy RM: Creatine and the creatine transporter: A review. Mol Cell Biochem 224: 169–181, 2001
18. Tarnopolsky MA: Gender differences in substrate metabolism during endurance exercise. Can J Appl Physiol 25: 312–327, 2000
19. Parise G, Mihic S, MacLennan D, Yarasheski KE, Tarnopolsky MA: Effects of acute creatine monohydrate supplementation on leucine kinetics and mixed-muscle protein synthesis. J Appl Physiol 91: 1041–1047, 2001
20. Esbjornsson-Liljedahl M. Human muscle response to sprint exercise in a gender perspective. PhD thesis. Huddinge University Hospital, Karolinska Institute, Stockholm, 2000
21. Balsom PD, Soderlund K, Ekblom B: Creatine in humans with special reference to creatine supplementation. Sports Med 18: 268–280, 1994
22. Moller P, Bergstrom J, Furst P, Hellstrom K: Effect of aging on energy-rich phosphagens in human skeletal muscles. Clin Sci 58: 553–555, 1980
23. Forsberg AM, Nilsson E, Werneman J, Bergstrom J, Hultman E: Muscle composition in relation to age and sex. Clin Sci Colch 81: 249–256, 1991
24. Harris RC, Soderlund K, Hultman E: Elevation of creatine in resting and exercised muscle of normal subjects by creatine supplementation. Clin Sci 83: 367–374, 1992
25. Harris RC, Hultman E, Nordesjo LO: Glycogen, glycolytic intermediates and high-energy phosphates determined in biopsy samples of musculus quadriceps femoris of man at rest. Methods and variance of values. Scand J Clin Lab Invest 33: 109–120, 1974
26. Lowry OH, Passoneau JV: In: A Flexible System of Enzymatic Analysis. Academic Press, New York, 1972
27. McKenna MJ, Morton J, Selig SE, Snow RJ: Creatine supplementation increase muscle total creatine, but not maximal intermittent exercise performance. J Appl Physiol 87: 2244–2252, 1999
28. Staton JM, Thomson AM, Leedman PJ: Hormonal regulation of mRNA stability and RNA-protein interactions in the pituitary. J Mol Endocrinol 25: 17–34, 2000
29. Rajab P, Greenhaff PL, Constantin-Teodosiu D, Bennett A, Anderton KA, White DA, Mayer RJ, Gardiner SM, Bennett T: Dietary creatine supplementation increases creatine transporter mRNA content and 26S proteasome activity in rat myocardium. J Physiol 515: 64P, 1999
30. Walzel B, Boehm E, Speer O, Kristiansen S, Richter EA, T W. Novel creatine transporter (CRT) isoforms: A minor species in plasma membrane vesicles and two major species in mitochondria. In 6th International Conference on Guanidino Compounds in Biology and Medicine. Cincinnati, USA, 2001
31. Essen-Gustavsson B, Borges O: Histochemical and metabolic characteristics of human skeletal muscle in relation to age. Acta Physiol Scand 126: 107–114, 1986
32. Simoneau JA, Bouchard C: Human variation in skeletal muscle fiber-type proportion and enzyme activities. Am J Physiol 257: E567–E572, 1989
33. Staron RS, Hagerman FC, Hikida RS, Murray TF, Hostler DP, Crill MT, Ragg KE, Toma K: Fiber type composition of the vastus lateralis muscle of young men and women. J Histochem Cytochem 48: 623–629, 2000
34. Ogata T, Yamasaki Y: Ultra-high-resolution scanning electron microscopy of mitochondria and sarcoplasmic reticulum arrangement in human red, white, and intermediate muscle fibers. Anat Rec 248: 214–223, 1997
35. Brannon TA, Adams GR, Conniff CL, Baldwin KM: Effects of creatine loading and training on running performance and biochemical properties of rat skeletal muscle. Med Sci Sports Exerc 29: 489–495, 1997
36. Esbjornsson-Liljedahl M, Sundberg CJ, Norman B, Jansson E: Metabolic response in type I and type II muscle fibers during a 30-sec cycle sprint in men and women. J Appl Physiol 87: 1326–1332, 1999

*Molecular and Cellular Biochemistry* **244:** 159–166, 2003.


# Acute and moderate-term creatine monohydrate supplementation does not affect creatine transporter mRNA or protein content in either young or elderly humans

Mark Tarnopolsky,[1,2] Gianni Parise,[2] Min-Hua Fu,[1] Andrea Brose,[2] Andrew Parshad,[1] Oliver Speer[3] and Theo Wallimann[3]

*Departments of [1]Medicine (Neurology and Rehabilitation); [2]Kinesiology, McMaster University, Hamilton, ON, Canada; [3]Institute of Cell Biology (Swiss Federal Institute of Technology), Hoenggerberg, Zurich, Switzerland*

## Abstract

Animal studies have shown that supra-physiological creatine monohydrate (Cr-mH) supplementation for 3 months reduced skeletal muscle creatine transporter (CRT) content. The doses of Cr-mH (1–2 g/kg/day) used in these studies were between 5 and 10 times those usually used in human studies, and it is unclear whether a down-regulation of CRT would occur in humans at the recommended doses of 0.1–0.2 g/kg/day. We measured CRT, and citrate synthase (CS) protein content using Western blotting before and after 2 months of Cr-mH supplementation and weight training in young men (N = 11 Cr-mH (0.125 g/kg/day); N = 8 placebo). CRT and CS were also measured before and after 4 months of Cr-mH supplementation and weight training in elderly (> 65 years) men and women (N = 14 Cr-mH (0.075 g/kg/day); N = 14 placebo). Finally, CRT mRNA was measured using competitive RT-PCR before and after 8–9 days of Cr-mH loading in young men and women (N = 14, CR-mH (mean = 0.18 g/kg/day); N = 13, PL). Total creatine content was significantly elevated after the Cr-mH supplementation period as compared to placebo in each of the studies. Neither Cr-mH supplementation, nor exercise training resulted in measurable alterations in CRT protein content and acute Cr-mH loading did not alter CRT mRNA. There were no gender differences in CRT mRNA or total creatine content in the young subjects and no gender differences in total creatine content or CRT protein content in the elderly subjects. Weight training in young men did not increase CS protein content, however, in the elderly there was a significant increase in CS protein content after exercise training ($p < 0.05$). These results demonstrated that Cr-mH supplementation during weight training resulted in increases in skeletal muscle total creatine without reductions in CRT protein and acute Cr-mH loading did not decrease CRT mRNA content. (Mol Cell Biochem **244**: 159–166, 2003)

*Key words*: creatine transport, dietary supplements, drug safety, transporter regulation

## Introduction

The transport of creatine into tissues is mediated predominantly by a sodium dependant creatine transporter (CRT, SLC6A8) that has been sequenced, mapped, and heterologously expressed [1, 2]. The SLC6A8 gene is on the X chromosome with an open reading frame of 1,905 bp, with 13 exons and 12 transmembrane spanning regions [1, 2]. There are two main protein species in skeletal muscle with apparent molecular weights of 52- and 70-kDa, which appear to arise from alternative splicing of the mRNA [3]. Prior to the identification of the CRT, it was known that creatine uptake into muscle was predominantly a saturable process [4], and that extracellular creatine regulated creatine transport in muscle by a protein synthesis dependant mechanism [5]. This process was later linked to the CRT from the results of ex-

*Address for offprints*: M. Tarnopolsky, Department of Neurology, Room 4U4, McMaster University Medical Center, 1200 Main Street West, Hamilton, Ontario, Canada (E-mail: tarnopol@mcmaster.ca)

periments where the transient expression of CRT in COS-7 cells showed sodium dependence and attenuation by exogenous creatine [1]. More recent *in vivo* evidence demonstrated that the administration of high dose Cr-mH (4% diet; ~ 1 g/kg/day) to rats over a 3 month period resulted in a decrease in CRT protein content [3].

Creatine monohydrate (Cr-mH) supplementation has become a popular practice among athletes wishing to increase muscle mass and power [6]. Oral Cr-mH supplementation at dosages of 20 g per day (~ 0.2–0.3 g/kg/day) over 3–5 days resulted in an increase in muscle total creatine and phosphocreatine by 12–20% [6–8]. Other studies have shown a maintenance of elevated total creatine and phosphocreatine concentration following an acute load with the continued consumption of 2 g per day (~ 0.03 g/kg/day) for a month and similar increases were observed without the initial load at 3 g per day (~ 0.04 g/kg/day) for a month [7]. Several studies have shown that this increased high intensity power output particularly with repetitive bouts of activity [6, 9, 10]. Other studies have demonstrated that the longer-term ingestion of Cr-mH over weeks to months resulted in a greater increase in fat free mass, muscle power and muscle fiber area during resistance exercise training [11–13].

Based upon the above data, there has been some concern that the longer term administration of Cr-mH in athletes and patients may down-regulate creatine uptake [3, 6]. Ultimately, if Cr-mH supplementation were stopped, there could be an 'undershoot' of muscle creatine stores. In one longer-term study, in young men, there was a significant increase in muscle total creatine at 1 week of supplementation, yet this was not significantly elevated with 12 weeks of continued supplementation (~ 0.075 g/kg/day) [13]. Conversely, the muscle phospho-creatine content remained elevated above baseline following 10 weeks of supplementation and following 6 weeks of discontinuing the supplementation there was no 'undershoot' [11]. To date there have been no determinations of muscle CRT protein content in humans following short or longer Cr-mH supplementation. Given that the dose of Cr-mH given to rats in the aforementioned study, showing a reduction in CRT content [3], was approximately 4 fold higher than the acute loading dose in humans, and 30 fold higher than the dose shown to increased total creatine content after 30 days of continuous dosing [7], it is likely that the stimulus for down regulating the CRT would be much less or even non-existent in humans following creatine supplementation at recommended doses used in human studies.

Therefore, the purpose of the current study was to determine whether a moderate-term (i.e. months) Cr-mH supplementation protocol would down-regulate the total amount of CRT protein in young and elderly individuals participating in a resistance exercise training protocol. We also evaluated whether an acute Cr-mH supplementation strategy would down regulate the CRT mRNA content.

## Materials and methods

### *Subjects/design*

The data for the current study came from muscle samples obtained from 3 separate studies that were all approved by the McMaster University Ethics Committee and conformed to the principles of the Declaration of Helsinki.

### *Study #1 – acute creatine loading in young men and women*

This study was a randomized, double-blind study designed to examine the effect of an acute 'loading' dose of Cr-mH compared to placebo (PL-glucose polymer) on protein turnover in moderately active young men (N = 13) and women (N = 14). The protein turnover data, phosphocreatine and total creatine concentrations have been reported previously [14]. Each subject had a muscle biopsy taken from the *vastus lateralis* under local anaesthesia (which was immediately quenched in liquid nitrogen) before and after 8–9 days of each intervention. The Cr-MH group consumed 4 × 5 $g \cdot day^{-1}$ of Cr-mH (99% pure, ISA, Hamilton, ON, Canada) for 5 days, and then consumed a maintenance dose of 5 $g \cdot day^{-1}$ for 3–4 days (mean = ~ 0.18 g/kg/day over the 8–9 days) and the PL group consumed an equivalent amount of glucose polymer. Subjects were instructed to consume their supplement dissolved in juice, chocolate milk, or a carbohydrate-containing soda beverage and refrained from leg exercise for 3 days prior to the testing session.

### *Study #2 – creatine supplementation during 2 months of resistance exercise training in young men*

This study was a randomized, double-blind study designed to examine the effect of Cr-mH supplementation as compared to PL when consumed during an intensive 2 month resistance exercise training program in moderately active young men (N = 19). The strength gains, total and fat-free mass gains and muscle fiber characteristics, protein turnover data, phosphocreatine and total creatine concentrations have been previously reported [15]. Each subject had a muscle biopsy taken from the *vastus lateralis* under local anaesthesia (which was immediately quenched in liquid nitrogen) before and after the two month resistance exercise program. The Cr-CHO group consumed 10 g of Cr-mH and 75 g of dextrose within 30 min of completing each of the exercise sessions, (N = 11; mean = ~ 0.125 g/kg/day, Cell Tech, Muscle Tech. Research and Development, Mississauga, ON, Canada), and the PL group consumed 10 g of casein and 75 g of dextrose (N = 8) at the same time. The 8 weeks of exercise training was performed 6 days per week under direct supervision with each subject completing a split-routine weight training, 1 h·day$^{-1}$ @ 80% of one repetition maximum, 3 sets, 6 days·week$^{-1}$, on a 3day rotating schedule.

*Study #3 – four months of resistance exercise training in men and women over 65 years of age*

Fifteen men (67.8 ± 4.0 years) and 15 women (69.3 ± 6.3 years, none taking hormone replacement therapy) volunteered to participate in a 14-week resistance exercise training program. The subjects were randomly assigned in a double blind manner to receive either a Cr-mH (5 g Cr-mH (~ 0.075 g/kg/day) + 2 g dextrose/day (Neotine, Avicena, Cambridge, MA, USA) or a placebo (PL; 7 g dextrose/day) supplement during the training program. Training was conducted 3 times weekly on nonconsecutive days for 14 weeks. Each training session was preceded by a 5-min warm-up and followed by a cool down consisting of stretching of the muscle groups involved in the resistance exercises. Using weight training machines, twelve exercises were used to train the major muscle groups of the upper and lower body in a circuit set system: seated chest press, latissimus pull-down, leg press, military press, calf raise, arm extension, arm flexion, back extension, abdominal crunch, upright row, knee extension, and knee flexion. Subjects performed 10 repetitions of each arm exercise and 12 repetitions of the remaining exercises. Training progressed from one set of each exercise at 50% of the initial 1 repetition maximum (1 RM) to three sets at 80% of 1 RM over the training period. Each subject had a muscle biopsy taken from the *vastus lateralis* under local anaesthesia (which was immediately quenched in liquid nitrogen) before and after the 14-week resistance exercise program.

*Muscle analyses*

*Phosphocreatine, creatine and ATP*
Muscle samples were lyophylized and powdered and the metabolites were extracted using a perchloric acid method as previously described [16]. The resultant extract was assayed for phosphocreatine, free creatine and ATP using enzymatic methods as previously described [16].

*Western blotting*
Muscle samples were homogenized and prepared for electrophoresis using methods previously described [17]. A membrane preparation was also made from fresh human erythrocytes. The muscle samples from the training studies in the young males (Study #2) and from the elderly men and women (Study #3) using polyclonal rabbit antibodies directed against the C-terminus of the rat CRT-1 (synthetic 15mer peptide corresponding to amino acids 621–635 derived from the known sequence of CRT-1, SwissProt P48029). This antibody recognizes two major polypeptides with apparent molecular weights of ~ 52 kDa (CRT-52) and 70 kDa (CRT-70) [3, 17, 18]. In addition, we had enough muscle homogenate left from Study #3 that we were able to probe using a newer antibody a new sheep antibody against the sequence motive, GSAWERRGESTMSAH, of a synthetic peptide of CRT-2 (Swiss Prot P53796) [2]. By using this antibody, we obtained a more robust and consistent band at 52 kDa and we found a polypeptide band at ~ 56 kDa (Walzel *et al.*, 2001 (in this issue); [19]). We also probed the blots from Study #2 and #3 using a polyclonal rabbit, anti-human antibody raised against purified human citrate synthase protein (Roche Chemicals) which recognized one polypeptide with an apparent molecular weight of ~ 65 kDa (a kind gift of Dr. B.H. Robinson).

*RT-PCR*
Total RNA was extracted from frozen human muscle by using TRIZOL Reagent according to the manufacturer's instructions (GIBCO-BRL, Life Technologies). Tissue was hand homogenized (glass-Teflon) in TRIZOL Reagent (25–30 mg/ml). The final RNA pellet was re-suspended in 12 μL of RNase-free water. Total RNA concentration was determined spectrophotometrically at 260 nm, and samples were treated with DNaseI using the DNA-free™ Kit (Ambion). Primers complementary to selected regions of the gene encoding for the CRT (GenBank accession No. L31409) were designed to produce a single 86 base pair (bp) product using Primer Express software (PE Biosystems, Foster City, CA, USA). The forward primer sequence (5′-3′) was GCCGGC-AGCATCAATGTC, and the reverse primer sequence (5′-3′) was GGTGTTGCAGTAGAAGACGATCAC. 18S rRNA was used as an internal control according to the manufacturer's instructions using a QuantumRNA™ 18S Internal Standards Kit (Ambion). The Alternate 18S Internal Standards were used, which produced a 324 bp PCR product.

Reverse transcriptase polymerase chain reaction (RT-PCR) was performed with the use of Titan One Tube RT-PCR Kit (Roche Molecular Biochemicals, Germany) according to the instruction manual with the following minor modifications; 10 μL reaction volumes of 1 × RT-PCR buffer, 0.5 μM of each dNTP, 0.5 μM forward CRT primer, 0.5 μM reverse CRT primer, 1.2 μM Alternate 18S Internal Standards primer mixture (18S PCR Primer Pair to 18S PCR Competimers, 1:18), 3.0 mM $MgCl_2$, 0.4 μl enzyme mixture, and 0.2 μg total RNA template per tube. Reverse transcription was carried out at 50°C for 30 min. RT-PCR was run for 1 cycle @ 94°C for 2 min, 10 cycles (94°C for 30 sec, 53°C for 30 sec, 68°C for 45 sec), 18 cycles (94°C for 30 sec, 53°C for 30 sec, 68°C for 50 sec with additional 5 sec to each successive cycle), and 1 final annealing cycle @ 68°C for 7 min. Control experiments were performed under identical conditions to determine the ratio of the density of the CRT PCR product:18S rRNA PCR product was within the linear range at PCR cycle 28. At 28 cycles, the coefficient of variation with replicate samples was 13.8%. For each gel, male and female, pre- and post-supplementation samples were run simultaneously. Products were run on a 2% agarose gel, stained with ethidium bromide. Image was acquired and analyzed with the use of UVP UV Darkroom and LabWorks™ Image Acquisition and Analysis Software.

*Statistical analyses*

A two-way analysis of variance (ANOVA) was used to analyze the mRNA data from Study #1 and the protein data from Study #3, with gender as a within variable and Cr-mH vs. placebo as the other between variable. When no gender differences were found, the genders were collapsed and the data was run as a between-within split-plot design with Cr-mH and placebo as the between variable and PRE/POST treatment as the within (repeated measure) variable. The protein data from Study #2 was analyzed with a between-within split-plot design with CR-CHO and PRO-CHO as the between variables and PRE/POST treatment as the within (repeated measure) variable. A p value of $\leq 0.05$ was considered to be statistically significant.

## Results

*Study #1*

There were no gender differences in CRT mRNA/18-S ribosomal RNA content between men and women (p = N.S.). CRT/18-S mRNA content did not change as a result of the acute Cr-mH supplementation protocol (P = N.S.) (Fig. 1B). Furthermore, men and women showed similar increases in muscle total creatine in response to the Cr-mH supplementation and had similar baseline total creatine concentrations (Table 1).

*Study #2*

Two months of supplementation with Cr-mH and carbohydrate (CR-CHO) resulted in a significant increase in total muscle creatine concentration, whereas there was no increase in response to protein and carbohydrate (PRO-CHO) (Table 1). Neither supplementation nor resistance exercise training altered either the citrate synthase (not shown) or CRT-70 content (Fig. 1A). The CRT-52 band was not reproducibly apparent in all of the blots using the CRT-1 antibody, hence, we did not include this analysis (with the limited number of homogenate available for probing with the newer CRT-

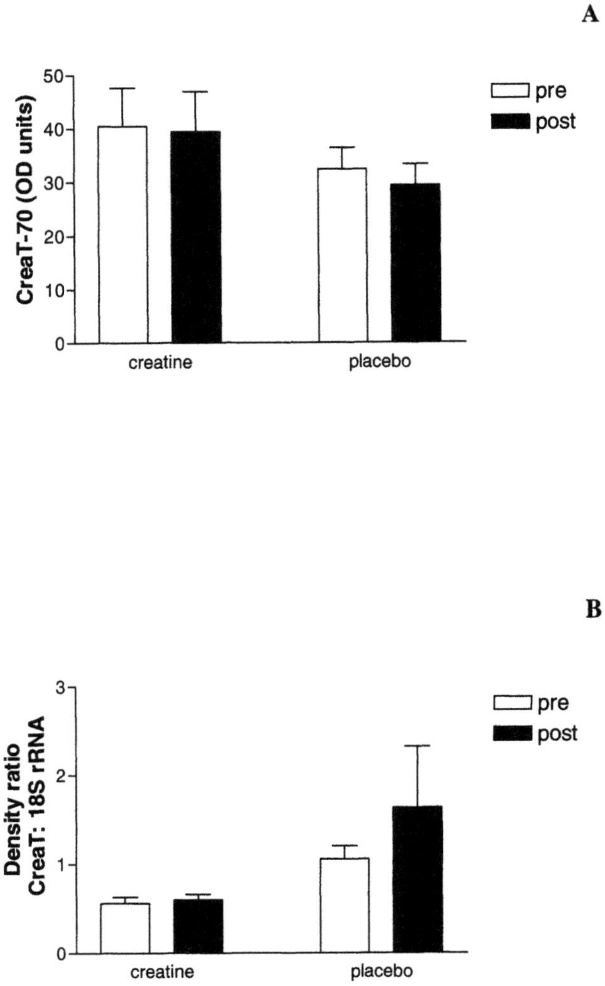


*Fig. 1.* (A) CreaT-70 protein content before (pre) and after (post) 2 months of resistance exercise training in young men consuming Cr-mH (10 g/day) + dextrose (70 g/day) or casein (10 g/day) + dextrose (70 g/day; 'placebo') (Study #2). (B) CreaT mRNA expressed relative to 18-S rRNA mRNA content in young men and women (collapsed for there were no gender differences) before (pre) and after (post) ~ 9 days of Cr-mH 'loading' (Study #1; see text for more details; values are mean ± S.D.).

2 antibody, we found that pre and post-supplementation analysis was possible on only 5 subjects and there was no evidence for down-regulation of the CRT protein, however with such a limited sample size, we feel that it is best to interpret these results as preliminary).

*Study #3*

Four months of Cr-mH supplementation (5 g/day, ~ 0.075 g/kg/day) in elderly subjects who were performing resistance exercise training, resulted in an increase in muscle total creatine concentration for both males and females (Table 1). CRT-70 protein content was similar in males and post-menopausal females and there was no effect of training (PRE/POST) or

*Table 1.* Total creatine concentration

| | Total creatine (mmol × $kg^{-1}$ dm) | | | |
|---|---|---|---|---|
| | Males | | Females | |
| | Cr-mH | Placebo | Cr-mH | Placebo |
| Study 1 | | | | |
| Pre | 125.5 ± 8.0 (n = 7) | 128.0 ± 16.0 (n = 7) | 134.0 ± 14.0 (n = 7) | 131.3 ± 13.0 (n = 7) |
| Post | 141.2 ± 12.0* | 130.0 ± 8.0 | 151.5 ± 25.0* | 134.7 ± 13.0 |
| Study 2 | | | | |
| Pre | 134.8 ± 12.0 (n = 11) | 125.6 ± 6.0 (n = 8) | N/A | N/A |
| Post | 142.9 ± 8.0* | 124.1 ± 7.0 | N/A | N/A |
| Study 3 | | | | |
| Pre | 116.8 ± 15.0 (n = 8) | 140.8 ± 21.0 (n = 7) | 129.7 ± 25.0 (n = 6) | 138.5 ± 14.0 (n = 7) |
| Post | 159.3 ± 24.0* | 125.5 ± 25.0 | 151.7 ± 19.0* | 147.5 ± 20.0 |
| Basal | | 130.1 ± 13.0 (n = 47) | | 127.2 ± 19.2 (n = 30) |

Values are mean ± S.D. *Denotes a significant change ($p < 0.05$) after supplementation. Study #1. Acute (9 day) creatine loading in young men and women. Study #2. Two months of resistance exercise training in young men with Cr-mH (10 g) + dextrose (70 g) or caseine (10 g) + dextrose (70 g). Study #3. Four months of resistance exercise training in elderly (> 65 years) men and women with Cr-mH (5 g), or placebo (dextrose). (see text for further details). Basal refers to resting, non-supplemented values in males and females.

supplementation (Cr-mH vs. PL) (Fig. 2A). In contrast to the younger subjects performing resistance exercise, the elderly subjects showed an increase in citrate synthase protein content ($p < 0.05$), with no effect of supplementation (N.S., data not shown). There was a significant correlation ($r = 0.72$, $p < 0.01$) between the change in citrate synthase and CreaT (CRT-70 + CRT-52) before and after exercise training. Using the original polyclonal rabbit antibody directed against the COOH-terminus of CRT-1 (SwissProt P48029), we also found low signal and inconsistent bands for the CRT-52 band in the elderly, however, the CRT-52 band was consistently apparent using the newer CRT-2 sheep antibody (Swiss Prot P53796), and there was no effect of supplementation or exercise training (Fig. 2B). In about 50% of the samples a faint band was also observed with an apparent MW of ~ 56 kDa using the CRT-2 antibody, that has been proposed as the actual sarcolemmal transporter (Walzel *et al.* (abstract in this issue) and Walzel *et al.* 2002 (in press). To confirm this, we made a human red blood cell preparation (no mitochondria) and found that the CRT-2 antibody recognized a single band with an apparent MW of ~ 56 kDa (Fig. 3, upper). When human muscle samples were overloaded on the gels, this same band was observed at the same MW of ~ 56 kDa (Fig. 3, lower). At this point however, it would be too speculative to report on the data concerning the quantitation of this latter polypeptide from the limited number of observations before and after supplementation in each of the groups.

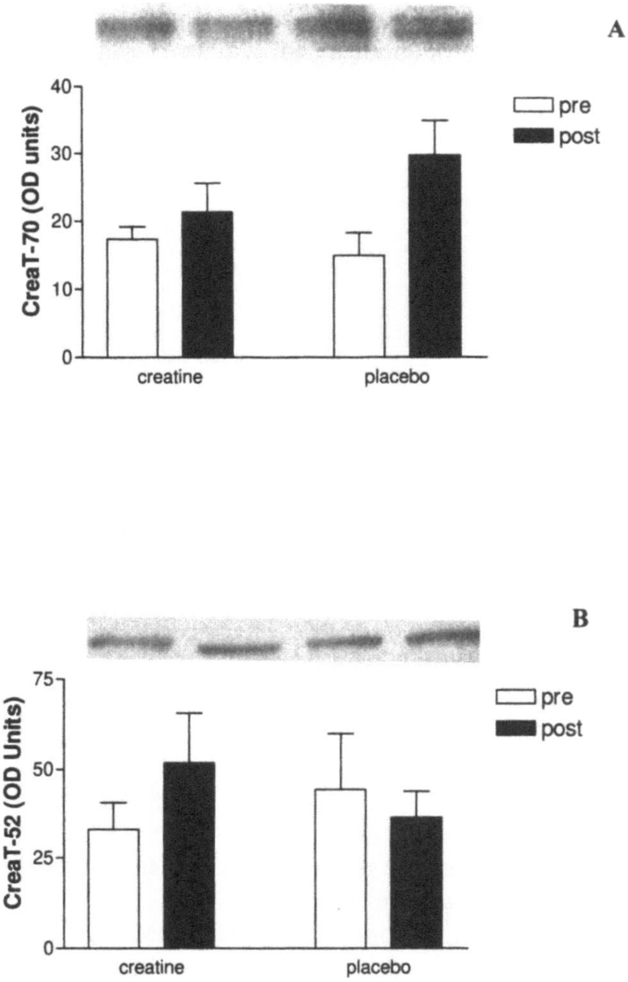


*Fig. 2.* (A) CreaT-70 protein content measured using the CRT-1 antibody before (pre) and after (post) 4 months of resistance exercise training in elderly (> 65 years) men and women (collapsed across gender for their were no gender differences) consuming Cr-mH (5 g/day) or placebo (dextrose 7 g/day) (Study #3). (B) Identical set-up but using the CRT-2 antibody to emphasize the 52 kDa band (Study #3).

## Discussion

The main findings from the current study are that two and four months of Cr-mH supplementation at doses of ~ 0.075–0.125 g/kg/day, did not result in a decrease in the amount of immunodetectable CRT protein. Furthermore, an acute Cr-mH 'loading' protocol for 8–9 days at a dose of 0.18 g/kg/day did not decrease muscle CRT/18-S mRNA content. Additionally, there does not appear to be a gender difference in basal CRT/18-S mRNA or CRT protein content. Finally, the increase in muscle total creatine in response to acute Cr-mH

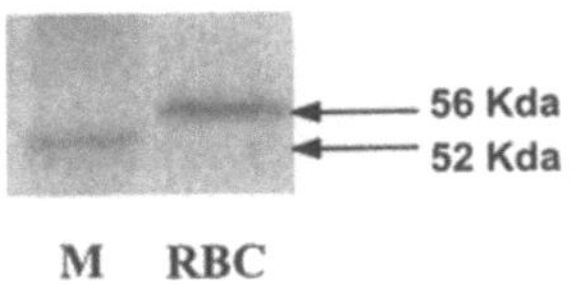


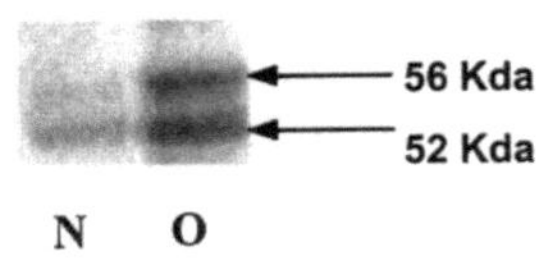


*Fig. 3.* (Upper) CreaT-52 protein in human skeletal muscle (M) using the CRT-2 antibody and the corresponding results in a purified human red blood cell (RBC) preparation highlighting the presumed plasma membrane isoform (56 kDa). (Lower) CreaT-52 protein in human skeletal muscle using the CRT-2 antibody loaded with the 'normal' (N) amount of protein and the same skeletal muscle sample with and overloaded amount of protein (O, 3 × the usual amount), highlighting that the 56 kDa band is visible in human skeletal muscle, however, to consistently observe the band ~ 3 times more protein must be loaded to obtain comparable optical density values to the 70 and 52 kDa bands.

loading or lower dose supplementation for four months was similar in men and women.

The addition of exogenous creatine (5 mM) to rat L6 myoblasts resulted in 33% of the creatine transport activity as compared to incubation in a creatine-free medium [5]. Furthermore, these changes were apparent within 24 h of incubation and 24 h of cyclohexamide treatment (inhibitor of protein synthesis) partially blocked the decrease in creatine transport induced by 5 mM of exogenous creatine [5]. Earlier research demonstrated that creatine transport into skeletal muscle was reduced with anaerobiosis, uncoupling of oxidative phosphorylation (2,4 dinitophenol), and cooling [4]. Together, these results demonstrate that creatine transport is modulated by an energy dependant process that involves functional protein synthetic capacity. The transport of creatine and it's attenuation by exogenous creatine was subsequently shown to involve the human CRT from the results of studies where the human CRT was transiently expressed in COS-7 cells [1]. Interestingly, the transport of creatine into COS-7 cells transfected with the CRT, was also acutely inhibited by β-guanidinopropionic acid [1]. Medium-term creatine feeding studies showed that skeletal muscle total CRT protein content was reduced that when rats received a diet containing 4% creatine and 50 mM in drinking water for 3 months [3]. In addition, the inhibition of creatine transport with β-guanidinopropionic acid in rats for 3–6 months resulted in a significant increase in CRT protein in skeletal muscle [3]. Therefore, the medium-term administration of exogenous creatine to rats in doses that were an order of magnitude higher than the doses used in humans [7], can decrease CRT protein content. Given the divergent effects of β-guanidinopropionic in the acute *in vitro* experiments (decrease creatine transport [1]), as compared to the longer-term administration studies *in vivo* (increase CRT protein [3]), it would appear that there are likely different regulatory mechanisms involved in the processes.

Our data shows that in response to Cr-mH supplementation strategies that increased muscle total creatine content in humans there was no decrease in CRT protein content. Clearly, the fact that we found no reduction in CRT protein or mRNA content does not exclude that creatine transport could have been inhibited by a different process such as post-translational modification of the protein (phosphorylation, glycosylation, etc.), or differences in the intra-cellular localization of the transporter such as the transmigration and recruitment seen in with the GLUT-4 glucose transporter [20]. Recently, it was shown that cyclosporin A by inhibiting, among others, the mitochondrial protein cyclophilin also inhibits creatine uptake by apparently altering the targeting of CRT to the plasma membrane by inhibiting the mitochondrial protein cyclophilin [21]. Furthermore, modification of CRT transport activity has been shown by site directed mutagenesis of a critical cysteine-144 residue [22]. In spite of the evidence that alterations in CRT transport can occur by mechanisms unrelated to the total CRT protein content, there did not appear to be an attenuation of muscle total creatine accumulation after acute, two month or four month Cr-mH supplementation. These observations are supported by the fact that the cessation of Cr-mH supplementation after 10 weeks (~ 0.125 g/kg/day), did not result in an 'undershoot' of muscle phosphocreatine concentrations 5 weeks later [11]. Thus, it appears that elevations of skeletal muscle total creatine concentration can be maintained for 2 months in men at a dose of 0.125 g/kg/day, for 4 months in elderly men and women at a dose of 0.075 g/kg/day, and for 10 weeks in women at a dose of ~ 0.0125 g/kg/day [11]. Whether or not these elevated concentrations can be maintained for longer periods of time awaits further investigation. These results are of importance for the long-term treatment of patients in whom creatine supplementation may prove to be effective.

Another issue raised by the results of the current study is the location of the CRT protein that is being detected by the antibodies. The first antibody (CRT-1) that was generated appears to recognize two distinct bands of apparent MW 52 and 70 kDa [3, 23], whereas, a newer antibody shows a faint band inconsistently in skeletal muscle crude homogenates at an apparent MW of ~ 56 kDa. The fact that there was only a single band in human erythrocytes at ~ 56 kDa suggests that this represents the true sarcolemmal transporter. Furthermore, the results of Murphy and colleagues who demonstrated that the ~ 52 and 70 kDa isoforms were in higher abundance in type I muscle fibers, yet the total creatine content was higher in type II muscle fibers, suggested that the 52 and 70 kDa isoforms are mitochondrial, in spite of the fact that immunofluoresence staining suggested that the transporter was located at the sarcolemma [23]. A difficulty with the immunofluorescence data is the fact that an apparent sarcolemmal localization may be due to the presence of sub-sarcolemmal mitochondria. Our data support that these isoforms are mitochondrially located for there was a positive correlation between the increase in the citrate synthase protein content (mitochondrial) and the total CRT protein (52 + 70 kDa) content in the 4 month training study in the elderly. By biochemical fractionation and immuno-localization, using the same COOH-terminal anti-CRT-1 antibody, both of these polypeptides were shown recently to be highly enriched in mitochondrial preparations and specifically localized in mitochondria, respectively (Walzel *et al.* this issue; [19]).

Finally, our group has found that there are gender differences in the increase in fat-free mass [24] and in the regulation of protein turnover [14] in response to Cr-mH supplementation. These findings are consistent with the results of Murphy and colleagues (in the current issue) and together, provide very strong evidence against a gender difference in total creatine concentration, CRT protein, or CRT mRNA content. Therefore, the apparent gender differences in FFM accumulation and protein breakdown and amino acid oxidation, must be attributable to processes distal to the accumulation of creatine in skeletal muscle.

Given the discovery of genetic defects in the CRT [25] and in creatine biosynthesis [26], the understanding of the control of the CRT in response to Cr-mH supplementation is of great importance. Furthermore, the potential for Cr-mH to be a therapeutic agent in neurological disorders [27–30], provides further reason to evaluate the long-term effects of Cr-mH on the regulation of the CRT. A major issue that is still not resolved is the exact intra-cellular localization of the various CreaT isoforms and their role in health and disease states. The recent characterization of a genetic defect in the SLC6A8 gene highlights the importance of this transporter to cellular function [25].

## Acknowledgements

Part of this study was funded by the Hamilton Health Sciences Corporation, Department of Rehabilitation and part by MuscleTech® Research and Development. Some of the equipment used in the analysis of the study came from a Canadian Foundation for Innovation Grant to Dr. M. Tarnopolsky. Part of this work was supported by the Swiss Society for muscle diseases and the German Benni and Co parents organization, as well as by the Swiss Natl. Science Foundation (grant # 31-62024.00 to T.W.).

## References

1. Sora I, Richman J, Santoro G, Wei H, Wang Y, Vanderah T, Horvath R, Nguyen M, Waite S, Roeske WR *et al.*: The cloning and expression of a human creatine transporter. Biochem Biophys Res Commun 204: 419–427, 1994
2. Iyer GS, Krahe R, Goodwin LA, Doggett NA, Siciliano MJ, Funanage VL, Proujansky R: Identification of a testis-expressed creatine transporter gene at 16p11.2 and confirmation of the X-linked locus to Xq28. Genomics 34: 143–146, 1996
3. Guerrero-Ontiveros ML, Wallimann T: Creatine supplementation in health and disease. Effects of chronic creatine ingestion *in vivo*: Down-regulation of the expression of creatine transporter isoforms in skeletal muscle. Mol Cell Biochem 184: 427–437, 1998
4. Fitch CD, Shields RP: Creatine metabolism in skeletal muscle. I. Creatine movement across muscle membranes. J Biol Chem 241: 3611–3614, 1966
5. Loike JD, Zalutsky DL, Kaback E, Miranda AF, Silverstein SC: Extracellular creatine regulates creatine transport in rat and human muscle cells. Proc Natl Acad Sci USA 85: 807–811, 1988
6. Terjung RL, Clarkson P, Eichner ER, Greenhaff PL, Hespel PJ, Israel RG, Kraemer WJ, Meyer RA, Spriet LL, Tarnopolsky MA, Wagenmakers AJ, Williams MH: American College of Sports Medicine roundtable. The physiological and health effects of oral creatine supplementation. Med Sci Sports Exerc 32: 706–717, 2000
7. Hultman E, Soderlund K, Timmons JA, Cederblad G, Greenhaff PL: Muscle creatine loading in men. J Appl Physiol 81: 232–237, 1996
8. Harris RC, Soderlund K, Hultman E: Elevation of creatine in resting and exercised muscle of normal subjects by creatine supplementation. Clin Sci (Colch) 83: 367–374, 1992
9. Greenhaff PL, Bodin K, Soderlund K, Hultman E: Effect of oral creatine supplementation on skeletal muscle phosphocreatine resynthesis. Am J Physiol 266: E725–E730, 1994
10. Greenhaff PL, Casey A, Short AH, Harris R, Soderlund K, Hultman E: Influence of oral creatine supplementation of muscle torque during repeated bouts of maximal voluntary exercise in man. Clin Sci (Colch) 84: 565–571, 1993
11. Vandenberghe K, Goris M, Van Hecke P, Van Leemputte M, Vangerven L, Hespel P: Long-term creatine intake is beneficial to muscle performance during resistance training. J Appl Physiol 83: 2055–2063, 1997
12. Kreider RB, Ferreira M, Wilson M, Grindstaff P, Plisk S, Reinardy J, Cantler E, Almada AL: Effects of creatine supplementation on body composition, strength, and sprint performance. Med Sci Sports Exerc 30: 73–82, 1998
13. Volek JS, Duncan ND, Mazzetti SA, Staron RS, Putukian M, Gomez AL, Pearson DR, Fink WJ, Kraemer WJ: Performance and muscle fiber adaptations to creatine supplementation and heavy resistance training. Med Sci Sports Exerc 31: 1147–1156, 1999
14. Parise G, Mihic S, MacLennan D, Yarasheski KE, Tarnopolsky MA: Effects of acute creatine monohydrate supplementation on leucine kinetics and mixed-muscle protein synthesis. J Appl Physiol 91: 1041–1047, 2001
15. Tarnopolsky MA, Parise G, Yardley NJ, Ballantyne CS, Olantunji S, Phillips SM: Creatine-dextrose and protein-dextrose induce similar strength gains during training. Med Sci Sports Exerc 33: 2001

16. Tarnopolsky MA, Parise G: Direct measurement of high-energy phosphate compounds in patients with neuromuscular disease. Muscle Nerve 22: 1228–1233, 1999
17. Tarnopolsky MA, Parshad A, Walzel B, Schlattner U, Wallimann T: Creatine transporter and mitochondrial creatine kinase protein content in myopathies. Muscle Nerve 24: 682–688, 2001
18. Neubauer S, Remkes H, Spindler M, Horn M, Wiesmann F, Prestle J, Walzel B, Ertl G, Hasenfuss G, Wallimann R: Downregulation of the Na(+)-creatine cotransporter in failing human myocardium and in experimental heart failure. Circulation 100: 1847–1850, 1999
19. Walzel B, Speer O, Boehm E, Kristiansen S, Chan S, Clarke K, Richter EA, Wallimann T: New creatine transport assay with giant plasma-membrane vesicles and identification of distinct creatine transporter isoforms in the sarcolemma and within mitochondria. Am J Physiol, 2002 (in press)
20. Martin S, Tellam J, Livingstone C, Slot JW, Gould GW, James DE: The glucose transporter (GLUT-4) and vesicle-associated membrane protein-2 (VAMP-2) are segregated from recycling endosomes in insulin-sensitive cells. J Cell Biol 134: 625–635, 1996
21. Tran TT, Dai W, Sarkar HK: Cyclosporin A inhibits creatine uptake by altering surface expression of the creatine transporter. J Biol Chem 275: 35708–35714, 2000
22. Dodd JR, Christie DL: Cysteine-144 in the third transmembrane domain of the creatine transporter is close to a substrate binding site. J Biol Chem, 2001
23. Murphy R, McConell G, Cameron-Smith D, Watt K, Ackland L, Walzel B, Wallimann T, Snow R: Creatine transporter protein content, localization, and gene expression in rat skeletal muscle. Am J Physiol Cell Physiol 280: C415–C422, 2001
24. Mihic S, MacDonald JR, McKenzie S, Tarnopolsky MA: Acute creatine loading increases fat-free mass, but does not affect blood pressure, plasma creatinine, or CK activity in men and women. Med Sci Sports Exerc 32: 291–296, 2000
25. Salomons GS, van Dooren SJ, Verhoeven NM, Cecil KM, Ball WS, Degrauw TJ, Jakobs C: X-linked creatine-transporter gene (SLC6A8) defect: A new creatine-deficiency syndrome. Am J Hum Genet 68: 1497–1500, 2001
26. Stockler S, Hanefeld F: Guanidinoacetate methyltransferase deficiency: A newly recognized inborn error of creatine biosynthesis. Wien Klin Wochenschr 109: 86–88, 1997
27. Walter MC, Lochmuller H, Reilich P, Klopstock T, Huber R, Hartard M, Hennig M, Pongratz D, Muller-Felber W: Creatine monohydrate in muscular dystrophies: A double-blind, placebo- controlled clinical study. Neurology 54: 1848–1850, 2000
28. Tarnopolsky M, Martin J: Creatine monohydrate increases strength in patients with neuromuscular disease. Neurology 52: 854–857, 1999
29. Klivenyi P, Ferrante RJ, Matthews RT, Bogdanov MB, Klein AM, Andreassen OA, Mueller G, Wermer M, Kaddurah-Daouk R, Beal MF: Neuroprotective effects of creatine in a transgenic animal model of amyotrophic lateral sclerosis. Nat Med 5: 347–530, 1999
30. Matthews RT, Yang L, Jenkins BG, Ferrante RJ, Rosen BR, Kaddurah-Daouk R, Beal MF: Neuroprotective effects of creatine and cyclocreatine in animal models of Huntington's disease. J Neurosci 18: 156–163, 1998

*Molecular and Cellular Biochemistry* **244:** 167–176, 2003.


# Antagonistic reactions of arginine and lysine against formaldehyde and their relation to cell proliferation, apoptosis, folate cycle and photosynthesis

Lajos Trézl,[1] Lehel Hullán,[2] Zsuzsa M. Jászay,[3] Tibor Szarvas,[4] Imre Petneházy,[1] Béla Szende,[5] József Bocsi,[5] Zoltán Takáts,[6] Károly Vékey[6] and László Töke[1]

[1]*Department of Organic Chemical Technology, Budapest University of Technology and Economics, Budapest;* [2]*National Oncological Institute, Budapest;* [3]*Organic Chemical Technological Research Group of the Hungarian Academy of Science;* [4]*Institute of Isotopes Co. Ltd., Budapest;* [5]*1st Institute of Pathology and Experimental Cancer Research, Semmelweis University, Budapest;* [6]*Chemical Research Institute of Hungarian Academy of Sciences, Budapest, Hungary*

## Abstract

$^{1}H$, $^{13}C$ NMR, ESMS and MS/MS investigations proved that there is an antagonism in the spontaneous reaction of formaldehyde with L-lysine and L-arginine. L-Arginine can only be hydroxymethylated on the guanidino group in a very fast reaction forming mono-, di-, and trihydroxymethyl arginines (HMA). L-Lysine can be methylated on the ε-amino group forming mono-, di-, and trimethyl lysine on physiological pH. Hydroxymethyl arginines are relative stable, isolable products, and can also be formed in biological systems, especially in plants. Significant amounts of hydroxymethyl arginines were identified in the aqueous extract of lyophilized kohlrabi, which can be formed in photosynthesis during $CO_2$ fixation. $^{14}C$-Formaldehyde formed in a short-term (10, 30 sec) $^{14}CO_2$ fixation reaction in *Zea mays* L. (early maturity variety: Szegedi TC 277) was captured by L-arginine, which occurs in leaves in large amount. Formaldehyde formed during photosynthesis can react not only with the arginine, but with ribulose-1,5-diphosphate present in leaves. In model reactions formaldehyde can react with the 'ene diole' group of ribulose-1,5-diphosphate in the absence of Rubisco enzyme, which is a similar reaction to the addition of formaldehyde to L-ascorbic acid. Hydroxymethyl arginines (HMA) are endogenous formaldehyde carrier molecules transferring the bound formaldehyde to thymidylate synthase enzyme system incorporating it into the folate cycle. HMA can also carry the bound formaldehyde to the cells especially to the tumorous cells (HT29 adenocarcinoma), and cause significant inhibition of cell proliferation and causes apoptosis. (Mol Cell Biochem **244**: 167–176, 2003)

*Key words*: formaldehyde, methyl lysine, hydroxymethyl arginine, photosynthesis, D-ribulose-1,5-biphosphate, cell proliferation

## Introduction

In the last 20 years, a great number of data have been accumulated proving, that there is an endogenous formaldehyde level in human and animal organisms, and also in different plants (accumulated in fruits) [1–9]. Endogenous formaldehyde had been first determined in human blood (0.4–0.6 μg/ml) in 1983 by our research group [2, 3]. The latest determination of endogenous formaldehyde of human blood plasma is in good agreement with our results [4]. The questions arise how endogenous formaldehyde could be formed, and what its role is in biological systems.

*Address for offprints*: L. Trézl, Department of Organic Chemical Technology, Budapest University of Technology and Economics, H-1521, Budapest, Hungary (E-mail: ltrezl@mail.bme.hu)

Formaldehyde can be liberated from different precursors in biological systems via permanent demethylation processes starting from $-N-CH_3$, $O-CH_3$, and $-S-CH_3$ groups, respectively [8–12]. Substrate specific demethylase enzymes can liberate formaldehyde only from specific endogenous precursors, e.g. formaldehyde can be liberated only from $N^\varepsilon$-methyl L-lysines by ε-alkyl lysinase, but from neither $N^\alpha$-methyl L-lysine or $N^\varepsilon$-methyl D-lysine [12].

Formaldehyde liberated in this way has a specific biochemical pathway with an elaborate enzymatic system already in place for handling it safely, e.g. introducing the methyl groups into a one-carbon pool via formaldehyde [10–18].

To the contrary, formaldehyde liberated from exogenous precursors (e.g. from N-methyl nitrosamines) by means of non-substrate specific enzymes (e.g. cytochrome P-450 isoenzymes) [15, 17] has no elaborate biochemical pathways, therefore the liberated formaldehyde can take part *in situ* random reactions in cellular components, such as formylation and methylation that leads to miscoding of DNA and RNA [2, 14, 16, 17]. If there are not enough acceptor molecules such as cysteine, glutathione and ascorbic acid present, these reactions are potentially dangerous, even carcinogenic.

Formaldehyde can also be formed permanently in a steady state reaction from serine by serine transhydroxymethylase enzyme [13]. In this case the formaldehyde will not be liberated, but an intermediate enzyme-formaldehyde complex occurs, which transfers the formaldehyde to tetrahydrofolic acid forming $N^5$,$N^{10}$-methylene tetrahydrofolic acid, the coenzyme of thymidylate synthase.

And last, we need to mention that, according to our results [7, 20], there is a new source of formaldehyde, which is formed in photosynthesis. We proved that formation of $^{14}C$-formaldehyde occurs from $^{14}CO_2$ in a short-term (10, 30 sec) $^{14}CO_2$ fixation reaction, and the liberated formaldehyde could bind to L-arginine [7, 22]. We have to emphasize that, although Willstätter and Stoll suspected the existence of formaldehyde in photosynthesis as early as in 1917 [19], Calvin omitted this fact in his cycle [21].

Our research dealing with the specific reactions of formaldehyde and its role in biological systems for once 20 years has uncovered some surprising chemical reactions of this molecule with amino acids, nucleic bases, and ascorbic acid, respectively, which can take place in biological systems. The most important reactions are spontaneous methylation, formylation and hydroxymethylation. The amino groups of lysine can be spontaneously methylated and formylated on the amino groups, while arginine can only be hydroxymethylated on the guanidino group [2, 3, 23]. Since the hydroxymethyl group is bound to the guanidino group of the arginine in a releasable form within biological systems, arginine molecules can carry and pass it on.

In this paper presented to the 6th International Conference on Guanidino Compounds in Biology and Medicine [22], we give account our latest results modeling the antagonistic reactions of arginine and lysine against formaldehyde with $^1H$, $^{13}C$-NMR, ESMS and HPLC-MS investigations, and review their formation in photosynthesis and their specific biological effects of these compounds.

## Materials and methods

### *Materials*

L-Lysine hydrochloride (Reanal, Budapest, Hungary), $N^\varepsilon$-monomethyl-L-lysine hydrochloride (Sigma Chemical Co., St. Louis, MO, USA), $N^\varepsilon$,$N^\varepsilon$-dimethyl-L-lysine hydrochloride (Vega, Columbia, USA), $N^\varepsilon$,$N^\varepsilon$,$N^\varepsilon$-trimethyl-L-lysine glutamate (Richter Pharm. Co., Budapest, Hungary), $N^\varepsilon$-formyl-L-lysine (Sigma), L-arginine, L-arginine hydrochloride (Reanal), $N^G$-monomethyl-L-arginine acetate (Calbiochem, San Diego, CA, USA), $N^G$, $N^G$-dimethyl-L-arginine hydrochloride (Calbiochem), formaldehyde solution p.a. (Fluka, Budapest, Hungary), tetrahydrofolic acid (Sigma), 2′-deoxyuridine-5′-monophosphate (Sigma), D-glucose (Sigma), D-xylose (Sigma), (5-$^3$H)-deoxyuridine monophosphate (Radiochemical Centre, Amersham, UK), D-ribulose 1,5-diphosphate sodium salt (Sigma), deuterium oxide (Fluka), acetonitrile HPLC grade (Fluka).

### *NMR measurements*

$^1H$ and $^{13}C$ NMR spectra were recorded on a Bruker AC 250 spectrometer equipped with a 5 mm probe, and operating at 62.896 MHz in case of $^{13}C$ NMR spectra. Chemical shifts are quoted relative to the internal standard HDO at 4.6 ppm in the case $^1H$ experiments, and to $CH_2(OH)_2$ at 83.2 ppm in the case of $^{13}C$ experiments, respectively.

### *Sample preparation*

Model mixtures were prepared by depolymerising paraformaldehyde (1 mmole, 30 mg) in 0.5 cm$^3$ boiling $D_2O$ in NMR tube, then allowing the formaldehyde to react with the appropriate amino acid ·HCl (L-lysine, $N^\varepsilon$-methyl L-lysine and L-arginine, 0.1 mmole) at 50°C (pH 5.5).

### *MS measurements*

Mass spectrometric measurements were carried out using a PE Sciex API 2000 LC-MS/MS system. The instrument was used with electrospray ion source in positive-ion mode. Ion spray voltage was 4000 V, orifice voltage was set at 50 volts, source temperature at 100°C. Flow injection analysis (0.1 ml/

min water flow) was applied. Mass spectrometer was used in simple MS1 scan mode, or in production scan mode.

*Sample preparation*
From L-lysine hydrochloride: A mixture of 400 µl of 100 mg/ml L-lysine hydrochloride solution and 160 µl formaldehyde solution was prepared (pH ~ 5.5), and to react for 4 days. From L-arginine hydrochloride: A mixture of 400 µl of 100 mg/ml L-arginine hydrochloride solution in 0.1 M K-phosphate buffer (pH 7.4) and 160 µl formaldehyde solution was allowed to react, and measurements were taken immediately. From L-arginine: A mixture of 400 µl of 100 mg/ml L-arginine solution and 160 µl formaldehyde solution (pH ~ 10.5) was allowed to react, and measurements were taken immediately.

From lyophilized kohlrabi extracts: Two g kohlrabi (5 weeks old separated leaves) were chopped and suspended in 15 ml water, after homogenized in an Ultra Turrax homogenizer and centrifuged (1000 × g, 10 min, 4°C). The supernatants were lyophilized and stored at –20°C.

*Photosynthetic experiments*

$^{14}CO_2$ was liberated from $Ba^{14}CO_3$ (specific radioactivity: 2.03 Gbq/mmol, product of Amersham) by means of lactic acid. The seedlings of *Zea mays* L. (single cross) were cultivated for 12 days for $^{14}CO_2$ fixation. It was carried out in an assimilation apparatus previously described [20]. The leaves were exposed to $^{14}CO_2$ (40 MBq) for 10 and 30 sec.

After the given exposure times the samples were frozen in liquid nitrogen, ground, and the powder was extracted with water, the aqueous extract was lyophilized for analysis.

*ESMS measurements of D-ribulose 1,5-biphosphate and its reaction with formaldehyde*

One hundred µM D-ribulose 1,5-biphosphate and 500 µM formaldehyde was allowed to react in 0.1 M phosphate buffer on pH 7.4 for 5 min at 20°C.

*HPLC-MS measurements*

The analysis was carried out using the above described mass spectrometer with a Perkin-Elmer 200 Micro LC system. A LiChrocart 250-4 HPLC cartridge was used with LiChosorb Diol 100 5 µl packing. Acetonitrile-water gradient was applied according the following: 80% acetonitrile-20% water (0 min), 80% acetonitrile-20% water (5 min), 0% acetonitrile-100% water (30 min), 0% acetonitrile-100% water (35 min), 80% acetonitrile-20% water (45 min).

*Thymidilate synthase enzyme assay*

Preparation of hydroxymethyl arginines: $N^G$-mono-, di-, and trihydroxymethyl-L-arginines (HMA) formed from the reaction mixture of L-arginine and formaldehyde were separated by column chromatography on Sephadex G-15 adsorbent. The $^1H$, $^{13}C$, and $^{13}CP$/MAS and other analyses were described earlier [17, 23, 26].

Thymidilate synthase activity was estimated, essentially as described by Roberts [24]. Reaction mixture (0.15 ml) contained 5-$^3H$-dUMP (22 kBq), 20 nmol tetrahydrofolate, 20 nmol $CH_2O$, 1.5 µmol 2-mercaptoethanol, 7.5 µmol NaF, 7.5 µmol tris-HCl buffer (pH 7.5) and 0.11 ml of cytosol prepared from P388 ascites tumor cells. Tumor transplantation, the check of tumor growth and cytosol preparation were carried out as previously described [25]. The reaction was started by adding the cytosol and after incubation (0–30 min) at 37°C was terminated by addition of 0.5 ml of 3.35% TCA and 0.5 ml of charcoal suspension (55 mg/ml). The samples were centrifuged at 1000 × g for 15 min. Radioactivity of supernatant was counted in LSC. Non-adsorbable and total radioactivity were determined in appropriate controls. Reaction velocity was linear with time under these conditions, and enzyme activities varied by < 5% in the control.

The same enzyme assay was followed with HMA prepared by the above described method.

*Measurement of the effect of HMA on cell proliferation of HT29 human colon adenocarcinoma*

*Cell cultures*
The human colon carcinoma cell lines HT-29 (ATCC HTB-38) were used. The monolayer cell cultures were maintained in 24-well and 6-well Greiner plates (Kremsmuenster, Austria) using RPMI supplemented with 10% fetal calf serum (Sigma-Aldrich, St. Louis, MO, USA). The number of cells at plating was $5.10^4$/ml.

*Treatment of cell cultures*
The colon carcinoma cells were treated with 1, 10 and 100 µg/ml HMA in triplicate samples, respectively for 48 h after plating. Samples for evaluation were taken 24, 48 and 72 h after treatment. Cell counts and viability were determined by Buerker chambre after Trypan blue staining with 200× magnification.

## Results and discussion

*Interpretation of the reaction of formaldehyde with lysine by $^1H$, $^{13}C$ NMR, ESMS and MS/MS spectra*

According to the $^1H$ NMR spectra obtained from the reaction mixture of L-lysine hydrochloride and formaldehyde after 5 min, 2 h and 16 h reaction times a growing singulet of the N-methyl protons can be observed at 2.45 ppm. The estimated conversion of lysine to $N^{\varepsilon}$-methyl-lysine is 6% calculated on the basis of the α-CH proton ratio of $N^{\varepsilon}$-methyl-lysine and lysine. A small signal at 2.6 ppm in the sample monitored after 16 h can be identified as traces of N,N-dimethyl lysine (Fig. 1). As a reference, $^{1}$H NMR spectra of the reaction of $N^{\varepsilon}$-monomethyl lysine and formaldehyde under the same circumstances was also taken (Fig. 2). After 16 h 15% conversion of monomethyl lysine to dimethyl lysine can be estimated from the ratio of the $N(CH_3)_2$ signal at 2.6 ppm, and the $NHCH_3$ signal at 2.48 ppm. The $^{13}$C NMR spectra taken from the reaction mixture of L-lysine hydrochloride and formaldehyde showed not only the signal of the $N^{\varepsilon}$-methyl carbon of the $N^{\varepsilon}$-methyl-lysine at 33 ppm and a small signal of the N,N-dimethyl carbon at 49 ppm, but the signal of a $NCH_2OH$ group at 71.8 ppm, which shows that lysine is

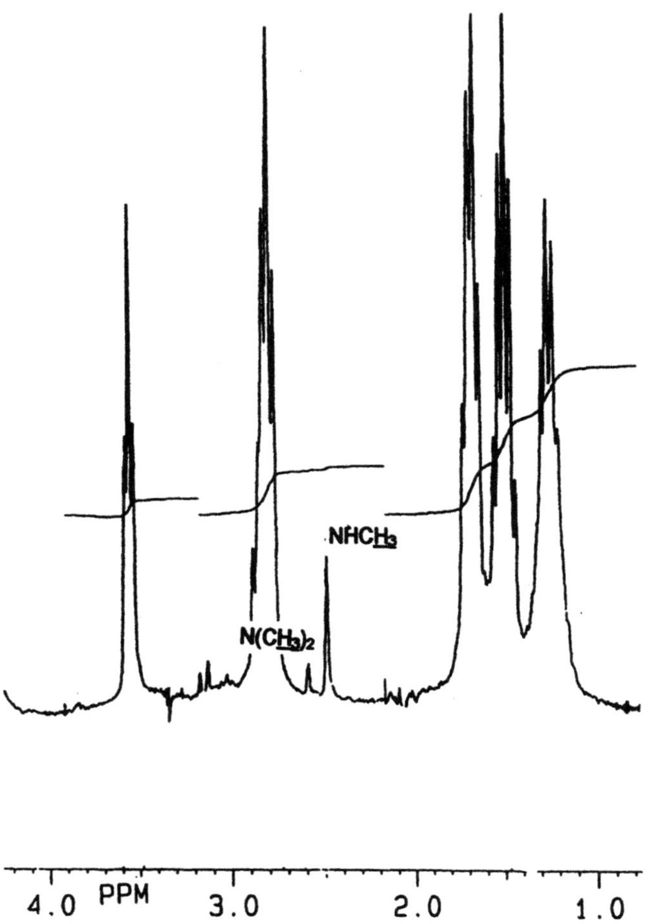


*Fig. 1.* Detail of $^{1}$H NMR spectrum taken from the reaction mixture of 1 mmol depolimerised formaldehyde and 0.1 mmol L-lysine in $D_2O$ at 50°C, after 16 h. The singulet of N-methyl proton at 2.45 ppm and a smaller singulet of N,N-dimethyl protons at 2.6 ppm can be assigned.

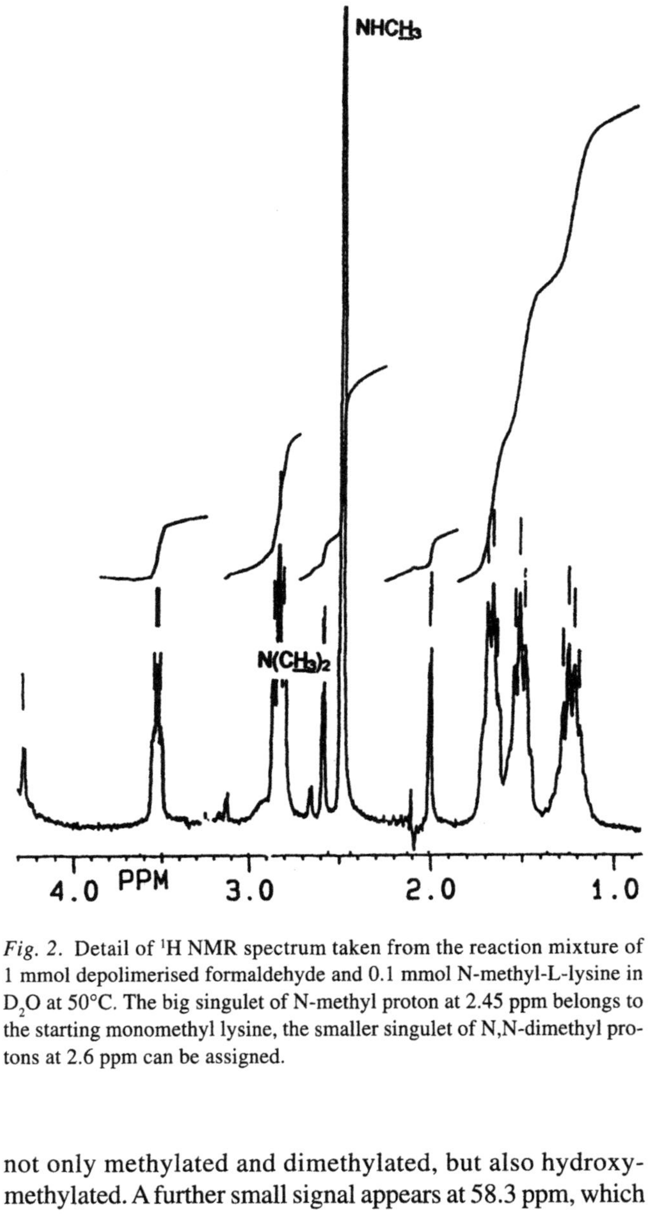


*Fig. 2.* Detail of $^{1}$H NMR spectrum taken from the reaction mixture of 1 mmol depolimerised formaldehyde and 0.1 mmol N-methyl-L-lysine in $D_2O$ at 50°C. The big singulet of N-methyl proton at 2.45 ppm belongs to the starting monomethyl lysine, the smaller singulet of N,N-dimethyl protons at 2.6 ppm can be assigned.

not only methylated and dimethylated, but also hydroxymethylated. A further small signal appears at 58.3 ppm, which can be assigned to the $CH_2$ group of a cyclic compound. The reaction scheme of formaldehyde with the $N^{\varepsilon}$-amino group of lysine is shown on Fig. 3.

According to the ESMS spectra taken from the reaction mixture after 4 days, a large monomethyl-lysine peak appeared at m/z 161.2 and a smaller dimethyl-lysine peak at m/z 175. A very small peak at m/z 189.1 was identified as trimethyl lysine, and at m/z 159 as the cyclic product. (Fig. 4). Mono-, di-, and trimethyl-lysines were identified from the splittings by the CID spectra. The structures of the identified lysine derivatives are in accordance with the results gained from $^{13}$C NMR spectra. There was no peak at 177 (hydroxymethyl lysine), because the carbon-nitrogen bond is too weak to appear in the circumstances of the MS measurements, and thus differs from arginine, which forms stable hydroxymethyl arginines (Fig. 6).

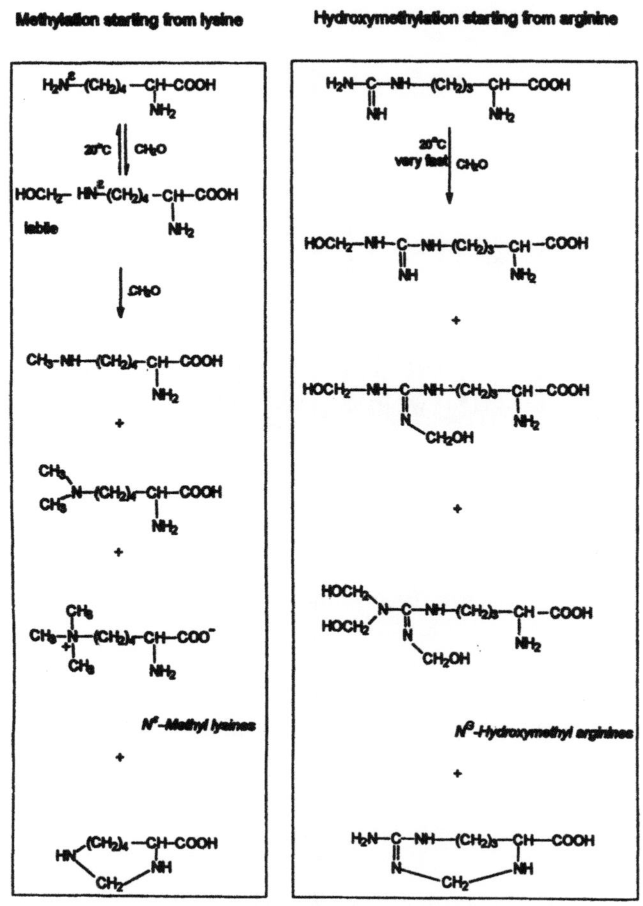


*Fig. 3.* Scheme of the reaction of formaldehyde with lysine and arginine, respectively.

## *Interpretation of in the reaction of formaldehyde with arginine, by $^{1}H$, $^{13}C$ NMR, ESMS and MS/MS spectra*

According to the $^{1}H$ NMR spectra obtained from the reaction mixture of L-arginine hydrochloride and formaldehyde, after a few minutes three singulets of the mono-, di-, and tri-N-$CH_2OH$ protons can be observed at 4.1, 4.2 and 4.35 ppm (Fig. 5). The signals of the mono-, di-, and tri-N-$CH_2OH$ groups are also assigned in the $^{13}C$-NMR spectra at 65.4, 66.7 and 68 ppm, respectively.

A typical ESMS spectra obtained from arginine hydrochloride and formaldehyde reaction mixture at pH 7 is shown in Fig. 6. Peaks at m/z 205, 235 and 265 can be identified as mono-, di-, and tri-hydroxymethyl arginines, respectively. We note that no methyl arginine peak was observed either in ESMS or in NMR spectra. In the spectrum taken at pH 10.5 peaks at m/z 187, and 199 can assigned to the appropriate cyclic compounds. In these circumstances hydroxymethyl arginine (m/z 205) is present only in traces.

Comparing the reactivity of lysine and arginine with formaldehyde according to the NMR and MS measurements, it

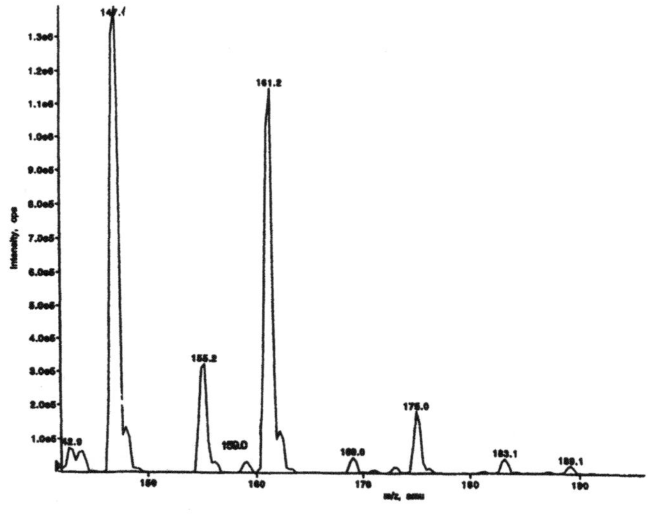

*Fig. 4.* The ESMS spectrum of the reaction mixture of formaldehyde with L-lysine. m/z 161.2: $N^{\varepsilon}$-monomethyl-L-lysine, m/z 175: $N^{\varepsilon}$,:$N^{\varepsilon}$-dimethyllysine, m/z 189.1: $N^{\varepsilon}$,:$N^{\varepsilon}$ $N^{\varepsilon}$-trimethyl lysine, m/z 159: cyclic product (supposed structure see in the reaction scheme, Fig. 3).

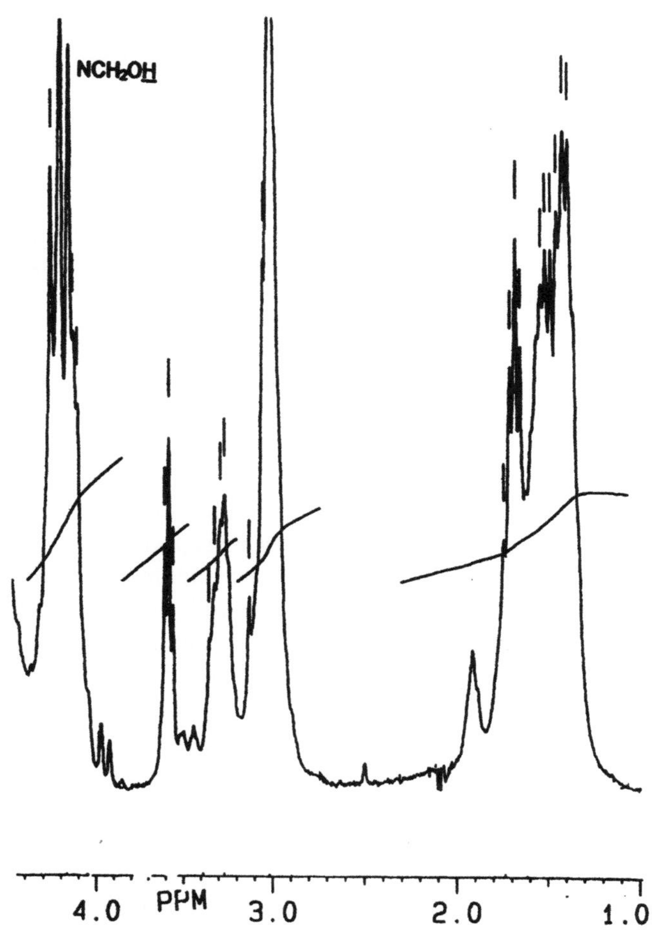


*Fig. 5.* Detail of $^{1}H$ NMR spectrum taken from the reaction mixture of 1 mmol depolimerised formaldehyde and 0.1 mmol L-arginine in $D_2O$ at 20°C, after a few minutes. Signals of mono, di, and trihydroxymethyl protons at 4.1, 4.2 4.35 ppm.

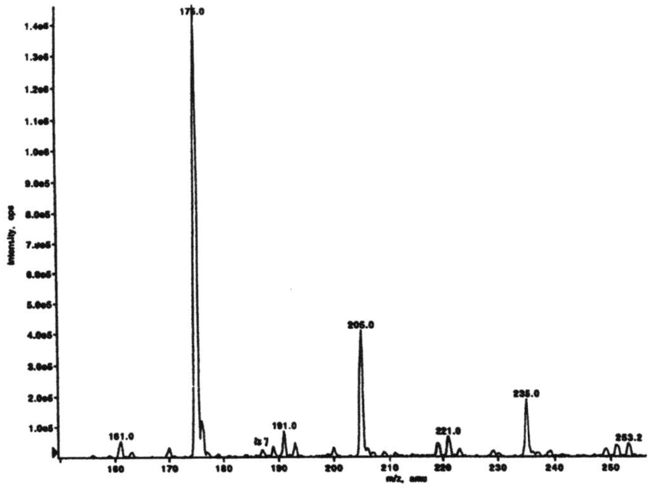

*Fig. 6.* ESMS spectrum of the reaction mixture of L-arginine.HCl and formaldehyde. Four hundred µl of 100 mg/ml L-arginine.HCl, 160 µl cc. formaldehyde was allowed to react in 0.1 M K-phosphate buffer (pH 7.4), at 20°C, and measurements were taken immediately. m/z 175: L-arginine, m/z 205: $N^G$-monohydroxymethyl L-arginine, m/z 235: $N^G$,$N^G$-dihydroxymethyl L-arginine, m/z 265: $N^G$, $N^G$,$N^G$-trihydroxymethyl L-arginine, m/z 187: cyclic product formed by reaction of formaldehyde with the a-amino group and the guanidino group of the L-arginine (supposed structure see in the reaction scheme, Fig. 3).

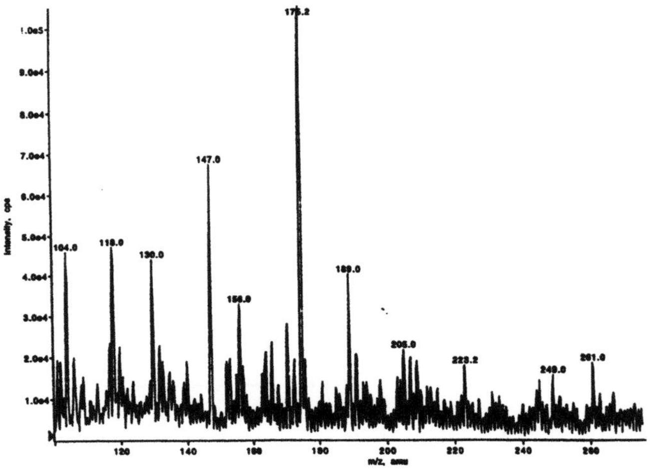

*Fig. 7.* ESMS spectrum of lyophilized young kohlrabi leaf extract. Five week old leaves, extracted with water, homogenized, centrifuged, and lyophilized m/z: 147 L-lysine, m/z 175: L-arginine, m/z: 189 $N^\varepsilon$, $N^\varepsilon$,$N^\varepsilon$-trimethyl-*L*-lysine, m/z: 205: $N^G$-monohydroxymethyl L-arginine.

has to be stressed that arginine reacts with formaldehyde in an immediate hydroxymethylation reaction, resulting in relative stable HMA-s, while lysine undergoes in a slower methylation, and although its hydroxymethyl derivative is also formed, it is very labile.

### *Identification of arginine and hydroxymethyl arginines in kohlrabi extracts*

ESMS spectra obtained from the aqueous solution of the extract is shown in Fig. 7. Besides the huge peak of arginine (m/z 175), peak of hydroxymethyl arginine (m/z 205), trimethyl lysine (m/z 189) and lysine (m/z 147) were observed, and identified by their CID spectra. Small peaks of di-, and trihydroxymethyl-arginines appeared in the ESMS spectra, too, but were not identified by CID. These results are in accordance with our earlier investigations [7], proving that there a number of fruits and vegetable contain high level of endogenous formaldehyde. We found the highest level in the ripe kohlrabi tuber (31 mg/kg) and in its young leaves (25.5 mg/kg), whereas there is a much lower value (6.3 mg/kg) in young kohlrabi tuber. This indicates that formaldehyde accumulates during the ripening time in tuber, but there is no significant difference in formaldehyde level between young and old leaves. Because the endogenous formaldehyde is not free, but bound in the form of hydroxymethyl arginine, it is essentially that arginine is present in leaves in significant amount, as verified by the ESMS results.

### *Role of formaldehyde in photosynthesis*

According to the HPLC-MS measurements of the lyophilized extracts of *Zea mays*, a m/z 207 peak could be identified as $^{14}$C-hydroxymethyl arginine (Fig. 8). In a control sample with non-labeled $CO_2$, we found only a m/z 205 peak (hydroxymethyl arginine), but not the m/z 207 peak. Earlier, Szarvas [20] identified $^{14}$C formaldehyde with dimedone liberated from $^{14}CO_2$ in photosynthetic experiment, and Trézl [7] identified $^{14}$C-hydroxymethyl arginines in $^{14}CO_2$ fixation experiments by radiochromatographic analysis. These results prove that formaldehyde is present in the photosynthetic system, because the only source of the hydroxymethyl group in hydroxymethyl arginine is formaldehyde, and hence arginine can be considered as the carrier molecule of formaldehyde in biological systems.

Our experiments verify the formation of formaldehyde in $CO_2$ fixation experiments, though Calvin excluded the role of formaldehyde in photosynthesis [21]. A formaldehyde fixation cycle was suggested by Quayle [27, 28] starting from D-ribulose 5-phosphate and formaldehyde resulting in 3-hexulose-6-phosphate catalyzed by hexululose phosphate synthase (Hu MP) enzyme in bacteria, such as in *Methylococcus capsulatus* and *Pseudomonas methanica,* and by means of other enzymes forming glyceraldehyde 3-phosphate, and other compounds. Quayle and other scientists did not suppose that the reaction can occur without an enzyme, therefore they did not adapt it for the photosynthesis. They considered

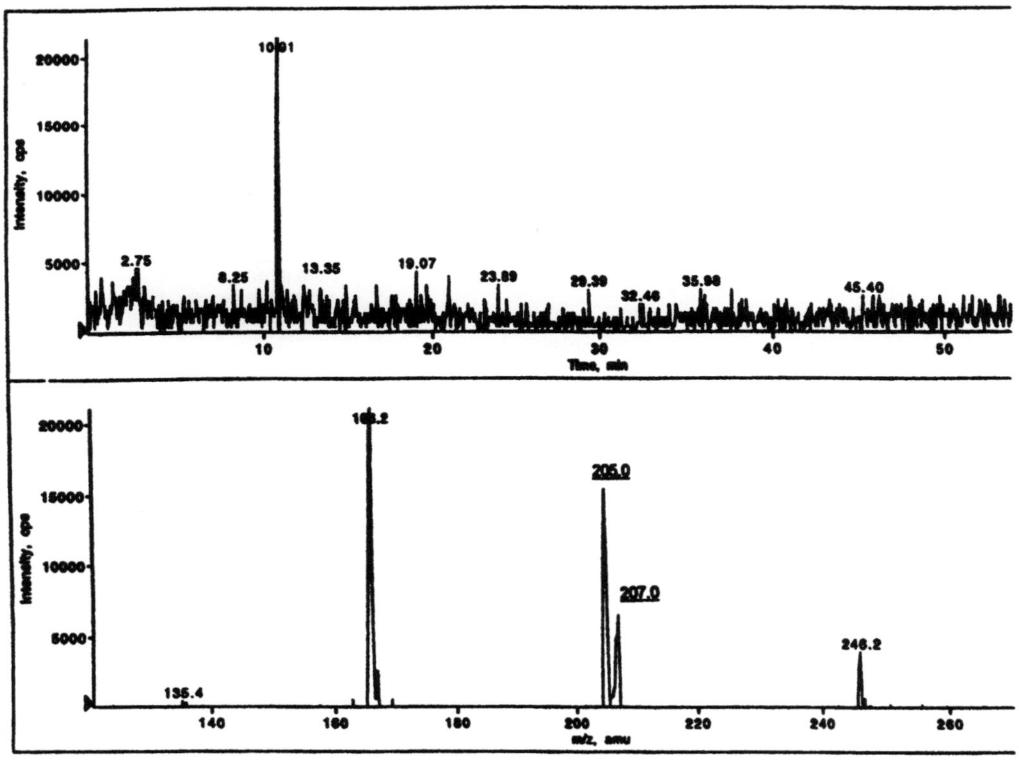


*Fig. 8.* HPLC diagram (upper) and ESMS spectrum (below) of lyophilized Zea mays leaves extract exposed to $^{14}CO_2$ for 10 sec. After 10 sec exposition time the sample was frozen in liquid nitrogen, then ground and extracted with water, and lyophilized for analysis. The sample was separated with acetonitrile-water eluent mixture (ratios see in 'Materials and methods'). A peak detected at 10.91 min retention time was examined by ESMS providing the following peaks: m/z 205: $N^G$-monohydroxymethyl L-arginine, m/z 207: $N^G$-$^{14}C$-monohydroxy methyl L-arginine.

the reaction as an aldol condensation of formaldehyde and D-ribulose 5-phosphate, which results in 3-hexulose-6-phosphate. We recognized that this biologically important reaction can take place without enzymes, and not in aldol condensation reaction, because the ribulose phosphate contains 'ene diol' group, which is being nucleophilic and therefore can react with electrophilic formaldehyde in a fast addition reaction. This type of addition reaction was demonstrated in the case of ascorbic acid by ourselves (Fig. 9) [17]. D-ribulose biphosphate, having also an 'ene diole' part, undergoes in a similar reaction with formaldehyde, resulting in 3-keto-2-hydroxymethyl arabitinol-1,5 biphosphate (Fig. 10). It is known from the literature that the enzymatic carboxylation of D-ribulose-biphosphate can undergo via an ene diole intermediate resulting in the keto acid [29], but to date nobody suspected the addition reaction of formaldehyde on ribulose biphosphate. This addition is fast at physiological pH, since the addition product can be identified at m/z 339 as a significant peak in the ESMS spectrum of the reaction mixture of ribulose biphosphate and formaldehyde (Fig. 11).

According to our preliminary investigations, the adduct formed is a substrate of D-ribulose-1,5-biphosphate carboxylase/oxygenase (Rubisco, EC 4.1.1.39), because the enzyme transforms the adduct to the same products as occur in $CO_2$ fixation, namely to 3-keto-2-carboxy arabitinol-1,5-biphosphate (m/z 353), and to glycerate-3-phosphate (m/z 185). These investigations indicate that there can exist a formaldehyde fixation cycle beside the Calvin cycle.

### *Thymidilate synthase (TS) enzyme activity*

The spontaneous reaction of formaldehyde with tetrahydrofolic acid (THF) as a control and TS assay were used. According to Fig. 12, both the reaction mixtures of HMA with THF and with formaldehyde, respectively, produced a considerable amount of labeled product during the TS reaction,

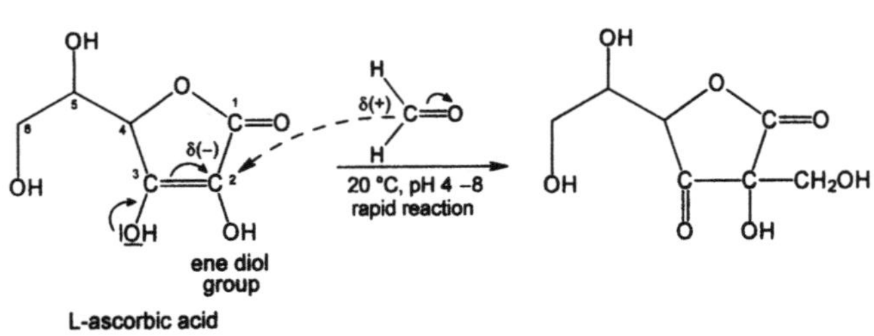


*Fig. 9.* The electrophilic addition reaction of formaldehyde to the 'ene diole' group of L-ascorbic acid.

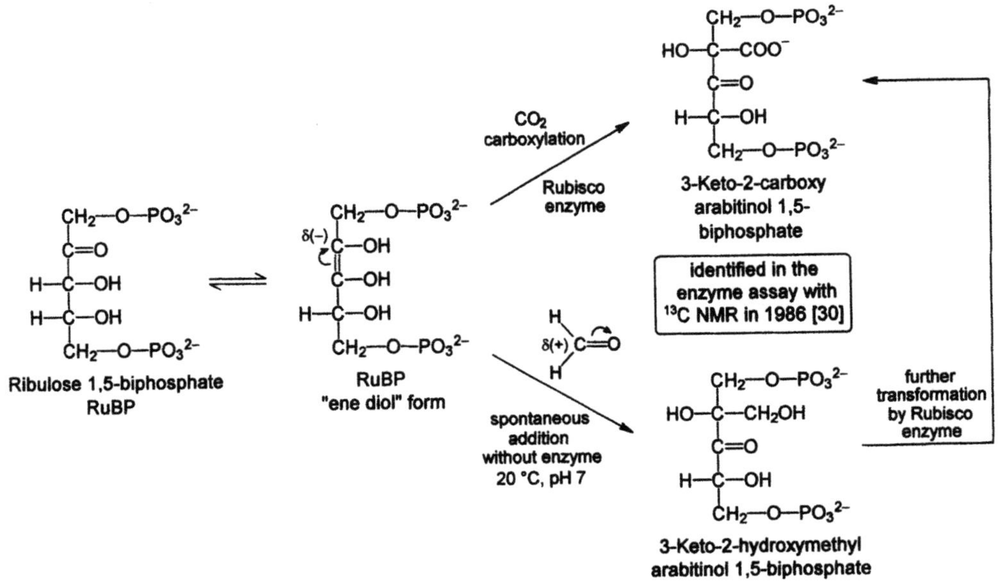


*Fig. 10.* The carbon dioxide fixation and formaldehyde addition reaction to the 'ene diole' group of ribulose 1,5-biphosphate.

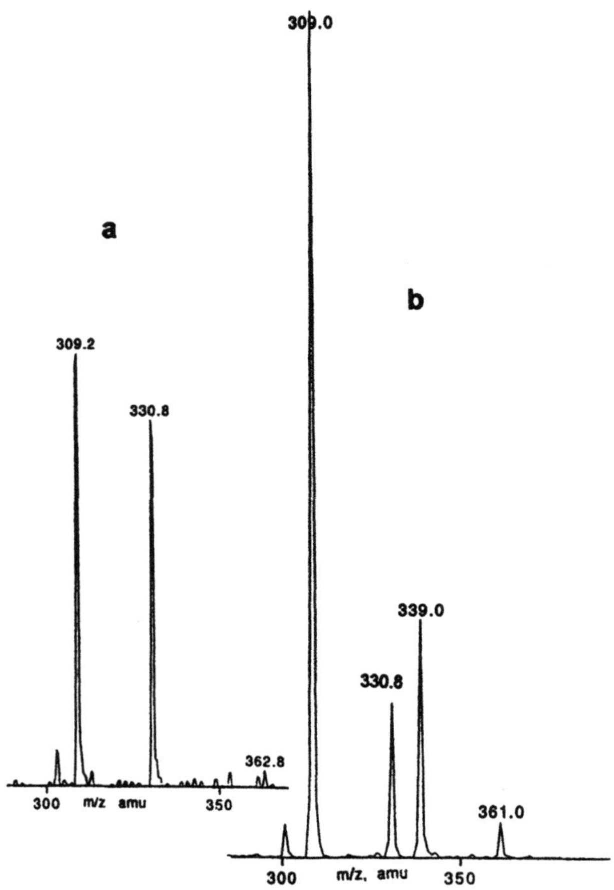


*Fig. 11.* ESMS spectra of D-ribulose 1,5-biphosphate (a), and its reaction with formaldehyde (b). One hundred μM D-ribulose 1,5-biphosphate and 500 μM formaldehyde was allowed to react in 0.1 M phosphate buffer on pH 7.4 at 20°C. m/z 309: D-ribulose 1,5-biphosphate, m/z 339: adduct of formaldehyde and D-ribulose 1,5-biphosphate, m/z 330.8: adduct of Na and D-ribulose 1,5-biphosphate.

and the radioactivity values were proportional to the reaction time. There was no radioactivity in the samples when trichloroacetic acid (TCA) was given prior to the cytosol, and consequently the results were similar, when either HMA or formaldehyde was preincubated with THF. Since the Robert's method is highly specific for TS, the results proved that HMA reacts with THF resulting in $N^5$,$N^{10}$-$CH_2$-THF, namely THF received a $C_1$-fragment from HMA. The source of the $C_1$-fragment must be the reversibly bound formaldehyde in the form of hydroxymethyl groups of the HMA, because arginine itself did not produce the coenzyme with THF.

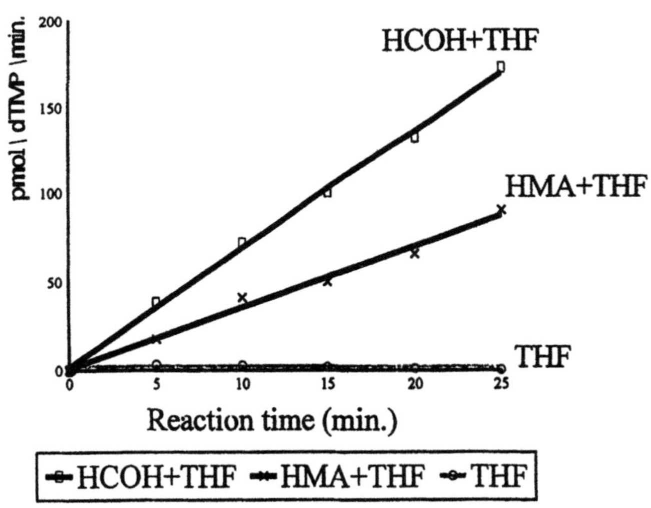


*Fig. 12.* Thymidylate synthase activity. Cytosol prepared from L 1210 ascites tumor cells. 1.2 μmol HMA, 1.2 μmol THF as well as 1.2 μmol HCHO, 1.2 μmol THF were incubated in 7.5 μmol tris.HCl buffer (pH 7.5) for 1 h at 37°C. Both reaction mixture were tested by TS assay.

As described earlier, HMA-s occur in plants during photosynthesis, especially in fruits. They are biomolecules responsible for the endogenous formaldehyde level, and may provide a direct supply of the $C_1$-fragment for the folate cycle.

In vitro *inhibition of cell proliferation of human HT29 colon adenocarcinoma cells with HMA*

Figure 13 shows clearly that HMA influences the cell proliferation in a dose and time dependent manner. A low dose (1 µg/ml) 'enhanced' cell proliferation, importantly, because it actually inhibited apoptotic activity (data not given). Higher doses (10, 100 µg/ml) caused significant inhibition on cell proliferation, especially in a dose of 100 mg/ml HMA. This dose induced a high apoptotic activity (data not given).

To interpret the effect of HMA, we stress that the main effect is that HMA-s contain formaldehyde in a releasable hydroxymethyl group; therefore formaldehyde can be slowly liberated from these groups, whereby a steady-state formaldehyde level can be maintained in cell culture. This formaldehyde level is not high enough to be toxic, and even high dose of HMAs are not toxic in contrast to formaldehyde.

These effects are verified by the investigations of Szende *et al.* [31], who documented that cell proliferation of HT29 colon adenocarcinoma and endothelial cells can be influenced by formaldehyde. Formaldehyde enhanced cell proliferation and decreased apoptotic activity in low doses (0.1 mM) in a similar manner as HMA, while formaldehyde at a higher doses (1 mM) slightly increased the apoptotic index and decreased the mitotic index. Formaldehyde in 10 mM concentration was very toxic causing nearly total cell loss.

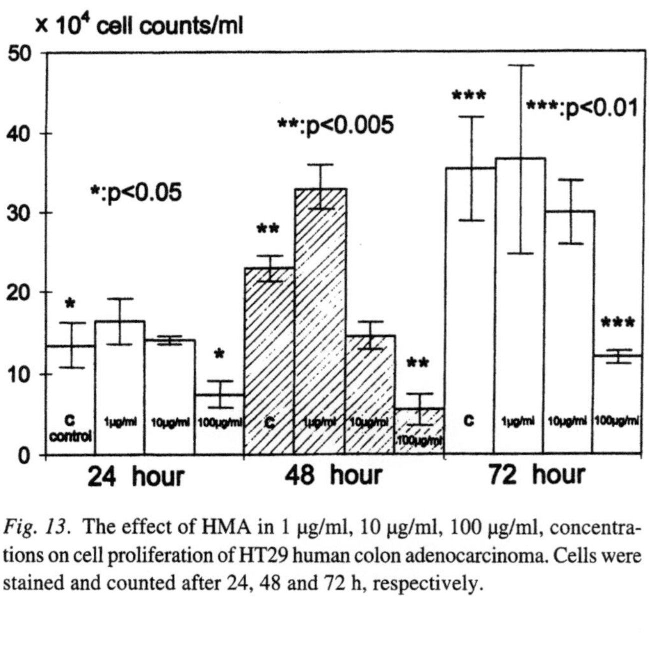


*Fig. 13.* The effect of HMA in 1 µg/ml, 10 µg/ml, 100 µg/ml, concentrations on cell proliferation of HT29 human colon adenocarcinoma. Cells were stained and counted after 24, 48 and 72 h, respectively.

It has to be supposed that formaldehyde can be captured by the side-chains of cell proteins, especially by arginine side-chains, and the bound formaldehyde can slowly be released from the hydroxymethyl groups in the same way when HMA is added to the cell culture. When a high dose of formaldehyde is present, the capacity of the 'formaldehyde capturer' will be exhausted, and the remaining free formaldehyde is toxic.

The main conclusion is that the guanidino group of L-arginine is a formaldehyde acceptor, which is able to control the endogenous free formaldehyde level.

*Effect of $N^{\varepsilon}$-methylated lysines on cell proliferation*

The biological effects of N-methylated lysines as a result of spontaneous methylation of formaldehyde on lysine were studied in detailed examinations, and published by us [32]. Methylated lysines (mono-, di-, and trimethyl) influenced the cell proliferation on animal and plant tissues (both healthy and tumorous tissues) very strongly. The effect of $N^{\varepsilon},N^{\varepsilon},N^{\varepsilon}$-trimethyl lysine (TML) is very especially significant, since it increases cell proliferation, and also increases the crop yield when sprayed on to culture plants.

## Acknowledgement

This work was supported by National Science Foundation OTKA Grant No. T034245.

## References

1. Heck H, White EL, Casanova-Schmitz M: Determination of formaldehyde in biological tissues by gas chromatograph/mass spectrometry. Biomed Mass Spectr 9: 347–353, 1982
2. Trézl L, Rusznák I, Tyihák E, Szarvas T, Szende B: Spontaneous $N^{\varepsilon}$-methylation and $N^{\varepsilon}$-formylation reactions between L-lysine and formaldehyde inhibited by L-ascorbic acid. Biochem J 214: 289–292, 1983
3. Szarvas T, Szatlóczky E, Volford J, Trézl L, Tyihák E, Rusznák I: Determination of endogenous formaldehyde in human blood and urine by dimedon-$^{14}$C-radiometric method. J Radioanal Nucl Chem Lett 106: 357–367, 1986
4. Luo W, Li H, Zhang Y, Ang CYW: Determination of formaldehyde in blood plasma by HPLC with fluorescence detection. J Chromatogr B: Biomed Sci Appl 753: 253–257, 2001
5. Bagchi D, Bagchi M, Hassoun EA, Kelly J, Stohs SJ: Adriamycin induced hepatic and myocardial lipid peroxidation and DNA damage, and enhanced excretion of urinary lipid metabolites in rats. Toxicology 95: 1–9, 1995
6. Trézl L, Csiba A, Juhász S, Szentgyörgyi M, Lombai G, Hullán L: Endogenous formaldehyde level of foods and its biological significance. Z Lebensm Unters Forsch A 205: 300–304, 1997
7. Trézl L, Hullán L, Szarvas T, Csiba A, Szende B: Determination of endogenous formaldehyde in plants (fruits) bound to L-arginine and

its relation to folate cycle, photosynthesis and apoptosis. Acta Biol Hung 49: 253–263, 1998

8. Tyihák E, Trézl L, Szende B: Formaldehyde cycle and the phase of stress syndrome. Ann NY Acad Sci 851: 259–270, 1998
9. Kalász H, Báthori M, Tyihák E: Formaldehyde generation by N-demethylation. Acta Biol Hung 49: 339–344, 1988
10. ElAmin B, Means GE: Introduction of methyl groups from ε-N-methyl lysine into the one-carbon pool. Biochem Biophys Res Commun 86: 407–417, 1979
11. Yu PH, Lai ChT, Zuo DM: Formation of formaldehyde from adrenaline *in vivo*: A potential risk factor for stress related angiopathy. Neurochem Res 22: 615–620, 1997
12. Kim S, Benoiton L, Paik WK: ε-Alkyllysinase. J Biol Chem 239: 148–154, 1964
13. Chen MS, Schirch L: Serine transhydroxymethylase. J Biol Chem 248: 3631–3635, 1973
14. Ashby J, Lefevre PA: Formaldehyde generators: relationship between stability, lipophilicity and carcinogenic potency. Carcinogenesis 3: 1273–1276, 1982
15. Lewis DFV: In: Cytochrome P-450, Structure, Function and Mechanism. Taylor and Francis Ltd., London, UK, 1996
16. Alderson T: Mechanism of formaldehyde induced mutagenesis. Nature 187: 485–489, 1960
17. Trézl L, Tyihák E, Lotlikar PB: Nonenzymatic protein methylation and its biological significance. In W.K. Paik, S. Kim (eds). Protein Methylation, ch. 22. CRC Press, Boca Raton, FL, 1990
18. Szende B, Tyihák E, Trézl L, Szoke E, László I, Kátay G, Király-Véghely ZS: Formaldehyde generators and captures as influencing factors of mitotic apoptotic processes. Acta Biologica Hung. 49: 323–329, 1998
19. Willstätter R, Stoll A: Über die Baeyersche assimilations hypothese. Ber 50: 1777–1791, 1917
20. Szarvas T, Pozsár B, Simon LP: Detection of endogenous free formaldehyde in plants and its relationship with photosynthesis. Acta Agron Acad Sci Hung 32: 313–329, 1983
21. Calvin M, Benson AA: The path of carbon in photosynthesis. Science 107: 476–480, 1948
22. Trézl L, Hullán L, MJászay Zs, Csiba A, Petneházy I, Szende B, Bocsi J, Vékey K, Toke L: Antagonistic reactions of arginine and lysine against formaldehyde and their relation to cell proliferation (apoptosis) folate cycle and photosynthesis. (Abstract of the 6th International Conference on Guanidino Compounds in Biology and Medicine, Cincinnati, Ohio, USA Aug. 31–Sept. 3, 2001) Cell Biol Int 25: 948, 2001
23. Csiba A, Trézl L, Tyihák E, Graber H, Vári É, Téglás G, Rusznák I: Assumed role of L-arginine in mobilization of endogenous formaldehyde. Acta Physiol Acad Sci Hung 59: 35–43, 1982
24. Roberts D: An isotopic assay for thymidylate synthetase. Biochemistry 5: 3546–3551, 1966
25. Hullán L, Szikla K, Tusnádi G, Holczinger L: Thymidylate synthase and thymidylate kinase activities and methotrexate cytotoxicity during growth of L 1210 an Ehrlich ascites tumor. Cancer Biochem Biophys 10: 131–139, 1988
26. Trézl L, Rusznák I, Náray-Szabó G, Szarvas T, Csiba A, Ludányi Á: Essential differences in spontaneous reactions of L-lysine and L-arginine with formaldehyde and its quantum chemical interpretation. Period Polytech Chem Eng 32: 251–260, 1988
27. Strom T, Ferenci T, Quayle JR: The carbon assimilation pathways of *Methylococcus capsulatus*, *Pseudomonas methanica* and *Methylosinus trichosporium* (OB3B) during growth on methane. Biochem J 144: 465–476, 1974
28. Quayle JR: Aspects of the regulation of methylotrophic metabolism. FEBS Lett 117(suppl) K16–K27, 1980
29. Rabin BR, Trown PW: Mechanism of action of carboxy dismutase. Nature 202: 1290–1293, 1964
30. Pierce J, Andrews TJ, Lorimer GH: Reaction intermediate partitioning by ribulose biphosphate carboxylase with different substrate specificities. J Biol Chem 261: 10248–10256, 1986
31. Tyihák E, Bocsi J, Timár F, Rácz G, Szende B. Formaldehyde promotes and inhibits the proliferation of cultured tumor and endothelial cells. Cell Prolif 34: 135–141, 2001
32. Tyihák E, Szende B, Trézl L. Biological effects of methylated amino acids. In W.K. Paik, S. Kim (eds). Protein Methylation, ch. 20. CRC Press, Boca Raton, FL, 1990

*Molecular and Cellular Biochemistry* **244:** 177–185, 2003.


# Arginine deprivation and tumour cell death: Arginase and its inhibition

Denys N. Wheatley, Ruth Philip and Elaine Campbell
*Department of Cell Pathology, University of Aberdeen, Aberdeen, Scotland, UK*

## Abstract

Arginase treatment of cell cultures reduced arginine in the medium to ~ micromolar levels within 5–30 min, and proved as effective as arginine-free medium (AFM) prepared by formulation. The enzyme was heat stable and as active at pH 7.2 as at pH 9.9. It persisted in culture for at least 3 days with only a small diminution in its speed of action, and still actively destroyed arginine after 6 days, since arginine supplementation failed to rescue viable cells.

Addition of L-norvaline, an inhibitor of arginase, rescued cells from arginase-induced deprivation. Its efficacy at low concentrations was short-lived (probably < 1 day), while at higher concentrations it did not appear to inhibit completely the enzyme. However, L-norvaline at these same levels also slowed the growth of positive non-enzyme treated controls receiving the normal arginine level. Thus the difference in this growth indicated that arginase was more inhibitory than cursory examination of initial kinetic data suggested. It also agreed with the inhibition of arginase in the ornithine assay used to measure biochemically enzyme activity. We conclude that norvaline partially but not completely antagonises arginase activity, which allows cell rescue in a dose-dependent manner between 0.4 and 4 mM, but cannot be used above about 2 mM without exhibiting a general non-specific interference of cell growth of its own, although no evidence of cell toxicity was observed in either AFM or arginine-containing medium. L-ornithine, the product of arginase that inhibits the enzyme by a feedback mechanism, had no inhibitory effect on arginase over a similar concentration range. (Mol Cell Biochem **244**: 177–185, 2003)

*Key words*: cancer, arginine, arginase, norvaline, ornithine, deprivation, catabolism, cell proliferation, cell death

## Introduction

A major aim of our work has been to lower arginine levels *in vitro* and *in vivo* to arrest the growth of normal cells, and selectively bring about the death of cells of malignant phenotype [1]. On balance, the evidence shows that most types of tumour cells can be killed by this strategy, even if it fails to destroy every tumour cell. Lowering arginine levels is not difficult in culture, the choice being either formulating a deficient medium that contains dialysed serum, or degradation of arginine *in situ* with an enzyme such as arginase, which has been used in a number of investigations over nearly 50 years. The first was arginase treatment of fibroblasts and Jensen sarcoma cells in culture by Bach and Simon-Reuss [2], who showed that it inhibited mitosis. Subsequently Bach and Swaine [3] treated rats bearing Walker carcinoma 256 with arginase and obtained 30–70% reduction in relative growth of tumours compared with untreated controls within 3–4 days. This is a remarkably rapid effect, but few studies have been able to emulate these findings, although *in vitro* work has continued to show good growth arrest of many malignant cell types (e.g. 4–6). Equally interesting, liver extracts were also known to have strong growth inhibitory effects on tumour cell growth that not unexpectedly was due to their high arginase content [7]. It should be noted that there is a big difference between an agent that simply inhibits growth and one that causes cell death. In our earlier work, it was stressed that arginine deprivation arrested normal cells in G1 (G0), but since most tumour cell lines do not have this 'luxury' because they remain in cycle, the consequence is that their growth becomes unbalanced and the cells die relatively quickly, the exact mechanism(s) is (are) not being fully understood [8]. Our objective has been to lower free arginine levels *in vitro* to at least < 5 μM, and preferably to < 1 μM, for a sufficient

*Address for offprints*: D.N. Wheatley, Department of Cell Pathology, University of Aberdeen, MacRobert Building, 581 King Street, Aberdeen, AB24 5UA, Scotland, UK (E-mail: wheatley@abdn.ac.uk)

period of time to induce cell death in malignant cells [1, 8]. The present *in vitro* experiments were designed to give further insight into the use of arginase as opposed to AFM in order that we might eventually improve its application in the treatment of cancer-bearing animals and ultimately human cancer patients, but if possible to gain more control over the enzyme through the use of a specific inhibitor.

Reducing arginine to the required level is more difficult to accomplish and maintain *in vivo* than in culture, because the normal healthy body responds by switching on a powerful homeostatic mechanism to restore free amino acid levels in the blood. The main pathway is the release of free amino acids from skeletal muscle breakdown. In culture, amino acid deprivation will also release arginine from 'luxury' proteins at the expense of vital house-keeping proteins [9, 10], as well as from dying cells. This aspect of the *in vitro* exercise has been only cursorily addressed, and the little we know is that arginine reutilisation is less efficient than that of other essential amino acids such as leucine under otherwise similar deprivation conditions [8]. Thus the use of a catabolic enzyme may have advantages in that any scavengable arginine released during increased protein turnover in the deficient state should be pre-emptorily destroyed, ostensibly producing better deprivation.

Here, we examine cultures made arginine-free by enzymatic means in comparison with our medium formulation procedure omitting arginine (AFM). Once arginase has been added to cultures, its fate has been largely unknown in previous work, but it is a relatively stable enzyme at 37°C as well as being heat-stable at 55–60°C; indeed 10–20 min at this temperature is supposed to fully activate it [11]. It is simpler to make medium deficient in arginine by adding a catabolic enzyme to complete medium than to prepare deficient medium, especially now that biological supply companies are reluctant to provide such 'customised media' except at considerable expense.

Once arginase has been added to medium, however, it is difficult to curtail its continued action because of its stability without physically removing it, hence the interested in an inhibitor of arginase to manipulate its activity. Chang *et al.* [12] have reported that L-norvaline inhibits arginase, and the presentation below includes some of our new data with this agent.

## Materials and methods

*Cell culture*

L1210 cells were maintained in flasks at between 0.1 and 2 × $10^6$ cells/ml in exponential phase growth (generation time ~ 15 h) in RPMI 1640 medium (Life Technologies, Paisley, Scotland, UK) at 37°C in a humidified atmosphere of 5% $CO_2$. For experiment, they were set out in 2 ml of culture medium at 5 × $10^4$ cells/ml in 12-well plates. HeLa cells and human diploid fibroblasts were seeded at 50,000 per ml in flasks or plates in a similar manner (previously described in ref. [1]).

Basic RPMI 1640 medium (Sigma, Poole, Dorset, UK) lacking the following supplements was prepared by the addition of these compounds to the stated final concentrations: 4 mM L-glutamine, 0.4 mM L-arginine, 0.2 mM L-cystine, 11.1 mM D-glucose, 0.2 mM I-inositol, 0.38 mM L-leucine, 0.1 mM L-methionine. 10% foetal calf serum (Harlan, Crawley Down, UK) was added for stock growth, but 5% dialysed serum was used in experimental plates. Arginine-free medium (AFM) was made by omitting L-arginine in the final formulation of the medium added to experimental cultures.

Cells were counted electronically with a Coulter Counter. Nearly all experiments involved not less than 3 samples per time-point and the average values have been plotted with one standard deviation (S.D.) of the mean in the bars on the Figures. The probability of differences being significant was taken at 5%, and in many cases the S.D. bars were tight, such that they will often be found within the symbols for the data-points. Significance was tested by the chi-squared test at appropriate points in the time-course of the experiments, $p < 0.05$ being taken as significant.

*Enzyme preparations*

*Arginase*

Bovine liver arginase, with an given activity of 150–250 units/mg (Sigma), and a lot number of 128H 7027 was almost exclusively in these experiments. However, each separate ampoule was nevertheless subject to enzyme assay (see below) to obtain its precise specific activity before use. Prior to use, the enzyme was initially heat-activated at 55–60°C for 10 min. The recovered enzyme was adjusted to 100 units per ml, filtered through a 0.2 μm filter (Nalgene, Nalge Co., Rochester, NY, USA) to sterilise, and stored at 0–4°C for use. A stock solution of 200 mM MnCl was prepared to provide a 100 times concentration of the co-factor for addition at the same time as the enzyme. The inhibitors, L-norvaline and L-ornithine, were prepared in PBS at 200 mM, sterilised by filtration, and diluted appropriately to treat cultures.

*Enzyme activity*

Enzyme activity under culture condition was analysed *in vitro* over time at 37°C in a humidified atmosphere of 5% $CO_2$. Arginase was added to basic RPMI medium (without foetal calf serum) with L-arginine at 0.4 mM in order to be in accord with our previous experimental work [1, 8]. The reaction was stopped at selected time-points by the addition of 10% sulphosalicyclic acid (Sigma), which precipitated the protein at 4°C for 30 min. The resulting samples were centrifuged at 13,000 rpm for 10 min in a microcentrifuge and the supernatant filtered through a MillexR-GV 0.22 μm filter (Millipore, Bedford, MA, USA). Arginine and ornithine was measured using a Biochrome 20 Amino Acid Analyser (Amersham Pharmacia Biotech, Cambridge, UK). Enzyme activity was calculated in units, I unit being the amount of enzyme catabolising 1 μmole of arginine to ornithine (and urea) per min.

*Ornithine assay*

Our biochemical assay of arginase activity employed the method of Ochoa *et al.* [13]. It was carried out at pH 9.9 [14], with the arginine being present in excess at 4 mM. For the purpose of comparison with the activity in culture media, as described above, the same biochemical test was carried out at pH 7.2 in phosphate buffered saline instead of bicarbonate buffer. The specific activity of arginase was defined as the amount of ornithine in μmoles per min per mg protein. Ornithine concentrations were determined against known standards from 10 to 100 μM [15].

*Protein*

An adaptation of the Lowry method [16] was used in conjunction with the above ornithine assay to determine the protein concentrations of samples, and the specific activity of the different preparations of the enzymes.

## Results

*Comparison of arginase with AFM on the cell cultures*

Arginase was as effective as AFM in reducing cell growth and causing cell death in both HeLa and L1210 cultures (Figs 1a and 1b). The continued presence of the enzyme throughout ensured that no arginine could be scavenged by any remaining viable cells throughout this time, which meant that cell death should have been more complete in HeLa cultures by this time than in AFM cultures, but if any difference existed it was marginal.

Arginase at 1 unit per ml lowered arginine to micromolar levels within 60 min of addition to culture medium at 37°C, 50% being destroyed within ~ 20 min (Fig. 2); higher concentrations to 5 units per ml led to almost complete degradation in < 10 min (see below), with ornithine remaining asymptotic over the ensuing incubation period at slightly lower than the stoichiometrically obverse level of arginine (see Discussion). Addition of fresh boluses of arginine to bring the medium concentration immediately back to 400 μM at daily intervals produced a step-wise rapid degradation of arginine, with about 80–90% of the initial step being seen in 20 min following addition after the first day, and about 60–70% after day 2, although subsequently the degradation of arginine went to completion. Hence enzyme activity was only slightly reduced over 3 days of incubation of the cell cultures, and its action remained probably considerably longer for other reasons given below.

Prior heat activation of the enzyme at 5–60°C for 10 or 20 min made little difference to these kinetics at 37°C (data

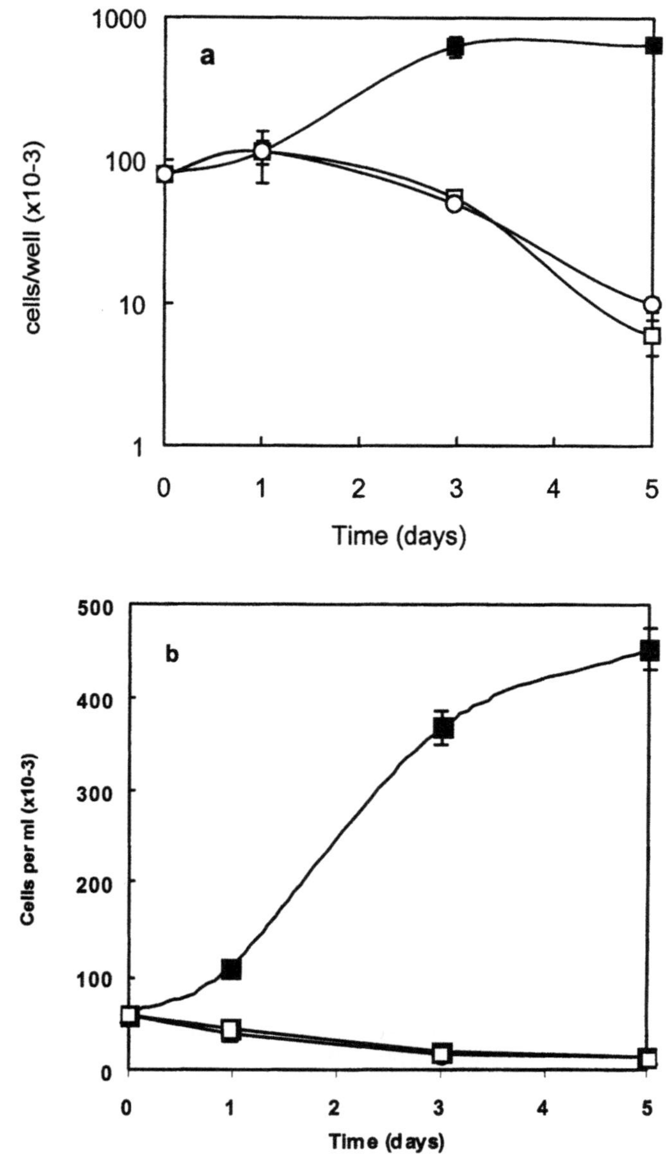


*Fig. 1.* (a) Growth kinetics of HeLa cells in (■) arginine containing medium (positive controls), (□) arginine-free medium (AFM), and (O) arginine-containing medium plus 1 unit per ml arginase. The total number of cells in each of 3 wells was averaged at each time-point, and the bars give 1 S.D. around the mean. The significance of the difference between (■) and (O) is very highly significant ($p < 0.001$), but the difference between (□) and (O) is not significant. (b) Growth kinetics as in (a), but for L1210 cells with a much faster generation time of ~ 15 h. The symbols of the negative control and the arginase-treated cultures superimpose of each other and the error bars are often within the symbols.

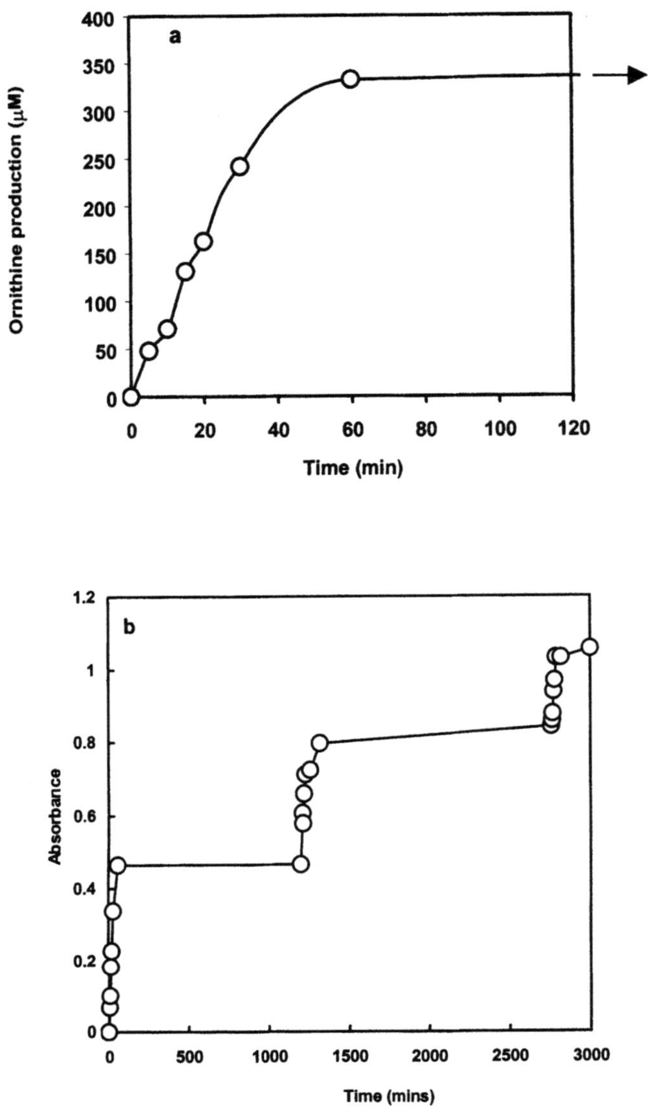


*Fig. 2.* (a) Degradation of 400 μM in 2 ml culture medium at 37°C, measured by the ornithine assay. In ~ 60 min all arginine had been converted to ornithine by 1 unit per ml arginase and none was detectable in the medium. Ornithine rose to a final concentration at 1000 min (arrow) of 360 μM, but the assay usually gave ~ 10% difference from the arginine value (which was probably due to technical difficulties with the slight colouration changes in the background). The data, however, were typical of all such test, and was taken as evidence that the reaction had gone to completion. (b) Similar data to (a), but at the times immediately prior to each step, 400 μM of fresh arginine were added to the ongoing incubation. Over 2 days, arginine was converted to ornithine within about 60 min each time, but the plateau levels were slightly reduced at each step (see text for more detail).

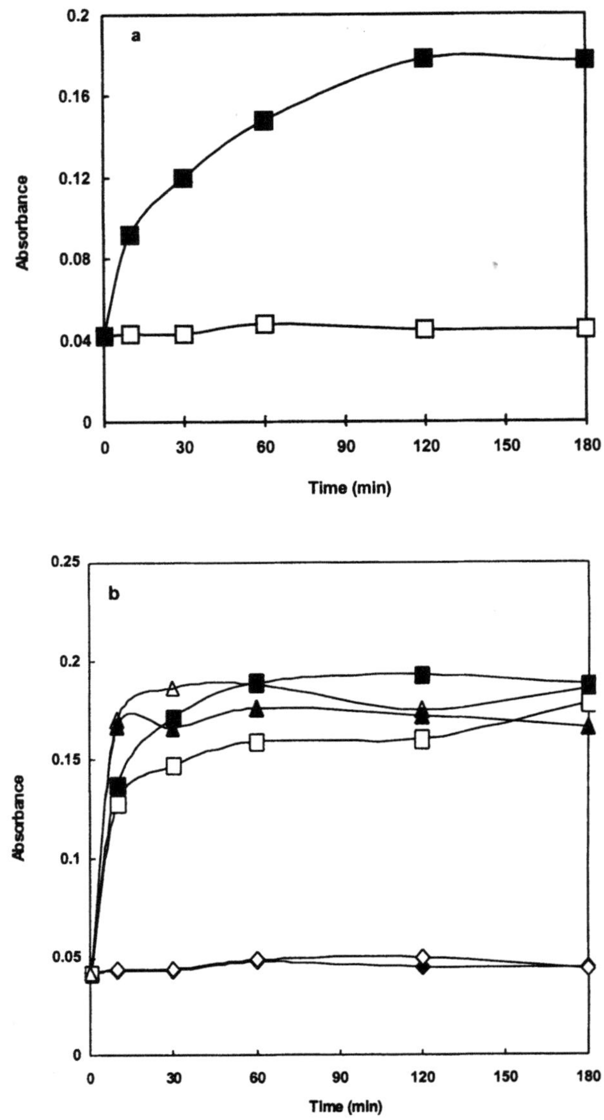


*Fig. 3.* (a) Biochemical assay of ornithine (ref. [13]) at pH 9.9 in bicarbonate buffer, showing incubation of 1 unit arginase per ml in tests containing 4 mM arginine (excess) per sample. After ~ 100 min, ornithine production has plateaued (■). The relatively high background of this assay is seen in the lower trace (□) where no enzyme was added may account for the discrepancy in not apparently attaining complete conversion of arginine to ornithine in Fig. 2a. (b) As for (a), but with 2.5 units arginase per ml in phosphate buffer pH 7.2 (□) or bicarbonate buffer (■) with similar data for 5 units per ml arginase, (Δ) and (▲), respectively. Backgrounds in both buffers where no arginase was present are shown with diamond symbols. The reactions have usually reached close to their asymptotic values in 10 min.

therefore not shown), and this procedure was omitted from the protocol thereafter. Similarly, further supplementation of $Mn^{2+}$ up to 20 μM made no significant difference compared with omitting it altogether (data therefore not shown), from which we presume that the manganous sulphate included in the Sigma preparation is more than adequate to sustain optimal enzyme activity.

Arginase assayed biochemically by the method of Ochoa *et al.* [13] involved using an excess of arginine (4 mM), but the conditions were alkaline (pH 9.9) and therefore did not allow comparison with the kinetics observed under the physiological (culture) conditions of Fig. 2. In the assay, the usual incubation period is 10 min, but if the reaction is allowed to go to completion as in Fig. 3, the quite small volume used meant that the arginine represented only a small number of micromoles which was completely degraded in ~ 120 min at 37°C at 1 unit per ml, ornithine rising stoichiometrically in near obverse proportion (Fig. 3a). At 2.5 and 5.0 units per ml, this arginine was mostly catabolised within the first 10–15 min. But more interestingly and contrary to the data on pH sensitivity of the enzyme reported by Roholt and Greenberg [13], degradation of arginine occurred as fast as (sometimes marginally *faster*) in phosphate buffered saline at pH 7.2 than in recommended bicarbonate buffer at pH 9.9 (Fig. 3b; see Discussion). From the kinetics, one can also calculate that in the 2 ml culture medium usually set out in 12 well-plates, almost all of the 0.8 micromoles of arginine initially present (at 400 μM) should be converted into ornithine in about 10 min at 1 unit of enzyme per ml, which agrees well with the observed rates in culture above.

*Rescue from arginase treatment*

*Addition of arginine*

Restoring arginine failed to rescue cultures at any time during a 6-day exposure to arginase, indicating that the enzyme was still capable of keeping arginine lower than that required for cell growth (i.e. well below 40 μM; see ref. [8]). Restoration of arginine to AFM cultures also gave the recovery expected from previous studies [8], (Fig. 4). The recovery of arginase-treated cultures by the addition of arginine only occurred when the enzyme was removed first, thus requiring a complete medium change. Under these circumstances, recovery was limited to cultures treated for 1–4 days before reversal, and not those that had had 6 days or more days of exposure, much in the same manner as AFM-treated cultures, except that in the latter a medium change was not necessary.

*Citrulline and ornithine*

Figure 4 shows that citrulline was fully capable of preventing AFM-treated cultures from arresting growth and the cells dying off when added from the outset, and in complete contrast to arginine this also took place *in the continued presence* of arginase (see Discussion). Ornithine failed on every occasion with both HeLa and L1210 cultures to prevent the effects of arginine deprivation due to arginase treatment (Fig. 4), just as it had no effect on AFM cultures, as previously noted [8].

*Norvaline*

Norvaline abrogated the effects of arginase on the cells, the experiments being controlled by cultures receiving AFM treatment (negative controls) as well as cultures given arginine but no enzyme along with the inhibitor (positive controls). The enzyme remained active and kept cultures free of arginine in the absence of norvaline, but in its presence a

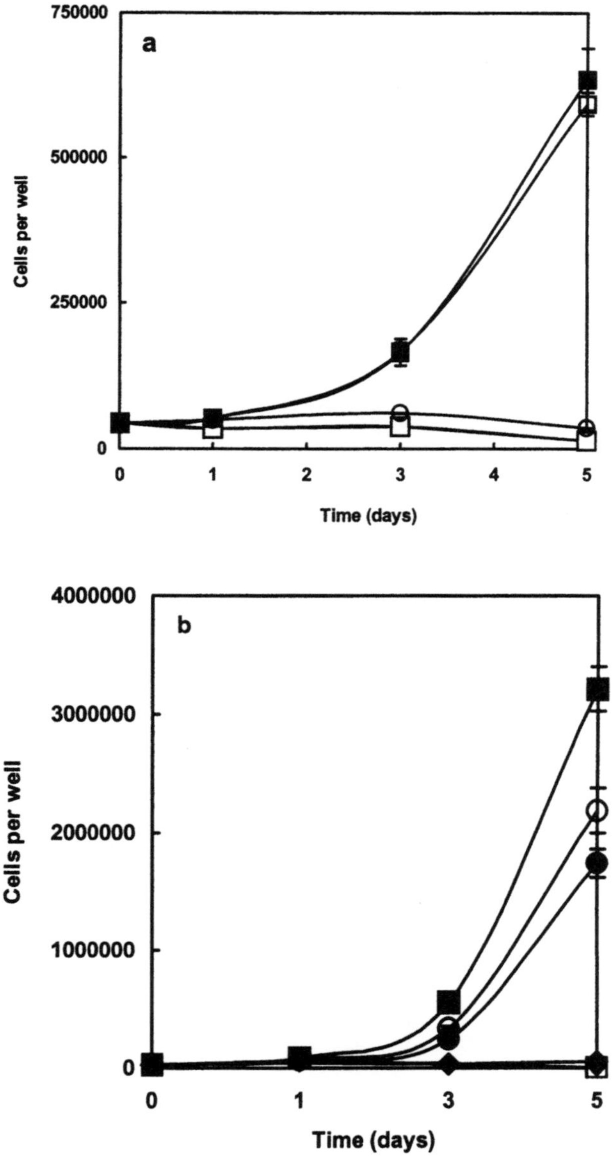


*Fig. 4.* (a) Effect of ornithine and citrulline on HeLa cell cultures inhibited by arginase. At time zero, the cultures received ornithine at 400 μM ornithine (□) or a placebo supplement of saline (O). Similarly treated cultures given 400 μM arginine (●) failed to thrive, whereas those given 400 μM citrulline (□) at time zero grew almost as well as positive control cultures in AFM given 400 μM arginine back at the start (■), and significant difference was detected at any time-point. (b) As for (a), but L1210 cells were incubated here in parallel cultures with arginase at 1 unit per ml. Again 400 μM ornithine (□) failed to rescue growth, and 400 μM arginine was equally ineffective (◆). Here the growth of cultures given 400 μM citrulline (●) recovered in the presence of the enzyme to almost the same extent as AFM-treated cultures similarly supplemented with 400 μM citrulline (O), the difference at 5 days not being significant. (Not all the data for the different controls and treatment of both AFM and arginase-treated cultures has been plotted here in order to avoid confusion of lines and points.)

dose-dependent effect was seen, reaching a maximum at the highest concentration (1.6 mM) used in this experiment (Fig. 5a). Although 1.6 mM or higher levels does not restore nor-

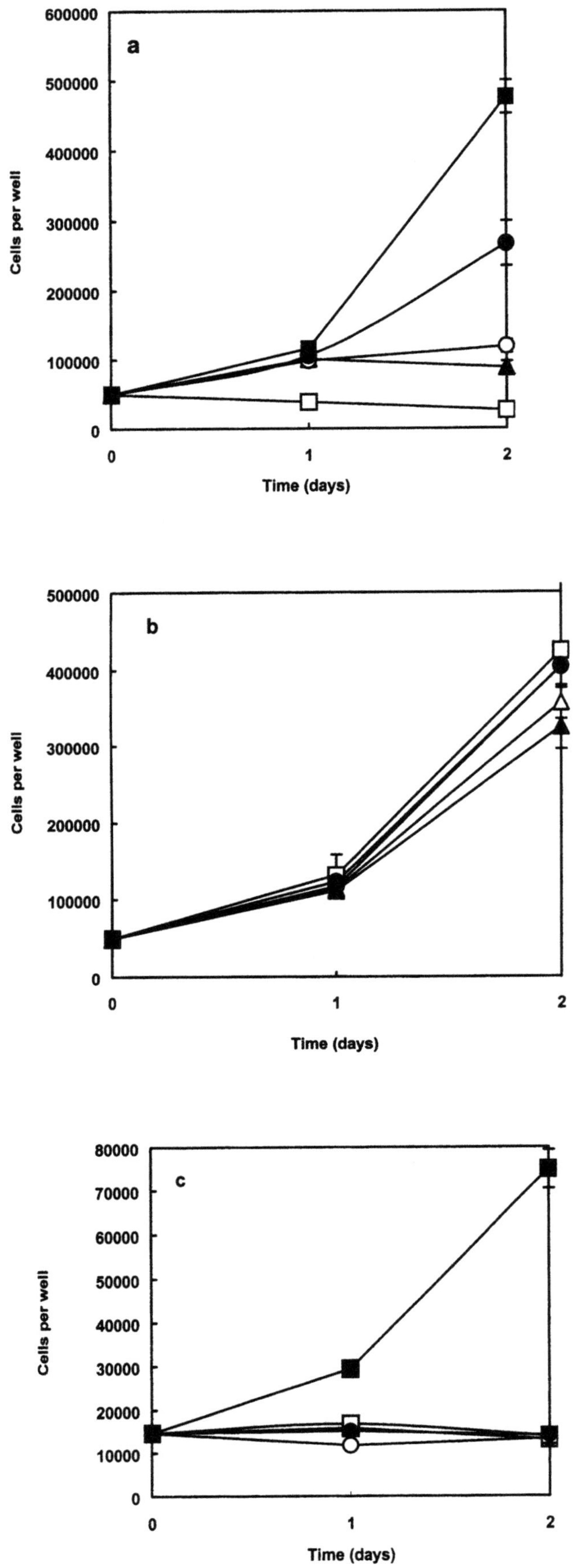


mal growth to arginase-treated cells, it should be noted that the analogue amino acid produced up to a 35% inhibition of growth in arginine-containing (positive) controls that had not received arginase (Fig. 5b). Taking this into account, the incomplete restoration of growth in the presence of the enzyme with 1.6 mM norvaline matches more closely with the growth level in cultures to which norvaline had been added on its own and suppressed growth by a mechanism (or mechanisms) other than a direct action on arginase (see Discussion). In the negative controls, L-norvaline did not exacerbated the arginine-free conditions (Fig. 5c).

In biochemically assaying of arginase activity, the presence of L-norvaline at 5 and 10 times the level of arginine brought down ornithine production by 45 and 65%, respectively (Fig. 6). Measurements were made up to 4 mM norvaline, which inhibited arginase more strongly than 1.6 mM, but the associated considerably greater arrest of positive (non-enzyme-treated) controls indicated that this level was too high to work with because of non-specific inhibition, and therefore we deemed that use above 2.0 mM was reasonable in these experimental circumstances. It is worth noting, however, that despite the growth suppression due to 1–4 mM norvaline, both positive and negative control cells remained perfectly viable and healthy in its presence, demonstrating very clearly that suppression of growth was not associated with any cytotoxicity of norvaline.

## Discussion

Arginase is a good enzyme to use to degrade arginine in culture because its product is ornithine, which cannot be recycled to arginine because the urea cycle is incomplete in most cultured cells. Our studies with that commercial preparations

←

*Fig. 5.* (a) In this panel, the controls received arginase at 1 unit per ml in arginine-containing medium, but no norvaline (□). The top curve (■) relates to positive control cells that received no arginase and had arginine present. The intermediate curves are for cultures like the arginase control given arginine, but they also received at the start norvaline at 0.4 mM (●), 0.8 mM (Δ) and 1.6 mM (▲), respectively. Recovery was seen in all, but fell back again at 0.4 mM, was small but continuing at 0.8 mM, and 55–60% effective in restoring growth at 1.6 mM. $p > 0.001$ for 1.6 mM group vs. its control given no norvaline. (b) Effect of norvaline on L1210 cells. This panel includes the responses of L1210 cells over two days of treatment in AFM to which 0.4 mM arginine was restored in all the groups. The positive controls (■) received no norvaline, whereas the other groups received norvaline at 0.4 mM (●), 0.8 mM (Δ) and 1.6 mM (▲), respectively, each progressively suppressing the control cell growth rate. (c) In the final panel, the negative control which received AFM and arginase is shown (O), and the three other superimposed curves were the same but with norvaline added at 0.4 mM (●), 0.8 mM (Δ) and 1.6 mM (▲), respectively, showing no further deleterious effects of norvaline upon deprived cells. A positive control is also included which received AFM and arginine at 0.4 mM (■).

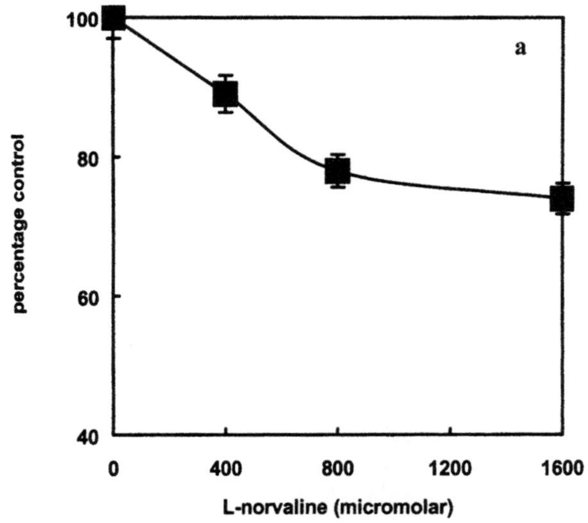


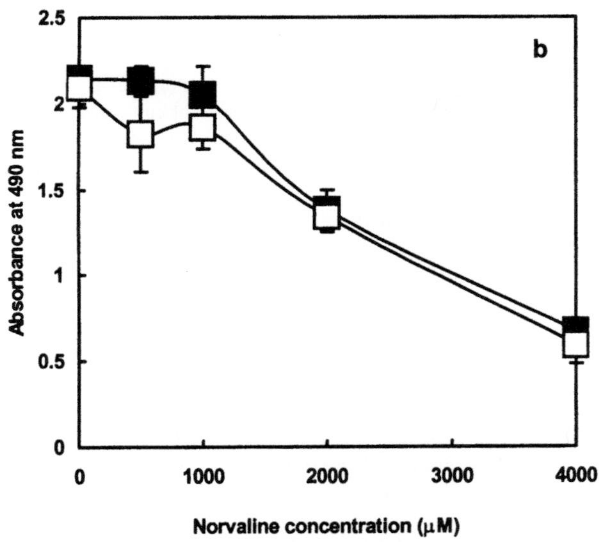


*Fig. 6.* (a) Depression of growth of L1210 cells in normal medium containing 0.4 mM arginine after incubation for 3 days in increasing concentrations of norvaline. (b) Two curves of the effect of increasing norvaline concentration on arginase in the Ochoa assay [13] where it is competing against 500 μM arginine at pH 7.2. The incubation time was 1 h. Norvaline has to be used at considerable excess (8-fold) to be effective, which suggests it is not a strong competitive inhibitor.

of the enzyme of the same enzyme as used by Bach's group [2, 3], i.e. bovine liver arginase, had similar effects on cancer cells. However, the Sigma preparations have a lower specific activity (by as much as an order of magnitude) than some preparations, such as human recombinant arginase expressed within transfected *E. coli* [17]. Nevertheless on a unit per ml basis, the conversion of arginine to ornithine was precisely that given by the producers and as estimated from our own assays [13, 18]. $Mn^{2+}$ availability being an integral part of the Sigma preparation meant that no further supplement was needed. Arginase is a remarkably heat-resistant enzyme, with a very long life on storage at 4°C in lyophilised form, but less so in solution, although it can be kept for 1–2 months before its activity falls < 50%. Once in solution and kept frozen, most of its activity will be retained for many months or years, but freeze-thawing cycle quickly diminishes its efficacy. The one feature not in keeping with previous experience related to its pH sensitivity. According to Roholt and Greenberg [14], arginase shows maximal activity at pH 9.9, falling steeply either side of this value. Our data (Fig. 3) shows unequivocally that it is as active, and occasionally marginally more active, at pH 7.2 than at pH 9.9. We can only suggest that the incubation conditions used by Roholt and Greenberg affected the conformation of the molecule, whereas in the Ochoa assay [13], it makes little difference whether the assay is carried out at pH 9.9 or 7.2. It also makes much greater sense that arginase released from the liver in the body is probably going to be selected to operate around neutral pH because few cells or tissues tolerate high pH. But this is important in our investigation of the catabolism of arginine, especially *in situ* in the body, because poor activity at pH 7.2 would have quickly eliminated the choice of this enzyme.

We conclude that arginase has given some highly encouraging results *in vivo* as well as *in vitro*, which clearly demonstrates its efficacy in both situations, its very rapid and considerable capacity to degrade significant quantities of substrate, its production of ornithine rather than (for example) citrulline as a direct product, and the fact that the former product requires more metabolic steps to be recycled to arginine. Cultured cells cannot normally use ornithine to substitute for arginine (Figs 4a and 4b). We are therefore surprised that the potential of arginase has not been more systematically explored and exploited following the original findings of Bach's group [2, 3]. Equally, we find it odd that much greater current interest is being shown in arginine deaminase [19, 20], whose product is the readily recyclable citrulline that rescues cells not only from AFM, but also from arginase-induced deficiency (Figs 4 and 5). On these grounds alone, arginase is a more logical and superior choice (see also below). In culture, we can add arginine back to AFM to rescue living cells, but this is ineffective when active arginase is present since it continues to destroy arginine for a long time, and fast enough, to prevent significant uptake and recovery of the deprived cells. Although this needs in depth analysis, it is at least 3 days and probably longer than 6 before the level of arginase given as a bolus at the start of the experiment falls below that which would permit sufficient arginine to persist for long enough to at least keep some of the cells alive. Clearly this does not happen (Figs 4 and 5).

We also know from Tanaka *et al.* [21] that arginine deaminase must remain active for at least 8 days in the murine body, and therefore arginine catabolising enzymes seem to be remarkably resilient to degradation. Since it is also known that pegylation of arginase further increases its stability [22], future work will compare the efficacy of the non-pegylated and pegylated forms. This is not as important with commercial arginase of relatively low specific activity as with recombinant human arginase prepared from *E. coli*, where the specific activity is likely to be about an order of magnitude greater. Increased stability coupled with greater specific activity means that much less protein (enzyme) is needed that will for longer *in vitro* and seemingly *in vivo* [21], provided that pegylation does not seriously diminish its activity. Another benefit is that human arginase is likely to be much less immunogenic than a bovine preparation.

Arginase-induced arginine deprivation was circumvented by the addition of citrulline, giving full recovery *without* the need to remove the enzyme. If we had to stop the action of the enzyme in circumstances where its physically removed was difficult, e.g. *in vivo*, inhibitors that can prevent arginine from being further catabolised will be of considerable importance in control. An arginine-deficient state *in vivo* is potentially life threatening because of the widespread involvement of arginine in metabolic pathways of vital importance to normal physiological functioning [23]. Norvaline may at first sight look less than adequate because considerable quantities were required for its action to be manifest (up to 1–4 mM). However, the higher millimolar levels also curb normal cell growth, possibly by non-specific interference with amino acid uptake. Thus the recovery seen with 1.6 mM norvaline in arginase-treated cultures must have more completely abrogated the action of arginase than would appear at first sight, because the amount of growth was quite closely in accord with the positive controls not receiving arginase but given 4 mM norvaline (Fig. 5). This is one of the more important issue emerging from these studies, since we have need of greater control over arginase in culture and in the body when the enzyme remains on the scene, and where rescue of normal cells from arrest is needed, even though tumour cells are by that time no longer recoverable. We are also encouraged by the fact that, although the positive control cells given norvaline simply grew relatively more slowly (Fig. 6a), the cells remained healthy and showed no signs of damage, as did the negative controls, which would have succumbed far faster in the presence of a cytotoxic agent. L-norvaline seems to be a safe amino acid analogue to use in culture, and therefore may also be non-toxic in the body. Furthermore, the fact that it seems to inhibit normal cell growth in a non-specific manner does not preclude its use at much higher concentration in the body to inhibit arginase activity since initially the cells will be suppressed from growing by the arginine deficient conditions induced by the enzyme. Its effect was much the same as seen after another straight-chained amino acid had been added to normal growing cells, i.e. norleucine [8], but which had no inhibitory action on arginase activity (data not shown).

Despite the fact that there are now hundreds of analogues and derivatives of arginine arising from the general interest in nitric oxide synthase, we have not so far come across any which is as active as L-norvaline in inhibiting arginase, although systematic has only just begun.

## Acknowledgements

We are grateful for the assistance of technical assistance of Lorraine Philip and Satvir Minhas. We thank Dr. Susan Smith for useful discussions on some of the data. The work was supported by funds from The Synos Foundation of Switzerland and Schlumberger International.

## References

1. Scott LA, Lamb JL, Smith S, Wheatley DN: Single amino (arginine) deprivation: Rapid and selective cells death of cultured transformed and malignant cells. Brit J Cancer 83: 800–810, 2000
2. Bach SJ, Simon-Reuss I: Arginase, an antimitotic agent in tissue culture. Biochim Biophys Acta 11: 396–402, 1953
3. Bach SJ, Swaine D: The effect of arginase on the retardation of tumour growth. Brit J Cancer 19: 379–386,1965
4. Terayama H, Koji T, Kontani M, Okumoto T: Arginase is an inhibitory principle in liver plasma membranes arresting the growth of various mammalian cells *in vitro*. Biochim Biophys Acta 720: 188–192, 1982
5. Storr JM, Burton AF: The effects of arginine deficiency on lymphoma cells. Brit J Cancer 30: 50–59, 1974
6. Sasada M, Terayama H: The nature of inhibitors of DNA synthesis in rat-liver hepatoma cells. Biochim Biophys Acta 190: 73–87, 1969
7. Holley RH: Evidence that a rat liver 'inhibitor' of the synthesis of DNA in cultured mammalian cells is arginase. Biochim Biophys Acta 145: 525–527, 1967
8. Wheatley DN Scott LA, Lamb JL, Smith S: Single amino (arginine) restriction: Growth and death of cultured HeLa and normal human diploid fibroblasts. Cell Physiol Biochem 10: 37–55, 2000
9. Auteri JS, Okada A, Bochaki V, Dice JF: Regulation of intracellular protein degradation in IMR-90 human diploid fibroblasts. J Cell Physiol 115: 167–174, 1983
10. Tanaka H, Zaitsu H, Onodera K, Kimura G: Influence of the deprivation of a single amino acid on cellular proliferation and survival in rat 3Y1 fibroblasts and their derivatives transformed by a wide variety of agents. J Cell Physiol 136: 421–430, 1988
11. Archibald RM: The synthesis of urea from glutamine by liver extract. J Biol Chem 157: 519–523, 1945
12. Chang CI, Liao JC, Kuo L: Macrophage arginase promotes tumor cell growth and suppresses nitric oxide mediated tumor cytotoxicity. Cancer Res 61: 1100–1106, 2001
13. Ochoa JB, Bernard AC, Mistry SK, Morris SM Jr, Figert PL, Maley ME, Tsuei BJ, Boulanger BR, Kearney PA: Trauma increases extrahepatic arginase activity. Surgery 127: 419–426, 2000
14. Roholt OA, Greenberg DM: Liver arginase: IV. Effect of pH on kinetics of manganese-activated enzyme. Arch Biochem Biophys 62: 454–470, 1956

15. Chinard FP: Photometric estimation of proline and ornithine. J Biol Chem 199: 91–95, 1952
16. Oyama VI, Eagle H: Measurement of cell growth in tissue culture with a phenol (Folin-Ciocalteau) reagent. Proc Soc Exp Biol Med 91: 305–307, 1952
17. Ikemoto M, Tabata M, Miyake T, Kono T, Mori M, Totani M, Murachi T: Expression of human liver arginase in *Escherichia coli*. Purification and properties of the product. Biochemical Journal 270: 697–703, 1990
18. Schimke RT: Enzymes of arginine metabolism in mammalian cell culture. 1. Repression of argininosuccinate synthetase and argininosuccinase. J Biol Chem 239: 136–145, 1964
19. Komada Y, Zhang XL, Zhou YW, Ido M, Azuma E: Apoptotic cell death of human T lymphoblastoid cells induced by arginine deaminase. Int J Hematol 65: 129–141, 1997
20. Gong H, Zölzer F, von Recklinghausen G, Havers W, Schweigerer L: Arginine deiminase inhibits proliferation of human leukaemia cells more potently than asparginase by inducing cell cycle arrest and apoptosis. Leukemia 14: 826–829, 2000
21. Takaku H, Takase M, Abe SI, Hayashi H, Miyaki K: *In vivo* anti-tumor activity of arginine deiminase purified from *Mycoplasma argini*. Int J Cancer 51: 244–249,1992
22. Savoca KV, Davis FF, Van Es T, McCoy JR, Palczuk NC: Cancer therapy with chemically modified enzyme: II. The therapeutic effectiveness of arginase modified by the covalent attachment of polyethylene glycol, on a Taper liver tumor and the L5178Y murine leukaemia. Cancer Biochem Biophys 7: 261–268, 1984
23. Wheatley DN: Arginine catabolism, liver extracts and cancer. Pathol Oncol Res

*Molecular and Cellular Biochemistry* **244:** 187–190, 2003.

# Index to Volume 244

## Developments in Molecular and Cellular Biochemistry

**Series Editor: Naranjan S. Dhalla, Ph.D., M.D. (Hon.), FACC**

1. V.A. Najjar (ed.): *Biological Effects of Glutamic Acid and Its Derivatives.* 1981 — ISBN 90-6193-841-4
2. V.A. Najjar (ed.): *Immunologically Active Peptides.* 1981 — ISBN 90-6193-842-2
3. V.A. Najjar (ed.): *Enzyme Induction and Modulation.* 1983 — ISBN 0-89838-583-0
4. V.A. Najjar and L. Lorand (eds.): *Transglutaminase.* 1984 — ISBN 0-89838-593-8
5. G.J. van der Vusse (ed.): *Lipid Metabolism in Normoxic and Ischemic Heart.* 1989 — ISBN 0-7923-0479-9
6. J.F.C. Glatz and G.J. van der Vusse (eds.): *Cellular Fatty Acid-Binding Proteins.* 1990 — ISBN 0-7923-0896-4
7. H.E. Morgan (ed.): *Molecular Mechanisms of Cellular Growth.* 1991 — ISBN 0-7923-1183-3
8. G.J. van der Vusse and H. Stam (eds.): *Lipid Metabolism in the Health and Diseased Heart.* 1992 — ISBN 0-7923-1850-1
9. Y. Yazaki and S. Mochizuki (eds.): *Cellular Function and Metabolism.* 1993 — ISBN 0-7923-2158-8
10. J.F.C. Glatz and G.J. van der Vusse (eds.): *Cellular Fatty-Acid-Binding Proteins, II.* 1993 — ISBN 0-7923-2395-5
11. R.L. Khandelwal and J.H. Wang (eds.): *Reversible Protein Phosphorylation in Cell Regulation.* 1993 — ISBN 0-7923-2637-7
12. J. Moss and P. Zahradka (eds.): *ADP-Ribosylation: Metabolic Effects and Regulatory Functions.* 1994 — ISBN 0-7923-2951-1
13. V.A. Saks and R. Ventura-Clapier (eds.): *Cellular Bioenergetics: Role of Coupled Creatine Kinases.* 1994 — ISBN 0-7923-2952-X
14. J. Slezák and A. Ziegelhöffer (eds.): *Cellular Interactions in Cardiac Pathophysiology.* 1995 — ISBN 0-7923-3573-2
15. J.A. Barnes, H.G. Coore, A.H. Mohammed and R.K. Sharma (eds.): *Signal Transduction Mechanisms.* 1995 — ISBN 0-7923-3663-1
16. A.K. Srivastava and J.-L. Chiasson (eds.): *Vanadium Compounds: Biochemical and Therapeutic Applications.* 1995 — ISBN 0-7923-3763-8
17. J.M.J. Lamers and P.D. Verdouw (eds.): *Biochemistry of Signal Transduction in Myocardium.* 1996 — ISBN 0-7923-4067-1
18. E.-G. Krause and R. Vetter (eds.): *Biochemical Mechanisms in Heart Function.* 1996 — ISBN 0-7923-4118-X
19. R. Vetter and E.-G. Krause (eds.): *Biochemical Regulation of Myocardium.* 1996 — ISBN 0-7923-4259-3
20. G.N. Pierce and W.C. Claycomb (eds.): *Novel Methods in Molecular and Cellular Biochemistry of Muscle.* 1997 — ISBN 0-7923-4387-5
21. F.N. Gellerich and S. Zierz (eds.): *Detection of Mitochondrial Diseases.* 1997 — ISBN 0-7923-9925-0
22. P.K. Singal, V. Panagia and G.N. Pierce (eds.): *The Cellular Basis of Cardiovascular Function in Health and Disease.* 1997 — ISBN 0-7923-9974-9
23. S. Abdel-aleem and J.E. Lowe (eds.): *Cardiac Metabolism in Health and Disease.* 1998 — ISBN 0-7923-8104-1
24. A.K. Srivastava and B. Posner (eds.): *Insulin Action.* 1998 — ISBN 0-7923-8113-0
25. V.A. Saks, R. Ventura-Clapier, X. Leverve, A. Rossi and M. Rigoulet (eds.): *Bioenergetics of the Cell: Quantitative Aspects.* 1998 — ISBN 0-7923-8118-1
26. G.N. Pierce, H. Rupp, T. Izumi and A. Grynberg (eds.): *Molecular and Cellular Effects of Nutrition on Disease Processes.* 1998 — ISBN 0-7923-8171-8
27. K. Ahmed, E. Chambaz and O.G. Issinger (eds.): *Molecular and Cellular View of Protein Kinase CK2.* 1998 — ISBN 0-7923-8208-0

## Developments in Molecular and Cellular Biochemistry

28. M.V. Cohen, J.M. Downey, R.J. Gelpi and J. Slezák (eds.): *Myocardial Ischemia and Reperfusion*. 1998 — ISBN 0-7923-8173-4
29. D.A. Bernlohr and L. Banaszak (eds.): *Lipid Binding Proteins within Molecular and Cellular Biochemistry*. 1998 — ISBN 0-7923-8223-4
30. R. Albarez-Gonzalez (ed.): *ADP-Ribosylation Reactions: From Bacterial Pathogenesis to Cancer*. 1998 — ISBN 0-7923-8235-8
31. S. Imai and M. Endo (eds.): *Muscle Physiology and Biochemistry*. 1998 — ISBN 0-7923-8264-1
32. D.K. Das (ed.): *Stress Adaptation, Prophylaxis and Treatment*. 1999 — ISBN 0-7923-8406-7
33. H. Rupp and B. Maisch (eds.): *Control of Gene Expression by Catecholamines and the Renin-Angiotensin System* — ISBN 0-7923-7981-0
34. X. Shi, V. Castranova, V. Vallyathan and W.G. Perry (eds.): *Molecular Mechanisms of Metal Toxicity and Carcinogenesis*. 2001 — ISBN 0-7923-7498-3
35. K. Ahmed, J.E. Allende and O.-G. Issinger (eds.): *Protein Kinase CK2 – From Structure to Regulation*. 2001 — ISBN 0-7923-7666-8
36. R.K. Sharma (ed.): *Guanylate Cyclase*. 2002 — ISBN 0-7923-7682-8
37. V. Vallyathan, V. Castranova and S. Shi (eds.): *Oxygen/Nitrogen Radicals: Cell Injury and Disease*. 2002 — ISBN 1-4020-7085-3
38. J.F.C. Glatz (ed.): *Cellular Lipid Binding Proteins*. 2002 — ISBN 1-4020-7220-1
39. E. Kardami, L. Hryshko and N. Mesaeli (eds.): *Cardiac Cell Biology*. 2003 — ISBN 1-4020-7296-1
40. J.F. Clark (ed.): *Guanidino Compounds in Biology and Medicine*. 2003 — ISBN 1-4020-7297-X

---

KLUWER ACADEMIC PUBLISHERS – DORDRECHT / BOSTON / LONDON

MIX
Papier aus verantwortungsvollen Quellen
Paper from responsible sources
FSC® C105338

If you have any concerns about our products,
you can contact us on
**ProductSafety@springernature.com**

In case Publisher is established outside the EU,
the EU authorized representative is:
**Springer Nature Customer Service Center GmbH**
**Europaplatz 3, 69115 Heidelberg, Germany**

Printed by Libri Plureos GmbH
in Hamburg, Germany